Clinical Research on
Diabetic Complications

Clinical Research on Diabetic Complications

Editor

Didac Mauricio

MDPI • Basel • Beijing • Wuhan • Barcelona • Belgrade • Manchester • Tokyo • Cluj • Tianjin

Editor
Didac Mauricio
Univesity of Vic
Spain

Editorial Office
MDPI
St. Alban-Anlage 66
4052 Basel, Switzerland

This is a reprint of articles from the Special Issue published online in the open access journal *Journal of Clinical Medicine* (ISSN 2077-0383) (available at: https://www.mdpi.com/journal/jcm/special_issues/Diabetic_Complications).

For citation purposes, cite each article independently as indicated on the article page online and as indicated below:

LastName, A.A.; LastName, B.B.; LastName, C.C. Article Title. *Journal Name* **Year**, *Volume Number*, Page Range.

ISBN 978-3-03943–825-9 (Hbk)
ISBN 978-3-03943-826-6 (PDF)

Contents

About the Editor . . . . **vii**

Didac Mauricio
Comments from the Editor of the Special Issue "Clinical Research on Diabetic Complications"
Reprinted from: *J. Clin. Med.* **2019**, *8*, 2193, doi:10.3390/jcm8122193 . . . . **1**

Andreea Ciudin, Enzamaria Fidilio, Angel Ortiz, Sara Pich, Eduardo Salas, Jordi Mesa, Cristina Hernández, Olga Simó-Servat, Albert Lecube and Rafael Simó
Genetic Testing to Predict Weight Loss and Diabetes Remission and Long-Term Sustainability
after Bariatric Surgery: A Pilot Study
Reprinted from: *J. Clin. Med.* **2019**, *8*, 964, doi:10.3390/jcm8070964 . . . . **3**

Laura Del Coco, Daniele Vergara, Serena De Matteis, Emanuela Mensà, Jacopo Sabbatinelli, Francesco Prattichizzo, Anna Rita Bonfigli, Gianluca Storci, Sara Bravaccini, Francesca Pirini, Andrea Ragusa, Andrea Casadei-Gardini, Massimiliano Bonafè, Michele Maffia, Francesco Paolo Fanizzi, Fabiola Olivieri and Anna Maria Giudetti
NMR-Based Metabolomic Approach Tracks Potential Serum Biomarkers of Disease Progression
in Patients with Type 2 Diabetes Mellitus
Reprinted from: *J. Clin. Med.* **2019**, *8*, 720, doi:10.3390/jcm8050720 . . . . **11**

Esmeralda Castelblanco, Lucía Sanjurjo, Mireia Falguera, Marta Hernández, José-Manuel Fernandez-Real, Maria-Rosa Sarrias, Nuria Alonso and Didac Mauricio
Circulating Soluble CD36 is Similar in Type 1 and Type 2 Diabetes Mellitus versus
Non-Diabetic Subjects
Reprinted from: *J. Clin. Med.* **2019**, *8*, 710, doi:10.3390/jcm8050710 . . . . **31**

Ming-Hsun Lin, Chien-Hsing Lee, Chin Lin, Yi-Fen Zou, Chieh-Hua Lu, Chang-Hsun Hsieh and Cho-Hao Lee
Low-Dose Aspirin for the Primary Prevention of Cardiovascular Disease in Diabetic
Individuals: A Meta-Analysis of Randomized Control Trials and Trial Sequential Analysis
Reprinted from: *J. Clin. Med.* **2019**, *8*, 609, doi:10.3390/jcm8050609 . . . . **45**

Chung-Hao Li, Feng-Hwa Lu, Yi-Ching Yang, Jin-Shang Wu and Chih-Jen Chang
Increased Arterial Stiffness in Prediabetic Subjects Recognized by Hemoglobin A1c with
Postprandial Glucose but Not Fasting Glucose Levels
Reprinted from: *J. Clin. Med.* **2019**, *8*, 603, doi:10.3390/jcm8050603 . . . . **59**

Juan J. Cabré, Teresa Mur, Bernardo Costa, Francisco Barrio, Charo López-Moya, Ramon Sagarra, Montserrat García-Barco, Jesús Vizcaíno, Immaculada Bonaventura, Nicolau Ortiz, Gemma Flores-Mateo, Oriol Solà-Morales and the Catalan Diabetes Prevention Research Group
Feasibility and Effectiveness of Electrochemical Dermal Conductance Measurement for the
Screening of Diabetic Neuropathy in Primary Care. Decoding Study (Dermal Electrochemical
Conductance in Diabetic Neuropathy)
Reprinted from: *J. Clin. Med.* **2019**, *8*, 598, doi:10.3390/jcm8050598 . . . . **71**

Javier Donate-Correa, Ernesto Martín-Núñez, Carla Ferri, Carolina Hernández-Carballo, V
íctor G. Tagua, Alejandro Delgado-Molinos, Ángel López-Castillo,
Sergio Rodríguez-Ramos, Purificación Cerro-López, Victoria Castro López-Tarruella,
Miguel Angel Arévalo-González, Nayra Pérez-Delgado, Carmen Mora-Fernández and
Juan F. Navarro-González
FGF23 and Klotho Levels are Independently Associated with Diabetic Foot Syndrome in Type
2 Diabetes Mellitus
Reprinted from: *J. Clin. Med.* **2019**, *8*, 448, doi:10.3390/jcm8040448 . . . . **85**

Esmeralda Castelblanco, Àngels Betriu, Marta Hernández, Minerva Granado-Casas,
Emilio Ortega, Berta Soldevila, Anna Ramírez-Morros, Josep Franch-Nadal,
Manel Puig-Domingo, Elvira Fernández, Angelo Avogaro, N úria Alonso and Dídac Mauricio
Ultrasound Tissue Characterization of Carotid Plaques Differs Between Patients with Type 1
Diabetes and Subjects without Diabetes
Reprinted from: *J. Clin. Med.* **2019**, *8*, 424, doi:10.3390/jcm8040424 . . . . **99**

Minerva Granado-Casas, Esmeralda Castelblanco, Anna Ramírez-Morros, Mariona Martín,
Nuria Alcubierre, Montserrat Martínez-Alonso, Xavier Valldeperas, Alicia Traveset,
Esther Rubinat, Ana Lucas-Martin, Marta Hernández, Núria Alonso and Didac Mauricio
Poorer Quality of Life and Treatment Satisfaction is Associated with Diabetic Retinopathy in
Patients with Type 1 Diabetes without Other Advanced Late Complications
Reprinted from: *J. Clin. Med.* **2019**, *8*, 377, doi:10.3390/jcm8030377 . . . . **111**

Noriko Ogama, Takashi Sakurai, Shuji Kawashima, Takahisa Tanikawa, Haruhiko Tokuda,
Shosuke Satake, Hisayuki Miura, Atsuya Shimizu, Manabu Kokubo, Shumpei Niida,
Kenji Toba, Hiroyuki Umegaki and Masafumi Kuzuya
Association of Glucose Fluctuations with Sarcopenia in Older Adults with Type 2
Diabetes Mellitus
Reprinted from: *J. Clin. Med.* **2019**, *8*, 319, doi:10.3390/jcm8030319 . . . . **125**

Olga Simó-Servat, Andreea Ciudin, Ángel M. Ortiz-Z úñiga, Cristina Hernández
and Rafael Simó
Usefulness of Eye Fixation Assessment for Identifying Type 2 Diabetic Subjects at Risk
of Dementia
Reprinted from: *J. Clin. Med.* **2019**, *8*, 59, doi:10.3390/jcm8010059 . . . . **141**

Young-Gun Kim, Ja Young Jeon, Hae Jin Kim, Dae Jung Kim, Kwan-Woo Lee,
So Young Moon and Seung Jin Han
Risk of Dementia in Older Patients with Type 2 Diabetes on Dipeptidyl-Peptidase IV Inhibitors
Versus Sulfonylureas: A Real-World Population-Based Cohort Study
Reprinted from: *J. Clin. Med.* **2019**, *8*, 28, doi:10.3390/jcm8010028 . . . . **153**

Pin-Pin Wu, Chew-Teng Kor, Ming-Chia Hsieh and Yao-Peng Hsieh
Association between End-Stage Renal Disease and Incident Diabetes Mellitus—A Nationwide
Population-Based Cohort Study
Reprinted from: *J. Clin. Med.* **2018**, *7*, 343, doi:10.3390/jcm7100343 . . . . **163**

Clara García-Carro, Ander Vergara, Irene Agraz, Conxita Jacobs-Cachá, Eugenia Espinel,
Daniel Seron and María José Soler
The New Era for Reno-Cardiovascular Treatment in Type 2 Diabetes
Reprinted from: *J. Clin. Med.* **2019**, *8*, 864, doi:10.3390/jcm8060864 . . . . **177**

About the Editor

Didac Mauricio, MD Ph.D., is the Director of the Department of Endocrinology & Nutrition, Hospital de la Santa Creu i Sant Pau in Barcelona, and an Associate Professor with the Faculty of Medicine, University of Vic (UVic/UCC). He is leading the Diabetes Research Group at this institution, that belongs to the excellence diabetes research network CIBER of Diabetes and Associated Metabolic diseases (CIBERDEM), Instituto de Salud Carlos III in Spain. He has published over 230 peer-reviewed articles and has contributed to multiple books. He has received the highest awards from the Spanish Society of Diabetes and the Spanish Society of Endocrinology & Nutrition. He has also served as a member of several local and international committees. D. Mauricio has been theprincipal investigator of several research projects funded by national and international agencies.

Journal of
Clinical Medicine

Editorial

Comments from the Editor of the Special Issue "Clinical Research on Diabetic Complications"

Didac Mauricio

Department of Endocrinology & Nutrition, CIBER of Diabetes and Associated Metabolic Diseases, Hospital de la Santa Creu i Sant Pau, Autonomous University of Barcelona, 08041 Barcelona, Spain; didacmauricio@gmail.com

Received: 4 December 2019; Accepted: 9 December 2019; Published: 12 December 2019

Abstract: With this Editorial, we are hereby presenting to the reader the Special Issue on "Clinical Research on Diabetic Complications". Chronic complications of diabetes mellitus have a major impact on the life of subjects with the disease, resulting in decreased quality of life and increased morbidity and mortality. This Special Issue includes contributions addressing different clinical aspects of the natural history, prevention and prediction, and characterization and management of diabetes-related complications.

Keywords: diabetic macroangiopathy; cardiovascular disease; heart disease; cerebrovascular disease; peripheral artery disease; diabetic foot disease; diabetic microangiopathy; diabetic retinopathy; diabetic kidney disease; diabetic neuropathy

---

Diabetes mellitus is associated with the development of chronic complications as a result of long-term exposure to hyperglycemia [1]. Diabetes-related complications are roughly divided between those affecting macrovessels (macroangiopathy or macrovascular complications), and those affecting microvessels (microangiopathy or microvascular complications). However, we currently know that we should have a more holistic view, beyond the traditional classification into micro- and macrovascular complications, as diabetes may affect any specific tissue or cell type. We have several articles of the latter concept in this Special Issue (e.g., diabetes-related cognitive impairment). The great burden of diabetes-related complications entails preventing or delaying its appearance as a main goal of diabetes management [2].

We have several articles dealing with diabetic macroangiopathy. One of the main pieces of research of this Special Issue, an article by Lin et al., deals with a very relevant topic in the field of cardiovascular disease prevention in diabetes. The authors performed a systematic literature search and a meta-analysis of randomized controlled trials of aspirin use for primary prevention of cardiovascular disease that points to a favorable balance between the risks and benefit of the use of this preventive treatment in individuals with diabetes. In another paper, Castelblanco et al. performed a study using ultrasound tissue characterization of carotid atherosclerotic plaques in which they found that plaques from type 1 diabetic subjects have a different pattern, with increased calcium plaque content, from that of non-diabetic individuals. Further, in another study by the same group, the circulating concentrations of soluble CD36, a potential cardiovascular risk biomarker, was only weakly associated with type 2 diabetes but not with type 1 diabetes. In addition, the work of Li et al. found that the increased arterial stiffness in prediabetes is mainly associated to postprandial glucose (i.e., subjects with impaired glucose tolerance). The proof-of-concept study of Donate-Correa and colleagues showed the potential involvement of the FGF23/Klotho system on the pathogenesis of diabetic foot disease. Finally, regarding diabetes-related cardiovascular disease, in a review paper García-Carro and colleagues wrote an overview of the current strategies and mechanisms of therapies that have demonstrated reno- and cardioprotective effects.

Two original articles contribute to greater insight in the field of diabetic microangiopathic complications. First, Cabré et al. showed that the measurement of dermal electrochemical conductance may be a useful tool to screen diabetic peripheral neuropathy. Further, a paper by Granado-Casas et al. showed that in subjects with type 1 diabetes, those with retinopathy had a poorer quality of life in the absence of other major diabetic complications.

As pointed out above, diabetes may have a deleterious effect on any cell type. A clear example of this is the impact of diabetes on cognitive function, indicating the metabolic damage on the central nervous system. We have several original articles addressing this issue. In a study by Simó-Servat et al., the authors found gaze fixation abnormalities in subjects with type 2 diabetes (T2D) associated with cognitive status; they concluded that microperimetry to measure parameters of fixation may be a method to detect prodromal stages of dementia. Further, Ogama et al. found that subjects with type 2 diabetes and sarcopenia, among those with cognitive impairment, showed higher glucose variability. Finally, Kim and coworkers provide us with a very interesting piece of work where they describe that the use of dipeptidyl peptidase-IV inhibitors, compared to sulfonylureas, in elderly subjects with type 2 diabetes is associated with a lower risk of dementia in real-world settings.

The rest of the articles of this Special Issue also address relevant questions. A pilot study by Ciudin et al. show the potential of a genetic tool to predict the response to bariatric surgery of subjects with type 2 diabetes. In search of new biomarkers, Del Coco et al. describe a new metabolic serum signature associated with the progression and burden of complications of type 2 diabetes. Finally, Wu et al. use a nationwide database to show that individuals treated with hemodialysis from Taiwan hold a lower risk of developing type 2 diabetes.

To conclude, we should underline that we are far from the optimal knowledge of diabetes-related complications. We need much more research efforts to provide clinicians with new tools to prevent and manage chronic diabetic complications.

Conflicts of Interest: The author declares no conflicts of interest.

References

1. Zheng, Y.; Ley, S.H.; Hu, F.B. Global aetiology and epidemiology of type 2 diabetes mellitus and its complications. *Nat. Rev. Endocrinol.* **2018**, *14*, 88–98. [CrossRef]
2. Gregg, E.W.; Sattar, N.; Ali, M.K. The changing face of diabetes complications. *Lancet Diabetes Endocrinol.* **2016**, *4*, 537–547. [CrossRef]

Journal of
Clinical Medicine

Article

Genetic Testing to Predict Weight Loss and Diabetes Remission and Long-Term Sustainability after Bariatric Surgery: A Pilot Study

Andreea Ciudin [1,2,*,†], Enzamaria Fidilio [1,†], Angel Ortiz [1], Sara Pich [3], Eduardo Salas [3], Jordi Mesa [1,2], Cristina Hernández [1,2], Olga Simó-Servat [1,2], Albert Lecube [2,4] and Rafael Simó [1,2,*]

[1] Institut de Recerca Vall d'Hebron, Universitat Autònoma de Barcelona (VHIR-UAB), 08035 Barcelona, Spain
[2] CIBER de Diabetes y Enfermedades Metabólicas Asociadas, Instituto de Salud Carlos III, 08950 Barcelona, Spain
[3] Scientific Department, Gendiag.exe, Joan XXIII, 10, Esplugues de LLobregat, 08950 Barcelona, Spain
[4] Endocrinology and Nutrition Department, Hospital Universitari Arnau de Vilanova, IRBLleida, Universitat de Lleida, 25198 Lleida, Spain
* Correspondence: aciudin@vhebron.net (A.C.); rafael.simo@vhir.org (R.S.); Tel.: +34-934894172 (A.C. & R.S.); Fax: +34-934894032 (A.C. & R.S.)
† Theses authors have contributed equally to this work.

Received: 26 June 2019; Accepted: 28 June 2019; Published: 3 July 2019

Abstract: Introduction: The aim of this pilot study was to assess genetic predisposition risk scores (GPS) in type 2 diabetic and non-diabetic patients in order to predict the better response to bariatric surgery (BS) in terms of either weight loss or diabetes remission. Research Design and Methods: A case-control study in which 96 females (47 with type 2 diabetes) underwent Roux-en-Y gastric by-pass were included. The DNA was extracted from saliva samples and SNPs were examined and grouped into 3 GPS. ROC curves were used to calculate sensitivity and specificity. Results: A highly sensitive and specific predictive model of response to BS was obtained by combining the GPS in non-diabetic subjects. This combination was different in diabetic subjects and highly predictive of diabetes remission. Additionally, the model was able to predict the weight regain and type 2 diabetes relapse after 5 years' follow-up. Conclusions: Genetic testing is a simple, reliable and useful tool for implementing personalized medicine in type 2 diabetic patients requiring BS.

Keywords: diabetes; obesity; bariatric surgery

---

1. Introduction

Obesity represents a major public health problem and it is associated with a significant economic burden on the health systems of developed countries, mainly due to the associated co-morbidities. Among these co-morbidities, type 2 diabetes (T2D) is one of the most important.

Bariatric surgery (BS) is a successful treatment for morbid obesity and leads to a dramatic improvement in obesity-related comorbidities [1]. The remission rate of T2D after BS is around 60–70% after 1 year of follow-up [2]. Therefore, there is a significant proportion of non-responders to BS in terms of diabetes remission. Additionally, after 5 years, there is about a 20–35% relapse of T2D after Y-de-Roux gastric by-pass (RYGB) [3–5]. A score based on clinical variables for the pre-operative prediction of T2D remission following RYGB surgery (DiaRem) was proposed [6]. However, this model has several limiting factors [7] and it has not been generally adopted in clinical practice. At present, there are no reliable predictors of T2D remission and relapse after BS.

In recent years, interest in the genetic influence on the response of different treatments for obesity has increased. Two retrospective studies [8,9] showed that several single nucleotide polymorphisms

(SNPs) were associated with a poor response to BS. However, in these studies the discrimination capacity of the GPS was not significant, and the role of T2D in the response to BS and the impact of these genetic factors on diabetes remission were not evaluated.

On this basis, the aim of the present study was to evaluate whether genetic markers can be used for the prediction of adequate weight loss and diabetes remission after BS.

2. Material and Methods

A single-center, retrospective observational pilot study in a third-level university hospital (Vall d´Hebron University Hospital, Barcelona, Spain) was conducted following the Strengthening the Reporting of Observational Studies in Epidemiology guidelines. The study comprised patients that underwent RYBG surgery between January 2010 and December 2012. The inclusion criteria were women, stable weight in the prior 6 months before BS, and minimum of 5 years of follow-up after BS. In order to avoid heterogeneity and given that the vast majority of the patients under bariatric surgery were women, we decided to rule out the inclusion of men in this pilot study. The patients were informed about the study and they all signed the written informed consent form.

The exclusion criteria were male, marked mobility problems, a different BS technique apart from RYBG, and severe psychiatric or eating disorders. For the genetic study, a sample of saliva was collected. The characteristics of RYBG were food loop length: 150–180 cm, and bilio-pancreatic loop length: 120 cm, gastric pouch 30 cc^3. The technique was the same in all cases, performed by the same surgical team in our hospital.

Excess body weight (EBW) was defined as the amount of weight that was in excess of the ideal body weight (IBW). The percentage of excess weight loss (EWL) was calculated according to the formula: %EWL = (weight before BS (kg) − weight after BS (kg)/EBW(kg)) × 100. The post-BMI weight regain was defined as a 10% regain of the minimal weight after BS. The minimal weight after BMI was achieved at 2 years follow-up for all of the patients.

Diabetes remission was defined according to American Diabetes Association (ADA) criteria [10]. Relapse of T2D was defined as one or more of the following conditions: (a) restarting diabetes medication; (b) one or more HbA1c measures ≥ 6.5%; and/or (c) one or more fasting glucose measures ≥ 126 mg/dL [11].

The study was approved by the Local Ethics Committee and registered at Clinical.Trials.gov, NCT02405949.

2.1. Genotyping and Sequencing

The DNA was extracted from saliva samples and processed by GoldenGate® Genotyping Assay for VeraCode. The genetic predisposition was assessed using Nutri inCode (NiC) (Ferrer inCode) and selecting the 57 SNPs associated with susceptibility to diabetes, obesity, appetite regulation, weight loss in response to hypocaloric diet, and the response to BS. The details about the SNPs are reflected in the Supplementary Materials. The selected SNPs were grouped into three genetic predisposition risk scores (GPS): diabetes remission, weight loss in non-diabetic subjects, and weight loss in subjects with diabetes.

2.2. Statistical Analysis

In order to assess the best predictive GPS, patients were distributed into 4 subgroups according to the BS response (%EWL) and the presence of T2D: (1) %EWL < 40% without diabetes (n = 15); (2) %EWL < 40% with diabetes (n = 16); (3) %EWL > 75% without diabetes (n = 35); and (4) T2D and %EWL > 75% with diabetes (n = 31). Univariate and multivariate logistic regressions were used to establish associations. Akaike Information Criterion (AIC)-based backward selection was used to remove insignificant terms from an initial model containing all the candidate predictors. The calibration of the model's adequacy was determined by the Hosmer–Lemeshow test. The area under the ROC curve (AUROC) was used for evaluating the prediction performance of the models. The cut-offs for the developed algorithms were selected as the point which maximizes the Youden index.

3. Results

The clinical characteristics of the patients included in the study are shown in Table 1. Apart from age, we did not find any significant differences between diabetic and non-diabetic subjects before BS. The diabetic treatment received by subjects with diabetes is displayed in Figure 1. No other medication apart from AINEs occasionally and vitamin supplements as per protocol after bariatric surgery (ciancobalamin 1000 mcg/month, colecalciferol 25.000–100.000 UI/month) were administered.

Table 1. Baseline characteristics of the patients included in the study.

	Non-Diabetic Patients	Type 2 Diabetic Patients	p
N	50	47	
Age (years)	48.0 (37.5; 55.0)	52.0 (46.0; 58.8)	0.0016
Initial BMI (Kg/m^2)	45.2 (43.0; 48.5)	42.5 (40.1; 46.4)	0.008
2 y post-BS BMI (Kg/m^2)	31.8 (26.1; 35.6)	30.9 (26.8; 35.7)	n.s.
5 y post-BS BMI (Kg/m^2)	32.63 (21; 52.14)	33.68 (21; 46.43)	n.s.
Hypertension (%)	48.3	49.5	n.s.
Dyslipidemia (%)	43.2	45.7	n.s.
Sleep apnea (%)	27.2	29.7	n.s.

In the subgroup of the non-diabetic patients, the multivariate logistic regression equation for predicting positive weight loss response (%EWL > 75%) after the BS (NiC-Bariatric-ND) includes SNPs associated with weight loss in response to a hypocaloric diet and SNPs associated to appetite regulation. The model showed an AUROC of 0.763 (95% CI 0.605 to 0.920; $p < 0.001$), a sensitivity of 86.49%, and a specificity of 57.14%. The calibration of the adequacy of the model determined by the Hosmer–Lemeshow test was 0.679.

The continuous variables were median (1st quartile; 3rd quartile) and the categorical data were percentages. BMI: body mass index. EWL: excess of weight loss. BS: bariatric surgery. Hypertension was defined by increased systolic ($\geq$140 mmHg) or increased diastolic ($\geq$90 mmHg) blood pressure or by the use of antihypertensive drugs, according to current guidelines. Dyslipidemia was defined by the use of lipid-lowering drugs, decreased values of HDL cholesterol (men < 0.9 mmol/L, women < 1.0 mmol/L) or by at least one increased value of total cholesterol (>5.2 mmol/L), LDL cholesterol or triglycerides (>1.7 mmol/L).

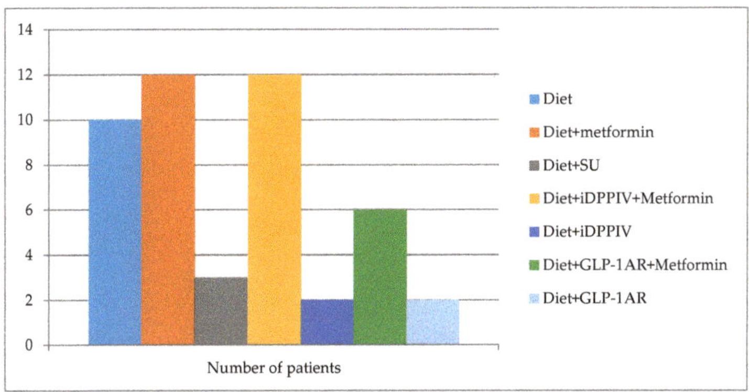

Figure 1. The diabetic treatment received before BS by the subjects with diabetes included in the study. SU: sulphonylurea, iDPPIV: DPPIV enzyme inhibitor, GLP-1AR: GLP-1 receptor agonists.

Weight regain after 5 years' follow-up was seen in 9.6% of the patients. The model to identify the patients who had presented weight regain after 5 years' follow-up showed an AUROC of 0.834 (95% CI 0.705 to 0.923; $p < 0.0001$), a sensitivity of 100%, and a specificity of 70.21%. The calibration of the adequacy of the model determined by the Hosmer–Lemeshow test was 0.5148.

In T2D patients, the multivariate logistic regression equation for the prediction of weight loss response (%EWL > 75%) after BS (NiC-Bariatric-D) included SNPs associated with weight loss in response to hypocaloric diet, SNPs associated to response to BS [9], and SNPs associated to response to lifestyle interventions [11]. The model showed an AUROC of 0.929 (95% CI 0.850 to 0.99; $p < 0.001$), a sensitivity of 87.10% and a specificity of 93.33%. The calibration of the model's adequacy determined by the Hosmer–Lemeshow test was 0.291. Weight regain in subjects with diabetes was observed in 17.5% of them. The model to identify patients with diabetes who will regain weight after a follow-up of 5 years after bariatric surgery showed in this case an AUROC of 0.781 (95% CI 0.623 to 0.896; $p < 0.04$), a sensitivity of 71.43%, and a specificity of 84.85%. The calibration of the model's adequacy determined by the Hosmer–Lemeshow test was 0.8664. Figure 2 shows the AUROC corresponding to weight regain in the whole (Figure 2A) population, non-diabetic subjects (Figure 2B), and T2D patients (Figure 2C).

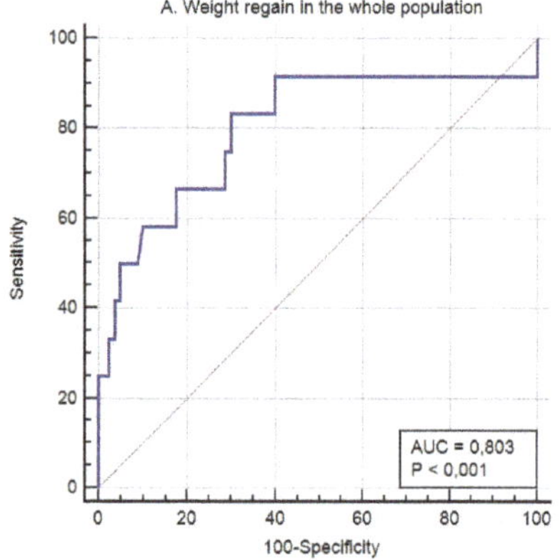

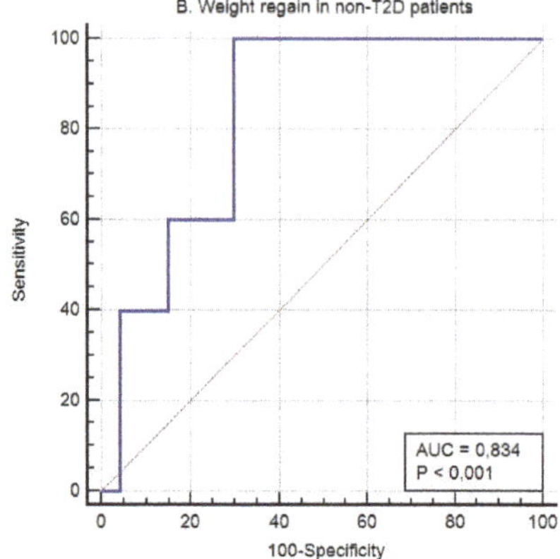

Figure 2. *Cont.*

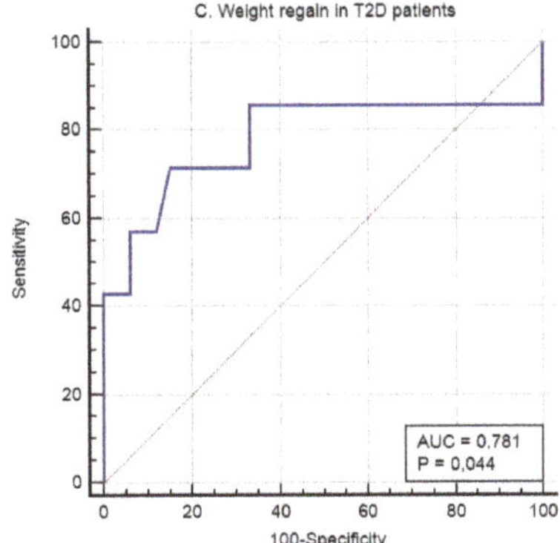

Figure 2. The predictive capacity of the genetic score for weight regain after 5 years' follow-up in the whole (**A**) population, non-T2D subjects (**B**), and T2D patients (**C**).

Diabetes remission was seen in 73.91% of the type 2 diabetic patients included in the study (66.67% in the group of %EWL < 40% and 77.42% in the group of %EWL > 75%). Diabetes relapse was seen in 25% of the patients.

The multivariate logistic regression equation for the prediction of diabetes remission and relapse after BS (NiC-Bariatric-DR) included SNPs associated with obesity, SNPs associated with weight loss in response to hypocaloric diet, SNPs associated with appetite regulation, and SNPs associated with genetic predisposition to diabetes. This prediction model showed an AUROC of 0.868 (95% CI 0.709 to 0.976; $p < 0.0001$) for diabetes remission, with a sensitivity of 76.47% and a specificity of 83.33%. In our population, the AEROC for DiaRem was lower than obtained by genetic testing, (0.69 versus 0.86), and when both scores were combined, the AUROC was 0.87, with a sensitivity of 88.49% and a specificity of 80% (Figure 3).

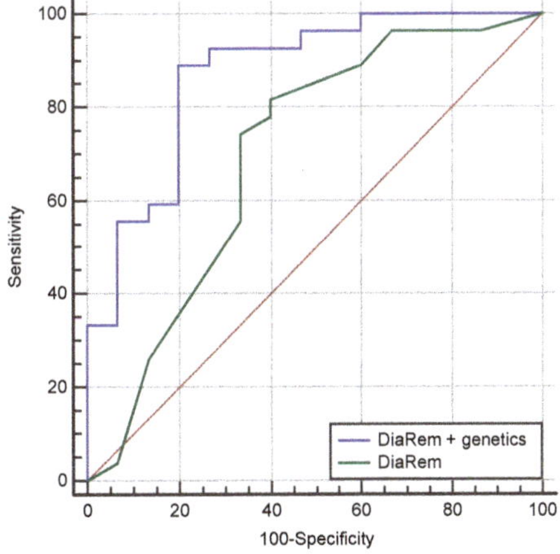

Figure 3. The predictive capacity of the DiARem score and the combination between DiARem and genetics in our study population. The AUROC for DiaRem was lower than obtained by genetic test (0.69 versus 0.86), and when both scores were combined the AUCROC was 0.87, with a sensitivity of 88.49% and a specificity of 80.00%.

Regarding diabetes relapse after 5 years, the model based showed an AUROC of 0.833 (95% CI 0.682 to 0.932; $p < 0.0001$), with a sensitivity of 90.00 and a specificity of 80.00 (Figure 4). The calibration of the adequacy of the model determined by the Hosmer–Lemeshow test was 0.280.

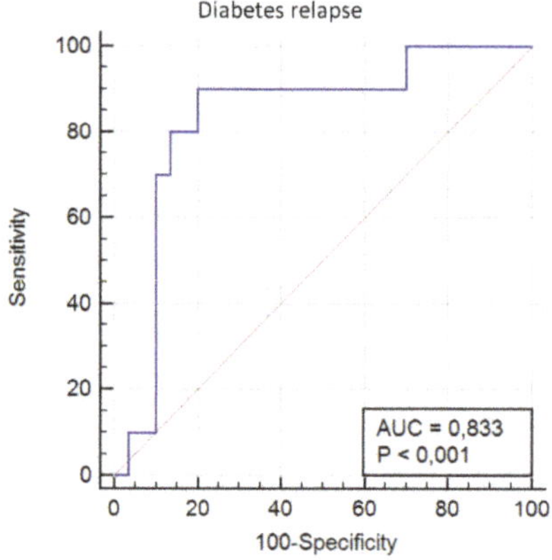

Figure 4. The predictive capacity of the genetic score for T2D relapse after 5 years' follow-up.

4. Discussion

Bariatric surgery provides adequate and sustainable weight loss and T2D remission, but 15–20% of the subjects do not reach these targets [12]. A recent study [13] showed a high inter-individual variability of the EWL response at mid-term after BS and that poor EWL could be illustrated by two different patterns: poor sustained weight loss or pronounced weight regain. At present, there are no reliable biomarkers for individual response to BS. Due to the increasing availability of BS around the world and the alarming prevalence of obesity and its associated co-morbidities such as T2D, the discovery of biomarkers that will permit us to identify the best candidates for BS are urgently needed.

In the present study, we developed genetic-based algorithms for the prediction of %EWL after BS and for T2D remission with high sensitivity and specificity. Still et al. [8] proposed a genetic score to predict the %EWL after BS, showing a non-statistically significant AUROC, a sensitivity of 48.39, and a specificity of 73.33. In addition, this score did not take into the account the presence of diabetes.

Regarding diabetes remission after BS, we analyzed for comparison purposes the predictive capacity of DiARem scores [8] in our study population. In our population, the AUROC for DiARem was lower than obtained by genetic test (0.69 versus 0.86), and when both scores were combined, the AUCROC was 0.87, with a sensitivity of 88.49% and a specificity of 80.00%. This finding supports the use of genetic testing in clinical practice.

It is worth mentioning that in diabetic patients, the rate of remission was not significantly different between the group with %EWL < 40% and the group with %EWL > 75% ($p = 0.674$). In addition, previous data showed that about 30% of T2D patients that are able to discontinue the medication after BS will present a relapse within the first 5 years [3–5]. Some studies found weak correlation between weight regain, younger age or lower BMI before BS as predictors of T2D relapse after BS [5,14], while other studies found no association [3]. Therefore, at present, there are no reliable predictors of T2D relapse after BS. In our study, the proposed score showed a high predictive value of T2D relapse after BS, thus, underlying the potential key role of genetic testing in precision medicine in order to assure better outcomes after BS. Interestingly, in our study, in the subgroup of T2D patients, the inclusion of SNPs associated to response to BS did not improve the prediction scores, suggesting that these genes are not critical or do not intervene in the remission and relapse of T2D. This finding suggests that the physiopathology of diabetes remission and relapse after RYGB might not be related with the %EWL in this population.

Overall, these results are intriguing and point to a genuine genetic background in the mechanisms involved in diabetes remission and relapse after BS, perhaps related to insulin resistance.

In conclusion, in this pilot study we have developed highly sensitive and specific genetic predictive scores of responses to BS in terms of weight loss and T2D remission and the long-term sustainability of these effects. These results would allow us not only to implement a more effective and personalized BS, but also to optimize healthcare resources. However, further studies with a larger sample size to confirm this pilot study are needed.

Supplementary Materials: The supplementary materials are available online at http://www.mdpi.com/2077-0383/8/7/964/s1.

Author Contributions: A.C. and E.F. contributed to the study design, obtained study data, assisted in data analysis, and drafted the initial manuscript. O.S. and A.O. obtained study data. S.P. and E.S. performed genetic and statistical analyses. J.M., C.H. and A.L. made substantial contributions to the acquisition, analysis and interpretation of data, and critically reviewed the manuscript. R.S. contributed to the study design, discussion, reviewed the manuscript, supervised the project, and is the guarantor of this work, and, as such, had full access to all the data in the study and takes responsibility for the integrity of the data and the accuracy of the data analysis. All authors approved the final version of the manuscript.

Funding: This study was supported by grants from the "Pla Estratègic de Recerca i Innovació en Salut" (PERIS) 2016-2020 (SLT002/16/00497), the Instituto de Salud Carlos III (Fondo de Investigación sanitaria, PI 18/00964), and the European Union (European Regional Development Fund). O.S. is a recipient of a "Río Hortega" fellowship grant from the Instituto de Salud Carlos III.

Conflicts of Interest: No potential conflict of interest relevant to this article were reported.

References

1. Rubino, F.; Kaplan, L.M.; Schauer, P.R.; Cummings, D.E. Diabetes Surgery Summit Delegates. The Diabetes Surgery Summit consensus conference: Recommendations for the evaluation and use of gastrointestinal surgery to treat type 2 diabetes mellitus. *Ann. Surg.* **2010**, *251*, 399–405. [CrossRef] [PubMed]

2. Esposito, K.; Maiorino, M.I.; Petrizzo, M.; Bellastella, G.; Giugliano, D. Remission of type 2 diabetes: Is bariatric surgery ready for prime time? *Endocrine* **2015**, *48*, 417–421. [CrossRef] [PubMed]

3. Schauer, P.R.; Bhatt, D.L.; Kirwan, J.P.; Wolski, K.; Aminian, A.; Brethauer, S.A.; Navaneethan, S.D.; Singh, R.P.; Pothier, C.E.; Nissen, S.E.; et al. Bariatric Surgery versus Intensive Medical Therapy for Diabetes-5-Year Outcomes. *N. Engl. J. Med.* **2017**, *376*, 641–651. [CrossRef] [PubMed]

4. Arterburn, D.E.; Bogart, A.; Sherwood, N.E.; Sidney, S.; Coleman, K.J.; Haneuse, S.; O'Connor, P.J.; Theis, M.K.; Campos, G.M.; McCulloch, D.; et al. A multisite study of long-term remission and relapse of type 2 diabetes mellitus following gastric bypass. *Obes. Surg.* **2013**, *23*, 93–102. [CrossRef] [PubMed]

5. DiGiorgi, M.; Rosen, D.J.; Choi, J.J.; Milone, L.; Schrope, B.; Olivero-Rivera, L.; Restuccia, N.; Yuen, S.; Fisk, M.; Inabnet, W.B.; et al. Re-emergence of diabetes after gastric bypass in patients with mid- to long-term follow-up. *Surg. Obes. Relat. Dis.* **2010**, *6*, 249–253. [CrossRef] [PubMed]

6. Still, C.D.; Wood, G.C.; Benotti, P.; Petrick, A.T.; Gabrielsen, J.; Strodel, W.E.; Ibele, A.; Seiler, J.; Irving, B.A.; Celaya, M.P.; et al. Preoperative prediction of type 2 diabetes remission after Roux-en-Y gastric bypass surgery: A retrospective cohort study. *Lancet Diabetes Endocrinol.* **2014**, *2*, 38–45. [CrossRef]

7. Aminian, A.; Brethauer, S.A.; Kashyap, S.R.; Kirwan, J.P.; Schauer, P.R. DiaRem score: External validation. *Lancet Diabetes Endocrinol.* **2014**, *2*, 12–13. [CrossRef]

8. Still, C.D.; Wood, G.C.; Chu, X.; Erdman, R.; Manney, C.H.; Benotti, P.N.; Petrick, A.T.; Strodel, W.E.; Mirshahi, U.L.; Mirshahi, T.; et al. High allelic burden of four obesity SNPs is associated with poorer weight loss outcomes following gastric bypass surgery. *Obesity* **2011**, *19*, 1676–1683. [CrossRef] [PubMed]

9. Mirshahi, U.L.; Still, C.D.; Masker, K.K.; Gerhard, G.S.; Carey, D.J.; Mirshahi, T. The MC4R(I251L) allele is associated with better metabolic status and more weight loss after gastric bypass surgery. *J. Clin. Endocrinol. Metab.* **2011**, *96*, E2088–E2096. [CrossRef] [PubMed]

10. American Diabetes Association. Standards of Medical Care in Diabetes 2012. *Diabetes Care* **2012**, *35*, S11–S63. [CrossRef] [PubMed]

11. Buse, J.B.; Caprio, S.; Cefalu, W.T.; Ceriello, A.; del Prato, S.; Inzucchi, S.E.; McLaughlin, S.; Phillips, G.L., II; Robertson, R.P.; Rubino, F.; et al. How do we define cure of diabetes? *Diabetes Care* **2009**, *32*, 2133–2135. [CrossRef] [PubMed]
12. Maggard, M.A.; Shugarman, L.R.; Suttorp, M.; Maglione, M.; Sugerman, H.J.; Sugarman, H.J.; Livingston, E.H.; Nguyen, N.T.; Li, Z.; Mojica, W.A.; et al. Meta-analysis: Surgical treatment of obesity. *Ann. Intern. Med.* **2005**, *142*, 547–559. [CrossRef] [PubMed]
13. de Hollanda, A.; Ruiz, T.; Jiménez, A.; Flores, L.; Lacy, A.; Vidal, J. Patterns of Weight Loss Response Following Gastric Bypass and Sleeve Gastrectomy. *Obes. Surg.* **2015**, *25*, 1177–1183. [CrossRef] [PubMed]
14. Chikunguwo, S.M.; Wolfe, L.G.; Dodson, P.; Meador, J.G.; Baugh, N.; Clore, J.N.; Kellum, J.M.; Maher, J.W. Analysis of factors associated with durable remission of diabetes after Roux-en-Y gastric bypass. *Surg. Obes. Relat. Dis.* **2010**, *6*, 254–259. [CrossRef] [PubMed]

Journal of
Clinical Medicine

Article

NMR-Based Metabolomic Approach Tracks Potential Serum Biomarkers of Disease Progression in Patients with Type 2 Diabetes Mellitus

Laura Del Coco [1], Daniele Vergara [1], Serena De Matteis [2], Emanuela Mensà [3], Jacopo Sabbatinelli [3], Francesco Prattichizzo [4], Anna Rita Bonfigli [5], Gianluca Storci [6], Sara Bravaccini [2], Francesca Pirini [2], Andrea Ragusa [1], Andrea Casadei-Gardini [7], Massimiliano Bonafè [6], Michele Maffia [1], Francesco Paolo Fanizzi [1,*], Fabiola Olivieri [3,8,*] and Anna Maria Giudetti [1]

[1] Department of Biological and Environmental Sciences and Technologies, University of Salento, via Monteroni, 73100 Lecce, Italy; laura.delcoco@unisalento.it (L.D.C.); daniele.vergara@unisalento.it (D.V.); andrea.ragusa@unisalento.it (A.R.); michele.maffia@unisalento.it (M.M.); anna.giudetti@unisalento.it (A.M.G.)

[2] Biosciences Laboratory, Istituto Scientifico Romagnolo per lo Studio e la Cura dei Tumori (IRST) IRCCS, 47014 Meldola, Italy; serena.dematteis@irst.emr.it (S.D.M.); sara.bravaccini@irst.emr.it (S.B.); francesca.pirini@irst.emr.it (F.P.)

[3] Department of Clinical and Molecular Sciences, DISCLIMO, Università Politecnica delle Marche, 60121 Ancona, Italy; emanuela.mensa@gmail.com (E.M.); jacopo.sabbatinelli@gmail.com (J.S.)

[4] IRCCS MultiMedica, 20099 Milan, Italy; francesco.prattichizzo@multimedica.it

[5] Scientific Direction, IRCCS INRCA, 60121 Ancona, Italy; a.bonfigli@inrca.it

[6] Department of Experimental, Diagnostic and Specialty Medicine, University of Bologna, 40126 Bologna, Italy; gianluca.storci@unibo.it (G.S.); massimiliano.bonafe@unibo.it (M.B.)

[7] Division of Medical Oncology, Department of Medical and Surgical Sciences for Children and Adults, University Hospital of Modena, 41125 Modena, Italy; casadeigardini@gmail.com

[8] Center of Clinical Pathology and Innovative Therapy, IRCCS INRCA, 60121 Ancona, Italy

* Correspondence: fp.fanizzi@unisalento.it (F.P.F.); f.olivieri@univpm.it (F.O.)

Received: 26 April 2019; Accepted: 15 May 2019; Published: 21 May 2019

Abstract: Type 2 diabetes mellitus (T2DM) is a metabolic disorder characterized by chronic hyperglycemia associated with alterations in carbohydrate, lipid, and protein metabolism. The prognosis of T2DM patients is highly dependent on the development of complications, and therefore the identification of biomarkers of T2DM progression, with minimally invasive techniques, is a huge need. In the present study, we applied a ^{1}H-Nuclear Magnetic Resonance (^{1}H-NMR)-based metabolomic approach coupled with multivariate data analysis to identify serum metabolite profiles associated with T2DM development and progression. To perform this, we compared the serum metabolome of non-diabetic subjects, treatment-naïve non-complicated T2DM patients, and T2DM patients with complications in insulin monotherapy. Our analysis revealed a significant reduction of alanine, glutamine, glutamate, leucine, lysine, methionine, tyrosine, and phenylalanine in T2DM patients with respect to non-diabetic subjects. Moreover, isoleucine, leucine, lysine, tyrosine, and valine levels distinguished complicated patients from patients without complications. Overall, the metabolic pathway analysis suggested that branched-chain amino acid (BCAA) metabolism is significantly compromised in T2DM patients with complications, while perturbation in the metabolism of gluconeogenic amino acids other than BCAAs characterizes both early and advanced T2DM stages. In conclusion, we identified a metabolic serum signature associated with T2DM stages. These data could be integrated with clinical characteristics to build a composite T2DM/complications risk score to be validated in a prospective cohort.

Keywords: branched-chain amino acids; metabolomics; NMR spectroscopy; type 2 diabetes mellitus

---

1. Introduction

Diabetes mellitus (DM) is a metabolic disorder characterized by chronic hyperglycemia associated with impairments in carbohydrate, lipid, and protein metabolism [1]. DM is classified in two main categories: type 1, due to cellular-mediated autoimmune pancreatic islet β-cells destruction which occurs in 5–10% of cases, and type 2 (T2DM), due to insulin resistance (IR) with a defect in compensatory insulin secretion, which affects 90% of diabetic patients [2]. Environmental and lifestyle changes in association with populations aging account for the rapid global increase in T2DM prevalence and incidence in recent decades [3]. A comprehensive summary of factors contributing to T2DM risk includes not only obesity and related aspects of diet quality and quantity, but also sedentary lifestyle and lack of physical activity, exposure to noise or fine dust, short or disturbed sleep, smoking, stress, depression, and a low socioeconomic status [4]. Chronic hyperglycemia leads to many long-term complications, such as cardiovascular disease (CVD), cerebrovascular disease, peripheral vascular disease, neuropathy, retinopathy, and renal failure, resulting in increasing disability, reduced life expectancy, and increased health costs [5]. The prognosis of patients with T2DM is highly dependent on the development of complications; the prevalence of patients with cardiovascular complication is growing exponentially and most T2DM patients die as a result of cardiovascular causes.

Despite the recently introduced therapies to manage T2DM, the progressive nature of IR and the inability of β-cells to cope with increased insulin demand still obligate insulin therapy for selected clusters of patients to achieve and maintain adequate glycemic control [6–8].

Because of the risks and benefits of treatments, the care of patients with T2DM involves complex decision-making [9]. Therefore, the identification of biomarkers of metabolic worsening in T2DM is of clinical relevance.

Recently, several metabolomic techniques were applied to the identification of metabolic signatures associated with T2DM [10–17]. Data in this research field were extensive, revealing new diagnostic/prognostic disease biomarkers [14], or metabolic markers of response to targeted therapies [15,16]. Metabolomics also provided tools for patient's stratification and recognizable metabolic patterns associated with organ dysfunction [17]. These studies focused on the diagnostic value of metabolic signatures in prospective cohorts, while the metabolic comparison of T2DM patients at different stages of the disease was less investigated. To do this, we took advantage of a selected cohort of T2DM patients that we characterized in our previous works [18–23].

Here, we aimed to identify, by a ^{1}H-Nuclear Magnetic Resonance (^{1}H-NMR)-based metabolomic approach coupled with multivariate data analysis, a serum metabolic signature associated with T2DM development and progression. The study design is characterized by the comparison of opposite phenotypes: non-diabetic subjects and two subsets of T2DM patients—T2DM patients at an early stage of the disease (without complications and treatments) and at a late stage (with diabetic complications on insulin monotherapy).

2. Materials and Methods

2.1. Patient Samples

Enrolled patients were accurately selected from a large cohort of Italian T2DM patients and control subjects, recruited from the Italian National Research Center on Aging (INRCA), Ancona. All subjects provided written informed consent, which was approved by the INRCA's Ethics Committee. The inclusion criteria for T2DM patients and the clinical information collected from each subject were as described by Testa et al. [24].

We selected 26 T2DM patients and seven control subjects with comparable age, body mass index (BMI), lipid plasma profile, and gender distribution. All studied subjects consumed a Mediterranean diet. Subjects were considered as controls if at the time of blood collection they did not have T2DM and any major acute and/or chronic age-related diseases such as acute myocardial infarction, chronic heart failure, Alzheimer's disease, or cancer. Among the 26 T2DM patients, we selected 13 patients without complications and 13 patients with documented complications on insulin monotherapy at the time of blood collection. The presence/absence of diabetic complications was established as follows:

(1) retinopathy was defined as dilated pupils detected on funduscopic and/or fluorescence angiography;
(2) incipient nephropathy was a urinary albumin excretion rate >30 mg/24 h and normal creatinine clearance;
(3) chronic renal failure was defined as an estimated glomerular filtration rate <60 mL/min per 1.73 m^2, based upon the four-variable modification of diet in renal disease (MDRD) formula;
(4) neuropathy was established by electromyography;
(5) ischemic heart disease was diagnosed by clinical history and/or ischemic electrocardiographic alterations; these patients had had ST- or non-ST-elevation myocardial infarction, which was defined as a major adverse cardiac event (MACE);
(6) Peripheral vascular disease, including arteriosclerosis obliterans and cerebrovascular disease, was diagnosed based on history, physical examination, and Doppler imaging.

Each patient could be affected by more than one complication (Table 1).

Table 1. Number and type of complications in T2DM-C patients.

	Number
Median Age (Range)	64 (55–72)
Gender	
Male	7/13
Female	6/13
Complications	
Neuropathy	9/13
Nephropathy	7/13
Retinopathy	12/13
Chronic renal failure	4/13
Lower limb arteriopathy	5/13
Mace	5/13

Uncomplicated patients were recruited within one month of diagnosis and had not received any specific pharmacological treatment for diabetes at the time of the blood collection. Other drugs prescribed to diabetic patients are shown in the Table 2. To avoid possible bias due to different treatments, T2DM patients taking glucose-lowering drugs others from insulin were excluded.

Table 2. Drug treatment prescribed to T2DM patients.

	T2DM-NC	T2DM-C
ACE inhibitors	1	8
Diuretics	-	5
Vasodilators	-	4
Beta blockers	-	4
Antiarrhythmic drugs	-	2
Calcium channel blockers	-	3
Statins	1	1
Antiplatelet drugs	-	5
NSAIDs	-	7
Proton-pump inhibitors	1	2
CNS agents	2	2

Abbreviations: ACE = angiotensin-converting-enzyme; Abbreviations: NSAIDs = nonsteroidal anti-inflammatory drugs; CNS = central nervous system.

2.2. Laboratory Assays

Overnight fasting venous blood samples were collected from 08:00 to 10:00 h. Blood concentration of glycosylated hemoglobin (HbA1c) was measured by a G8 HPLC analyzer (TOSOH BIOSCIENCE, Tokyo, Japan). Plasminogen activator inhibitor-1 (PAI-1) antigen was quantified with an immune-enzymatic method (Biopool, Sweden). Total and high-density lipoprotein (HDL) cholesterol, triacylglycerols (TAG), fasting insulin, fasting glucose, fibrinogen, apolipoprotein AI and B (ApoAI and ApoB), and creatinine were measured using commercially available kits on an automated clinical chemistry COBAS Modular Platform (C module) analyzer (Roche-Hitachi, Basel, Switzerland). Highly sensitive C-reactive protein (hsCRP) was determined by the particle-enhanced immune-turbidimetric assay (CRP High Sensitive, Roche-Hitachi) on a COBAS analyzer. The homeostasis model assessment (HOMA) index was calculated as glucose (mg/100 mL)*insulin (uIU/mL)/405. BMI was calculated as weight divided by height squared (kg/m^2). α-Fucosidase and β-galactosidase were quantified as described by Spazzafumo et al. [25]. Insulin growth factor 1 (IGF1) level was quantified with a commercially available ELISA kit.

2.3. Sample Preparation and NMR Measurements

Serum samples were stored at a temperature of -80 °C until the NMR measurements were performed. Prior to NMR analysis, serum samples were thawed and an aliquot of 200 µL was mixed with 400 µL saline buffer solution (in 100% D2O containing TSP as a chemical shift reference, $\delta = 0$ ppm, NaCl 0.9%, 50 mM sodium phosphate buffer and pH 7.4) to minimize the pH variation and transferred in a 5 mm NMR tube [26]. All measurements were performed on a Bruker Avance III 600 Ascend NMR spectrometer (Bruker, Ettlingen, Germany), operating at 600.13 MHz for 1H observation, equipped with a TCI cryoprobe (Triple Resonance inverse Cryoprobe) incorporating a z-axis gradient coil and automatic tuning-matching (ATM). Experiments were acquired at 300 K in automation mode after loading individual samples on a Bruker Automatic Sample Changer, interfaced with the IconNMR software (Bruker). For each sample, two types of 1D ^{1}H-NMR experiments were recorded: a standard (ZGCPPR Bruker standard pulse sequence) spectrum, with pre-saturation and composite pulse for selection, and a Carr–Purcell–Meiboom–Gill (CMPG) spin-echo sequence, with 32 transients, 16 dummy scans, 5 s relaxation delay, size of FID (free induction decay) of 64 K data points, spectral width of 12,019.230 Hz (20.0276 ppm), an acquisition time of 1.36 s, a total spin-spin relaxation delay of 1.2 ms, and solvent signal saturation during the relaxation delay. The resulting FIDs were multiplied by an exponential weighting function corresponding to a line broadening of 0.3 Hz before Fourier transformation, automated phasing, and baseline correction. Moreover, peak assignments were carried out using 2D NMR experiments (^{1}H-^{1}H J-resolved, ^{1}H-^{1}H COSY, Correlation Spectroscopy, ^{1}H-^{13}C HSQC Heteronuclear Single Quantum Correlation, ^{1}H-^{13}C HMBC, Heteronuclear Multiple Bond Correlation) and by comparison with published data [26,27].

2.4. Metabolic Pathway Analysis

The most relevant metabolic pathways potentially involved in the metabolomic study were identified using MeTPA MetaboAnalyst software [28]. The purpose is to investigate if certain metabolic pathways are significantly different for the two groups of patients, when compared with control subjects and also whit each other. Metabolites of interest previously quantified by selected distinctive unbiased NMR signals were used as the input matrix for the metabolic pathway analysis. The pathway impact is calculated as the sum of the importance measures of the matched metabolites normalized by the sum of the importance measures of all metabolites in each pathway [29].

2.5. Statistical Analysis

All clinical parameters were computed with Excel (Microsoft 7) and presented as mean $\pm$ SD. The comparison among the data was made using one-way repeated measures ANOVA. Further comparisons were made by using paired-sample *t*-test. The SPSS/PC computer program (SPSS, Chicago, IL, USA) was used to perform all statistical analyses. Statistical significance was set at $p \leq 0.05$.

The metabolic profile of serum samples from controls and T2DM patients at different stages were analyzed by NMR. The ^{1}H-NMR Carr–Purcell–Meiboom–Gill (CPMG) spectra were processed using Topspin 3.5 and Amix 3.9.13 (Bruker, Biospin, Italy), both for simultaneous visual inspection and the successive bucketing process. The full NMR spectra (in the range 9.0–0.5 ppm) were segmented in fixed rectangular buckets of 0.04 ppm width and successively integrated. The spectral region between 5.10 and 4.7 ppm was discarded because of the residual peak of water. The total sum normalization was applied to minimize small differences due to sample concentration and/or experimental conditions among samples. The data set (bucket table) resulted in a matrix, made of 204 variables, corresponding to the bucketed ^{1}H-NMR spectra values (in columns), measured for each sample (in rows). Multivariate statistical analysis was performed using MetaboAnalyst software [28]. Unsupervised principal component analysis (PCA), and partial least squares/supervised orthogonal partial least squares discriminant analysis (PLSDA and OPLSDA, respectively) were applied to examine the intrinsic variation in the data, and also to screen out potential biomarkers [30]. In particular, OPLSDA analysis focuses the predictive information in one component, so that the first OPLS component shows the between-class difference. The remaining systematic information is transferred in higher components, thus facilitating interpretation. Two parameters, R^2 and Q^2, describe the goodness of the statistical models. The former (R^2) explains the total variations in the data, whereas the latter (Q^2, calculated via 10-fold cross-validation, CV) provides an estimate of the predictive ability of the models [31]. By ^{1}H-NMR spectroscopy, metabolites of interest were quantified by analyzing the integrals of selected distinctive unbiased NMR signals [32–34]. Results, represented as mean intensities and standard deviation of the selected NMR signals, were validated by one-way ANOVA with Tukey's honestly significant difference (HSD) post-hoc test. To better visualize data, a heatmap was performed on metabolites and samples, using Euclidean for distance measure and Ward for the clustering algorithm. Then, to identify the potential biomarkers associated with T2DM disease in patients with complications (T2DM-C), the receiver operating characteristic (ROC) curve was applied, using the Biomarker Analysis module of the MetaboAnalyst software. Multivariate ROC curve exploratory analysis was used to identify the promising biomarkers with high sensitivity and high specificity. The ROC curves were generated using Monte–Carlo cross validation (MCCV) algorithm and linear Support Vector Machines (SVM) clustering to evaluate the feature importance of the selected metabolites [35]. Both univariate and multivariate statistical analyses were performed using MetaboAnalyst software [28].

3. Results

3.1. Clinical Characteristics

Three different selected groups of subjects were included for this investigation: non-diabetic subjects, referred as the control group (CG) and two different groups of T2DM patients, with or without complications and insulin treatment. Patients with complications on insulin monotherapy were indicated as T2DM-C; patients without complications and treatments were indicated as T2DM-NC. Clinical characteristics of CG and T2DM groups are summarized in Tables 1 and 3. There was no significant difference in age (range 60–68 years) and BMI between groups. All subjects were overweight as confirmed by BMI values exceeding 25, with prevalence in android obesity as indicated by the waist to hip ratio (WHR). Fasting plasma glucose level was significantly higher in both T2DM groups compared to CG and significantly different ($p < 0.05$) between T2DM groups. In both T2DM groups, HbA1c and HOMA were higher than in CG, especially in T2DM-C ($p < 0.001$ and $p < 0.005$, respectively, T2DM-C vs. CG). It is important to note that insulin treatment holds plasma insulin levels to values not significantly different from that of other groups, so that metabolic changes we measured in our study cannot be related to defects in insulin secretion.

Table 3. Clinical information for control group (CG) and type 2 diabetes mellitus (T2DM) patients (T2DM-NC and T2DM-C) enrolled in the study. (T2DM-C: patients with complications on insulin monotherapy; T2DM-NC: without complications).

	CG	T2DM-NC	T2DM-C
Number of subjects (n)	7	13	13
Male gender (n, %)	4, 57.1%	8, 61.5%	7, 53.8%
Age (years)	63 ± 2	64 ± 3	64 ± 4
BMI (kg/m^2)	26.94 ± 3.05	28.78 ± 4.05	29.31 ± 3.82
WHR	0.90 ± 0.07	0.94 ± 0.06	0.94 ± 0.06
Fasting glucose (mg/dL)	93 ± 6.2	165 ± 59.4 [***,b]	247 ± 65.0 [***,a]
HbA1c (%)	5.68 ± 0.38	6.92 ± 0.90 [***,b]	8.87 ± 1.73 [***,a]
Fasting insulin (mU/L)	5.2 ± 1.6	9.2 ± 7.8	8.5 ± 4
hsCRP (mg/L)	2.4 ± 2.1	3.5 ± 3.1	3.9 ± 2.8
PAI-1 (ng/mL)	18.9 ± 5.7	24.7 ± 11	16 ± 8*
Creatinine (mg/dL)	0.81 ± 0.17	0.91 ± 0.19	1.33 ± 0.70 *
Azotemia (mg/dL)	38 ± 7.9	38 ± 9.7 [b]	54 ± 30 [**,a]
Ferritin (ng/mL)	111 ± 71	236 ± 177 **	140 ± 113
Total cholesterol (mg/dL)	237 ± 25	230 ± 36	215 ± 43
HDL cholesterol (mg/dL)	59 ± 11	58 ± 15	51 ± 14
LDL cholesterol (mg/dL)	141 ± 24	136 ± 36	116 ± 34
Triglycerides (mg/dL)	106 ± 46	142 ± 128	160 ± 95
ApoB (mg/dL)	109 ± 2	112.8 ± 30	105.1 ± 30
IGF1 (ng/mL)	32.7 ± 6.5	38.5 ± 9.7	32.9 ± 5.8
β-Galactosidase (nM/ml/h)	5.47 ± 1.73	3.77 ± 3.11	6.65 ± 3.87
α-Fucosidase (nM/ml/h)	374 ± 179	328 ± 208	389 ± 219
HOMA-IR	1.2 ± 0.4	2.9 ± 2.0 *	4.3 ± 1.9 ***
eGRF (ml/min)	89.7 ± 28.1	82.8 ± 14.7	60.5 ± 25.3 *
Disease duration (years)	n/a	n/a	22 ± 12

Values are given as the mean ± SD. p values were obtained by ANOVA and post-Tukey. *, ** and *** refer to a p-value < 0.05, 0.005 and < 0.001, respectively, for the type 2 diabetes mellitus (T2DM) classes versus control group (CG). Different letters represent significant differences among different T2DM groups. T2DM-C = T2DM patients with complications and insulin treatment; T2DM-NC = T2DM patients without complications and treatment; BMI = body mass index; WHR = waist to hip ratio; PAI-1 = plasminogen activator inhibitor-1; IGF1 = Insulin growth factor; HOMA-IR = Homeostatic Model Assessment for Insulin Resistance; eGRF = estimated glomerular filtration rate; ApoB = apolipoprotein B; hsCRP = highly sensitive C-reactive protein.

PAI-1, azotemia and creatinine were significantly increased in T2DM-C. No significant changes in IGF1, β-galactosidase, and α-fucosidase levels were observed between groups. All patients were hyperlipidemic, with no significant differences in their plasma total cholesterol, low-density lipoprotein (LDL), HDL, ApoB, and TAG levels. Ferritin level was significantly higher in T2DM-NC compared to CG (Table 3).

3.2. ^{1}H-NMR Analysis of Serum Samples

Typical ^{1}H-NMR CMPG (600 MHz) spectra of serum, obtained from CG and T2DM samples are reported in Figure 1. Resonance assignments were performed according to the literature [36] and further confirmed by 2D NMR spectra. A complex pattern of signals ascribable to aromatic molecules (1), sugar moieties, and aliphatic metabolites (2) was shown (Figure 1), with some of the identified metabolites reported for each different group. Although the NMR spectra appeared similar among different serum samples, there were striking differences in peak intensities for the different groups. The ^{1}H-NMR spectra were dominated by high-intensity signals of sugars (α and β glucose) and some high molecular weight metabolites, such as lipoproteins (very low and low density lipoprotein, VLDL/LDL). As expected, a significant increase in sugar content was observed in T2DM patients (with or without complications) compared to the CG. Small molecules, as branched-chain amino acids (leucine, isoleucine, valine), aliphatic and aromatic amino acids (alanine, arginine, glutamine, glutamate, methionine, glycine, phenylalanine, tyrosine), organic acids (lactate, formate, pyruvate, citrate, acetate, acetoacetate, β-hydroxybutyrate), osmolytes (choline, TMA-N-oxide), and others, including methyl-histidine, *N*-acetyl-glycoproteins, dimethylamine (DMA), trimethylamine (TMA) and creatinine were also identified [27].

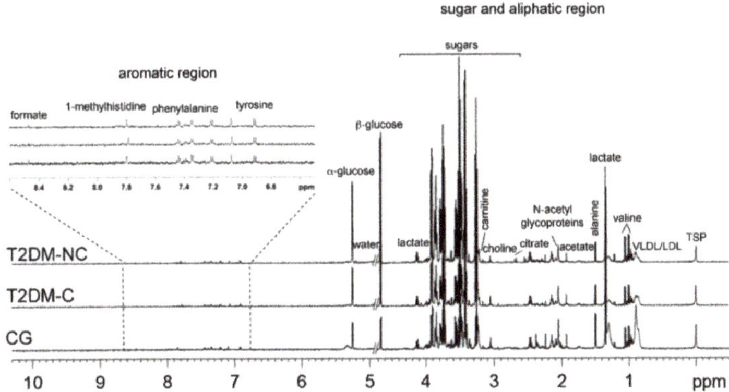

Figure 1. Representative ^{1}H CPMG (Carr–Purcell–Meiboom–Gill) NMR spectra of serum isolated from different groups of diabetic (T2DM-NC, T2DM-C) and control (CG) patients. Aromatic, and sugar and aliphatic regions with some identified metabolites were visualized.

3.3. Multivariate Analysis of NMR Data

Multivariate data analysis was applied to the NMR spectra for reducing the complexity and the volume of the data. After the pre-processing of the NMR spectra, including the bucketing process and total sum normalization to minimize small differences due to sample concentration and/or experimental conditions among samples [36], both unsupervised (PCA) and supervised (PLSDA, OPLSDA) multivariate statistical methods were applied. As a first attempt, PCA analysis was conducted to display natural groupings of samples without imposing any preconception about class membership, allowing the general trend display and the identification of potential outliers among samples. Supervised statistical methods, such as PLSDA and OPLSDA analyses, were also applied in order to search for discriminating features and potential biomarkers, responsible for the separation between groups. In particular, pairwise multivariate (PCA, PLSDA, OPLSDA) analyses were obtained comparing the CG vs. T2DM-NC (Figure 2) and the CG vs. T2DM-C groups (Figure 3). Moreover, T2DM patients were also compared with each other (T2DM-C vs. T2DM-NC, Figure 4). The PCA score plot obtained for CG and T2DM-NC (Figure 2A) revealed a good separation between the two groups, especially along the first principal component PC1 (PC1 and PC2 accounted for 51.3% and 23% of the total variance, respectively). A clear separation between the two groups was observed also in the corresponding PLSDA (Figure S1A in Supplementary Materials) and OPLSDA score plots (Figure 2B). The PLSDA model was obtained, with the first two components explaining 50% and 23.5% of the total variance, $R^2 = 0.86$ and $Q^2 = 0.78$, and the corresponding OPLSDA score plot was obtained with the first predictive and orthogonal components, accounting for 17.8% and 53% of the total variance, respectively. The variables (bucket reduced NMR signals) responsible for the CG and T2DM-NC separation were observed in the corresponding VIP plot ($p < 0.05$) (Figure S1B in Supplementary Materials). In addition to a relative higher α and β-glucose content that characterized T2DM-NC patients with respect to CG, a relative decrease of alanine, creatine/creatinine, glutamine, glutamate, leucine, lysine, methionine, *N*-acetylglycoproteins, phenylalanine, and tyrosine was observed in T2DM-NC patients with respect to the CG (Figure 2C). Moreover, the quantitative variation in discriminating metabolites among the observed groups was calculated by the integration of specific signals, identified by the NMR-based multivariate analysis (Table 4). Results, reported as the mean and standard deviation of integrals for each group, were validated by one-way ANOVA with the HSD post-hoc test. The level of statistical significance was at least at *p*-values < 0.05 with 95% confidence level. PCA, PLSDA, and OPLSDA analyses were also performed for CG and T2DM-C patients (Figure 3). PCA analysis showed an even better separation than that observed comparing CG and T2DM-NC, between CG and T2DM-C patients (PC1 and PC2 accounted for 73.7% and 10.5% of the total variance, respectively) (Figure 3A). This result appeared more evident when PLSDA (with the first two components explaining 73.6% and 9.3% of the total variance, $R^2 = 0.85$ and $Q^2 = 0.72$) and OPLSDA (with the first predictive and orthogonal components accounting for 22.9% and 48.6%) analyses were applied to the data (Figure S2 in Supplementary Materials and Figure 3B, respectively). Together with the expected higher level of sugars, several metabolites, such as alanine, carnitine, citrate, creatine, glutamate, glutamine, isoleucine, leucine, lactate, lysine, methionine, N-acetyl-glycoproteins, phenylalanine, tyrosine and valine were also quantified and statistically validated as significantly reduced in T2DM-C patients compared to the CG (Figure 3C and Table 4).

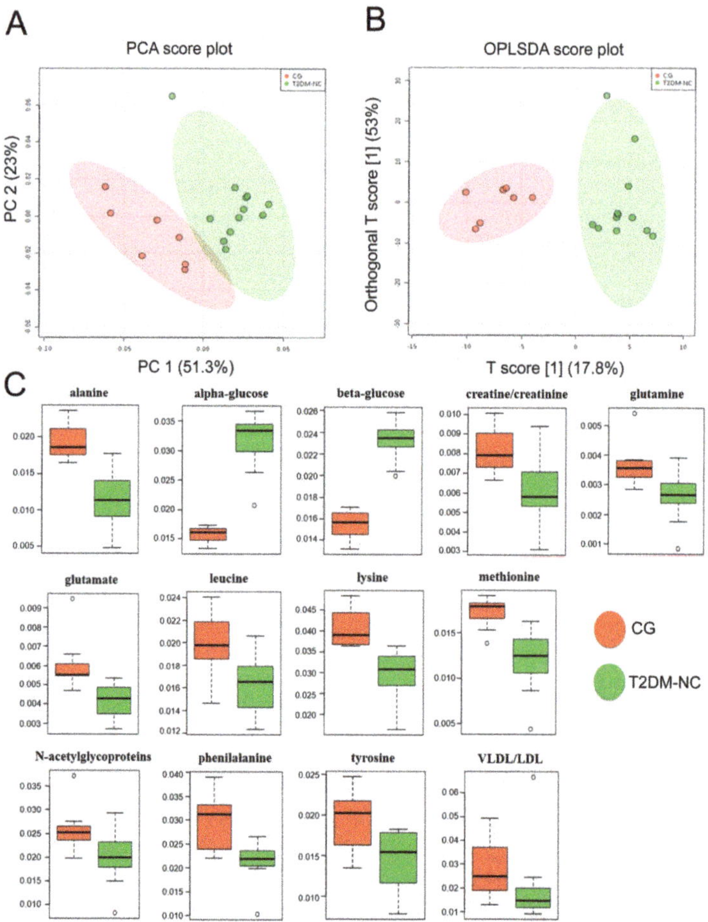

Figure 2. PCA and OPLSDA score scatter plots distinguishing between CG and T2DM-NC. (**A**) PCA (PC1/PC2, 51.3% and 23% of the total variance, respectively); (**B**) OPLSDA (obtained with the first predictive and orthogonal components, accounting for 17.8% and 53% of the total variance, respectively) analyses of NMR data; (**C**) metabolites significantly changed between T2DM-NC and CG patients. PCA = principal component analysis; OPLSDA = orthogonal partial least squares discriminant analysis.

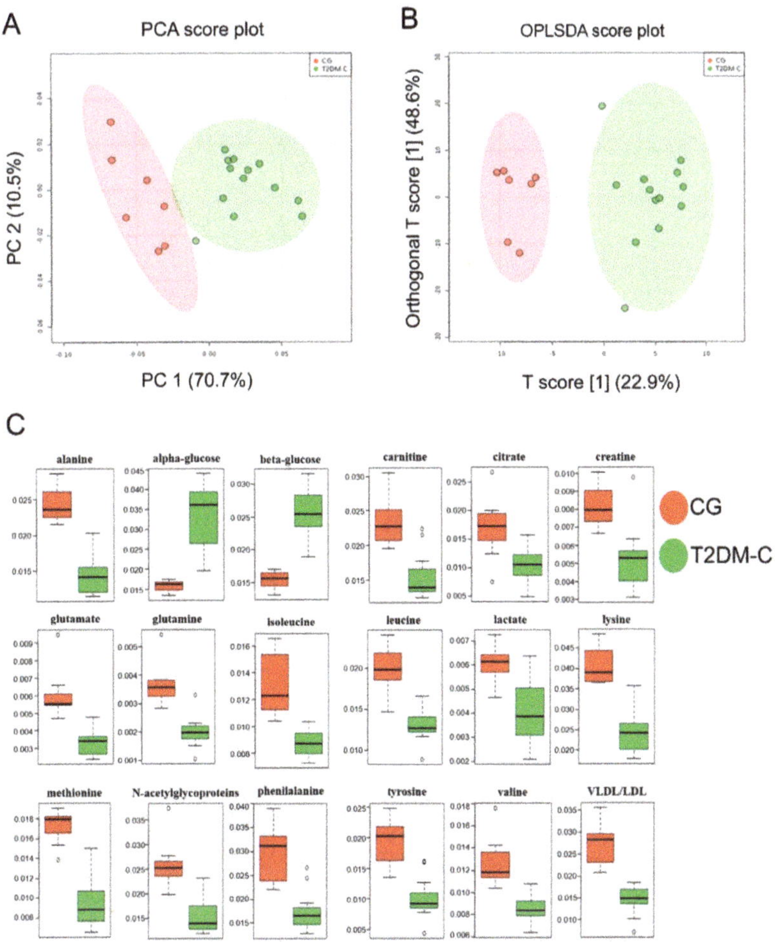

Figure 3. PCA and OPLS-DA score scatter plots distinguishing between CG and T2DM-C. (**A**) PCA (PC1/PC2, 73.7% and 10.5% of the total variance, respectively); (**B**) OPLSDA (obtained with the first predictive and orthogonal components, accounting for 22.9% and 48.6% of the total variance, respectively) analyses of NMR data; (**C**) metabolites significantly changed between T2DM-C and CG patients.

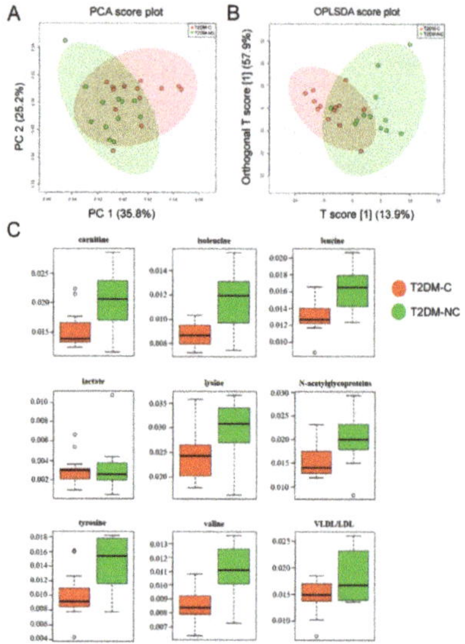

Figure 4. PCA and OPLS-DA score scatter plots distinguishing between T2DM-NC and T2DM-NC. (**A**) PCA (PC1/PC2, 35.8% and 25.2% of the total variance, respectively); (**B**) OPLSDA (obtained with the first predictive and orthogonal components, accounting for 13.9% and 57.9% of the total variance, respectively) analyses of NMR data; (**C**) metabolites significantly changed between T2DM-C and T2DM-NC patients.

Finally, T2DM patients were also compared with each other (T2DM-C vs. T2DM-NC). The statistical models obtained appeared well descriptive but weakly predictive, showing a partial overlap of samples (Figure 4). In fact, PCA model (Figure 4A) was built with PC1 and PC2 accounting for 35.8% and 25.2% of the total variance, while in the PLSDA score plot (Figure S3 in Supplementary Materials) the first two components explained 33.5% and 24.5% of the total variance, respectively ($R^2 = 0.46$ and $Q^2 = -0.27$); the OPLSDA score plot was obtained with the first predictive and orthogonal components accounting for 13.9% and 57.9% (Figure 4B). Metabolites showing a significant variation between the two groups were carnitine, isoleucine, leucine, lactate, lysine, *N*-acetylglycoproteins, tyrosine, and valine (Figure 4C and Table 4).

The metabolites listed in Table 4 were also used as input variables in order to investigate the distance of any type of samples, obtaining a heatmap (Figure 5). Moreover, to identify the potential biomarkers associated with T2DM disease, a multivariate ROC curve analysis was also performed on selected metabolites, considering the T2DM-C patients compared to controls (GC) (Figure 6). An overview of all ROC curves (created from six different biomarker models using different number of features) and predictive accuracies with different features are shown in Figure 6A,B, respectively. Considering the first top metabolites (selected frequency >~0.7%), the obtained AUC value was 0.999, with a 0.938–1 confidence interval (CI). Moreover, from the variable importance in projection (VIP) plot (Figure 6C) and prediction of T2DM-C patients and controls using MCCV analysis (Figure 6D), the most discriminating metabolites in descending order of importance were indicated. In addition to glucose, also leucine, glutamate, lysine, valine, and isoleucine resulted the top five important metabolites, based on their frequency selection during cross-validation (Figure 6C). In conclusion, the ROC curves analysis allowed to obtain a general overview of significant alterations of biomarkers in T2DM patients with complications. In particular, BCAAs such as leucine, valine, and isoleucine resulted significant in this analysis.

Table 4. Quantitative comparison of metabolites found in the serum of T2DM and CG patients.

Metabolite	Chemical Shift (ppm)	Integral in CG Group [a] (Mean $\pm$ SD) $\times 10^{-2}$	Integral in T2DM-NC [a] Group (Mean $\pm$ SD) $\times 10^{-2}$	Integral in T2DM-C [a] Group (Mean $\pm$ SD) $\times 10^{-2}$	p-Value	Tukey's HSD Adjusted p-Value (FDR) Cutoff: 0.05 [b]
Alanine	1.49 (d)	1.95 $\pm$ 0.27	1.14 $\pm$ 0.38 ($\downarrow$)	0.96 $\pm$ 0.25 ($\downarrow$)	6.89×10^{-7}	T2DM-C/CG; T2DM-NC/CG
α-glucose	5.25 (d)	1.57 $\pm$ 0.14	3.16 $\pm$ 0.45 ($\uparrow$)	3.37 $\pm$ 0.77 ($\uparrow$)	5.35×10^{-7}	T2DM-C/CG; T2DM-NC/CG
β-glucose	4.66 (d)	1.54 $\pm$ 0.14	2.33 $\pm$ 0.18 ($\uparrow$)	2.56 $\pm$ 0.33 ($\uparrow$)	6.26×10^{-9}	T2DM-C/CG; T2DM-NC/CG
Carnitine	3.22 (s)	2.35 $\pm$ 0.4	2.04 $\pm$ 0.45 ($\downarrow$)	1.54 $\pm$ 0.33 ($\downarrow$)	3.38×10^{-4}	T2DM-C/CG; T2DM-NC/T2DM-C
Citrate	2.65 (d)-2.56(d)	0.17 $\pm$ 0.006	0.14 $\pm$ 0.04 ($\downarrow$)	0.11 $\pm$ 0.035 ($\downarrow$)	1.2×10^{-2}	T2DM-C/CG
Creatine/creatinine	3.05 (s)	0.82 $\pm$ 0.13	0.6 $\pm$ 0.16 ($\downarrow$)	0.53 $\pm$ 0.17 ($\downarrow$)	2.1×10^{-3}	T2DM-C/CG; T2DM-NC/CG
Glutamate	2.38 (m)	0.61 $\pm$ 0.16	0.41 $\pm$ 0.08 ($\downarrow$)	0.34 $\pm$ 0.08 ($\downarrow$)	2.35×10^{-5}	T2DM-C/CG; T2DM-NC/CG
Glutamine	2.45 (m)	0.37 $\pm$ 0.008	0.26 $\pm$ 0.07 ($\downarrow$)	0.2 $\pm$ 0.05 ($\downarrow$)	5.34×10^{-5}	T2DM-C/CG; T2DM-NC/CG
Isoleucine	1.02 (d)	1.32 $\pm$ 0.25	1.15 $\pm$ 0.24 ($\downarrow$)	0.88 $\pm$ 0.1 ($\downarrow$)	1.15×10^{-4}	T2DM-C/CG; T2DM-NC/T2DM-C
Leucine	0.98 (d)	1.99 $\pm$ 0.31	1.64 $\pm$ 0.24 ($\downarrow$)	1.32 $\pm$ 0.2 ($\downarrow$)	9.48×10^{-6}	T2DM-C/CG; T2DM-NC/CG; T2DM-NC/T2DM-C
Lactate	1.33 (d)	6.05 $\pm$ 0.82	5.4 $\pm$ 1.6 ($\downarrow$)	4.1 $\pm$ 1.4 ($\downarrow$)	1.18×10^{-2}	T2DM-C/CG; T2DM-NC/T2DM-C
Lysine	1.74 (m)	0.41 $\pm$ 0.005	0.29 $\pm$ 0.05 ($\downarrow$)	0.24 $\pm$ 0.05 ($\downarrow$)	6.20×10^{-7}	T2DM-C/CG; T2DM-NC/CG; T2DM-NC/T2DM-C
Methionine	2.14 (m)	1.72 $\pm$ 0.19	1.21 $\pm$ 0.33 ($\downarrow$)	0.96 $\pm$ 0.26 ($\downarrow$)	1.17×10^{-5}	T2DM-C/CG; T2DM-NC/CG
N-acetylglycoproteins	2.05 (m)	2.61 $\pm$ 0.55	2.02 $\pm$ 0.55 ($\downarrow$)	1.53 $\pm$ 0.35 ($\downarrow$)	0.2×10^{-3}	T2DM-C/CG; T2DM-NC/CG; T2DM-NC/T2DM-C
Phenylalanine	7.34 (d)	0.29 $\pm$ 0.006	0.21 $\pm$ 0.04 ($\downarrow$)	0.17 $\pm$ 0.04 ($\downarrow$)	2.54×10^{-5}	T2DM-C/CG; T2DM-NC/CG
Tyrosine	7.18 (m)	1.92 $\pm$ 0.004	0.14 $\pm$ 0.04 ($\downarrow$)	0.1 $\pm$ 0.03 ($\downarrow$)	4.09×10^{-5}	T2DM-C/CG; T2DM-NC/CG; T2DM-NC/T2DM-C
Valine	1.04 (d)	1.28 $\pm$ 0.24	1.10 $\pm$ 0.19 ($\downarrow$)	0.86 $\pm$ 0.12 ($\downarrow$)	7.35×10^{-5}	T2DM-C/CG; T2DM-NC/T2DM-C

The arrows ($\uparrow$/$\downarrow$) were used to show the metabolite levels increase/decreased compared with CG. [a] Mean and relative standard deviation refers to the relative integrals of metabolites, determined from serum 1D ^{1}H-NMR spectra of each group. The chemical shifts used for calculating integrals were reported in the second column. [b] An FDR (False Discovery Rate) adjusted p-value of 0.05 was obtained from analysis of variance (one way-ANOVA) with Tukey's honestly significant difference (HSD) post hoc test. The levels of statistical significance were at least at p-values < 0.05 with 95% confidence level. Letters in parentheses indicate the peak multiplicities: s, singlet; d, doublet; m, multiplet).

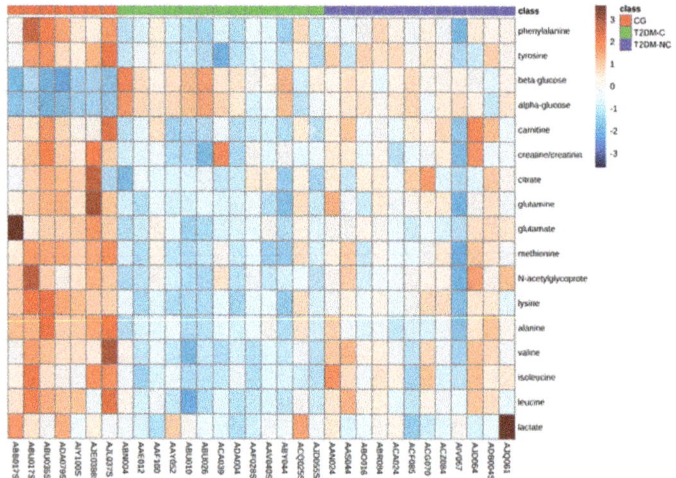

Figure 5. Heatmap visualization of group averages based on 17 biomarkers. Heatmap visualization based on 17 metabolites. Rows: biomarkers; columns: samples. Red: GC; green and blue: T2DM-C and T2DM-NC patients, respectively. Color key indicates metabolite values: dark blue: lowest; dark red: highest.

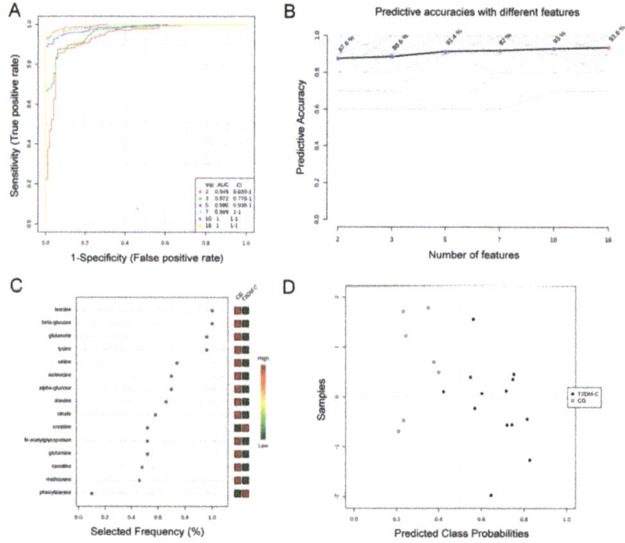

Figure 6. Comparison of variables based on ROC curve. (**A**) Multivariate receiver operating characteristic (ROC) analysis, showing the feature numbers the AUCs and the confidence intervals of the six models; (**B**) predictive accuracies with different features based on the ROC curves; (**C**) percentage selected frequency of metabolites based on ROC curves, with the variable importance in projection (VIP) plot indicating the most discriminating metabolite in descending order of importance; (**D**) prediction of T2DM-C patients and controls using Monte–Carlo cross validation (MCCV) analysis.

3.4. Metabolic Pathway Analysis

A metabolic pathway analysis using the MeTPA (Metabolomics Pathway Analysis) software [37] was performed in order to identify the most relevant pathways potentially involved in the observed changes of T2DM serum metabolites. Based on the observed quantitative variation for the identified metabolite content found in CG and T2DM patients at different stages, according to the p-value and the impact value (Tables S1–S3 in Supplementary Materials), at least six potential target pathways resulted altered in T2DM-NC patients (Figure 7A). These pathways included alanine, aspartate and glutamate metabolism, D-glutamine and D-glutamate metabolism, lysine degradation, aminoacyl-tRNA biosynthesis, and carbohydrate metabolism (starch and sucrose metabolism, glycolysis, or gluconeogenesis). The same metabolic pathways, with a different significance, appear to be compromised in T2DM-C patients with respect to CG (Figure 7B), with the exception of pathways involving carbohydrate metabolism. Finally, when the analysis was conducted by comparing the two diabetic groups, phenylalanine and branched-chain amino acid (BCAA) metabolism resulted perturbed and discriminant of the two diabetic stages (Figure 7C).

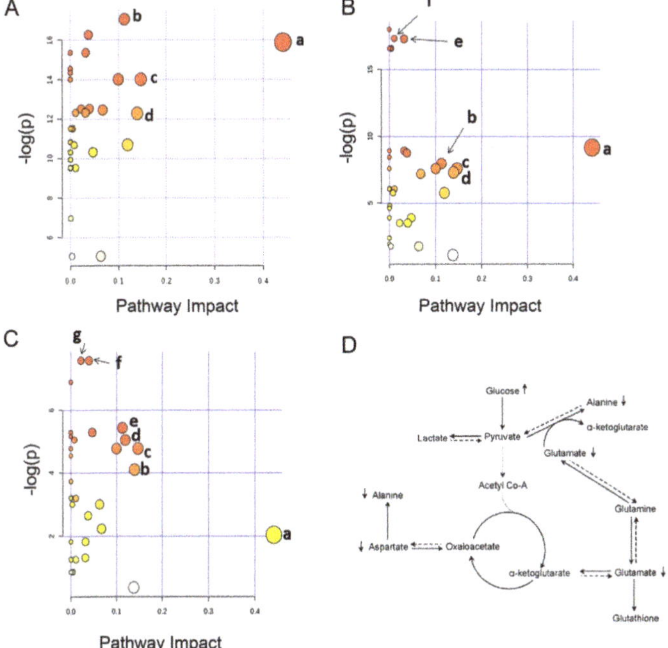

Figure 7. Summary of metabolic analysis conducted by MetPA. (**A**) CG vs. T2DM-C. (**a**) Alanine, aspartate, and glutamate metabolism; (**b**) aminoacyl-tRNA biosynthesis; (**c**) lysine degradation; (**d**) D-glutamine and D-glutamate metabolism. (**B**) CG vs. T2DM-NC. (**a**) Alanine, aspartate and glutamate metabolism; (**b**) aminoacyl-tRNA biosynthesis; (**c**) lysine degradation; (**d**) D-glutamine and D-glutamate metabolism; (**e**) starch and sucrose metabolism; (**f**) glycolysis or gluconeogenesis. (**C**) T2DM-NC vs. T2DM-C. (**a**) Alanine, aspartate and glutamate metabolism; (**b**) D-glutamine and D-glutamate metabolism; (**c**) lysine degradation; (**d**) phenylalanine metabolism; (**e**) aminoacyl-tRNA biosynthesis; (**f**) valine, leucine and isoleucine biosynthesis; (**g**) valine, leucine and isoleucine degradation. (**D**) Schematic diagram of the metabolic pathways supposed to be altered in T2DM detected by ^{1}H NMR analysis. Arrows (↑ ↓) represented the increase or decrease of metabolites in the serum.

4. Discussion

T2DM is a metabolic disorder characterized by chronic hyperglycemia due to the resistance of target tissues to the metabolic action of insulin and dysfunction of pancreatic β cells. According to the most recent guidelines, individuals with T2DM are considered at high to very-high cardiovascular risk, depending on the presence or not of established complications [38]. Insulin therapy is used for the treatment of T2DM when the progression of the disease overcomes the effectiveness of oral hypoglycemic drugs [39]. Due to the paucity of metabolic markers of disease progression and the suboptimal performance of risk equations, the characterization of biomarkers that can indicate specific and temporal metabolic and vascular disturbances in the progression of T2DM is an active field of investigation. Metabolomics has gained growing applications in the identification of circulating biomarkers thus contributing to the diagnosis, prognosis and risk estimation of diseases [40]. In the present study, the serum metabolic profiling of control and T2DM groups was investigated to explore the metabolic response to diabetic complications leading to the identification of a metabolic signature that might serve as predictor of disease progression.

To do this, we profiled a well-characterized diabetic cohort of patients who showed hyperglycemia, with increased levels of HbA1c and HOMA index with respect to CG closely matched for age and BMI. Among T2DM patients, retinopathy, nephropathy, and neuropathy were the most frequent observed complications. Analysis of clinical data revealed that a set of few metabolites including PAI-1, azotemia, and creatinine were increased in the T2DM-C group and potentially related to disease complications. Moreover, an increased ferritin level was observed in T2MD-NC patients compared to the CG (Table 3). This was consistent with previous studies that described a functional correlation between PAI-1, ferritin, and type II diabetes [41,42]. These conventional clinical and biochemical parameters clearly distinguished T2DM patients from CG but do not allow the detection of specific alterations useful for the clinical discrimination of T2DM patient cohorts. Application of metabolomics to these datasets has led to a broader evaluation of a large set of serum metabolites associated with each class of patients. As a result of disease complexity [43], several metabolites were identified in T2DM, most of which were involved in amino acid metabolism (Figure 7D). This result is in agreement with previously reported metabolomic studies highlighting the association of serum amino acids alteration with metabolic disorders, and T2DM [44–47].

More in detail, MetPA software analysis of NMR data showed a significant perturbation of alanine, glutamine and glutamate metabolism in both T2DM groups (T2DM-C and T2DM-NC) of patients (Figure 5, Tables S1–S3). Multivariate analyses pointed out a decreased serum level of these amino acids in both groups of diabetic (T2DM-C and T2DM-NC) with respect to control patients. This result confirms previous studies that identified alanine and glutamine as consistently associated with an increased risk of developing T2DM [36,48]. Alanine, glutamine, and glutamate are gluconeogenic amino acids precursors for glucose synthesis. Therefore, decreased serum level of these amino acids could be associated to a major hepatic utilization for glucose synthesis in agreement with previous reports of increased gluconeogenesis from amino acids in T2DM patients [49–51]. Indeed, our metabolic pathway analysis clearly suggests the involvement of the gluconeogenesis pathway, particularly at the earliest T2DM stage (Figure 7). Moreover, the reduced amount of serum lactate, measured in T2DM-C compared with T2DM-NC patients, could be seen as a consequence of the higher alanine utilization for glucose synthesis. In this case, pyruvate is shunted towards alanine and away from lactate synthesis.

The multivariate analysis performed on serum metabolic profiles also showed and allowed to quantify significant differences in the levels of isoleucine, lysine, tyrosine, and valine in T2DM-C with respect to CG patients and between the two diabetic groups (Table 4). Leucine, isoleucine, and valine are essential amino acids also known as BCAAs. Indeed, MetPA analysis highlighted a compromised BCAA metabolism as a discriminant of the two diabetes stages (Figure 7C). Thus, the strong decrease in BCAA serum level observed in the advanced with respect to T2DM early stage could be considered as a time-dependent perturbation of BCAA metabolism. Moreover, ROC curve analysis validated the clinical relevance of BCAAs as potential biomarkers in the diagnosis of T2DM complications (Figure 6).

In recent years, BCAA metabolism was found to be significantly altered in patients with several diseases including diabetes [52]. Both human and animal studies demonstrated a functional correlation between BCAA metabolism and a higher susceptibility to IR in T2DM patients [53–57]. Here, we showed lower levels of BCAA in T2DM patients compared to CG, raising questions regarding the mechanisms regulating BCAA metabolism in our cohort. In this context, we speculate that IR levels and disease complications may have a major impact on this result, as corroborated by other studies that utilize functional and NMR approaches. For instance, lower plasma levels, with respect to healthy controls, of BCAAs in T2DM patients were indicated as a first sign of kidney dysfunction [56]. Further, the decreased serum levels of BCAAs that we observed in T2DM are in line with the results of Shin et al. [57] who demonstrated that alteration in hypothalamic insulin signaling can decrease plasma BCAA levels by inducing the hepatic activity of branched-chain α-keto acid dehydrogenase, a rate-limiting enzyme in the BCAA degradation pathway. Overall, this indication well correlated with the HOMA-IR values that we measured in the diabetic patients with respect to CG, with the highest value of IR measured in the complicated T2DM patients (Table 3). Moreover, a higher BCAA level has been found in retinal Müller cells from diabetic compared to euglycemic rats [58]. This study demonstrated that BCAAs competitively inhibited glutamate transamination and induced glutamate excitotoxicity and neuronal cell death, providing an early sign of retinopathy. This data well correlated with the high occurrence (more than 90% of subjects) of retinopathy that we measured among T2DM-C patients (Table 1). Moreover, increased muscle BCAA oxidation has been reported to improve muscle glucose uptake in metabolic syndrome by enhancing the recycling of glucose via the glucose-alanine cycle [59].

Serum levels of lysine decreased from control subjects to T2DM patients at an early stage and to T2DM complicated patients. Interestingly, recent work reported a therapeutic effect of lysine administration in T2DM patients to counteract the production of glycated lysozyme [60]. Protein glycation is a well-established process associated with long-term hyperglycemia characterizing many pathophysiological conditions such as cancer, inflammation, metabolic dysfunctions and aging [61].

We also found that tyrosine serum level significantly decreased in both groups of diabetic patients with respect to CG and also differentiated T2DM-C from T2DM-NC patients (Table 4). It is important to note that a previous work, conducted on rat liver, demonstrated that insulin was capable to increase the activity of the tyrosine aminotransferase enzyme. Thus, by activating the transamination of tyrosine to p-hydroxyphenylpyruvate the serum level of tyrosine was reduced [62]. On the basis of these data, the high fasting insulin level measured in diabetic subjects (Table 3) could be linked to the decreased tyrosine serum level in diabetic patients. Importantly, it has been also reported that circulating tyrosine level could be correlated with the risk of developing various major complications of diabetes [56]. Moreover, other studies described an association of a low tyrosine level with the impairment of kidney function, which itself could predict future microvascular events [63,64]. Tyrosine is also linked to catecholamine synthesis, a mechanism that may be involved in the development of T2DM complications [65]. Increasing evidence suggested a correlation among tyrosine levels, obesity, and insulin concentration in both diabetic and non-diabetic subjects [65]. Moreover, tyrosine has been identified as the only metabolite significantly associated with HOMA in obesity-independent models [66,67].

The serum level of patients in the advanced T2DM stage was also characterized by a lower level of carnitine with respect to T2DM-NC. L-carnitine, primarily synthesized in the liver and kidneys from lysine and methionine, has an essential role in the transfer of activated long-chain fatty acids into the mitochondria where β-oxidation takes place [68]. Considering that impaired fatty acid transport inside mitochondria can result in triglyceride cytosolic accumulation in and IR [69], we can consider the reduced serum carnitine level measured in complicated T2DM patients closely associated to the highest HOMA-IR of these subjects.

5. Conclusions

Results from the present study identified a set of metabolites discriminating control from T2DM groups. We also demonstrated that specific metabolic alterations characterize T2DM groups at the systemic level with main differences in amino acids pathways. If this result reflects a way to decrease the pathological state by activating compensatory mechanisms or is a mere consequence of deranged metabolic pathways is yet to be understood. In this context, the metabolism of gluconeogenic amino acids and BCAAs may have a role. On the basis of these results, considering that BCAAs are essential amino acids introduced with the diet, changes in dietary habit, not only as carbohydrate intake, have to be taken into account for the prevention of T2DM complications. Finally, if this metabolic change represents the pathological basis for the onset of complications increasing, for instance, the risk of developing cancer should be also considered and evaluated. Our data will represent the basis for future studies involved patients affected by tumors related to metabolic dysfunctions such as hepatocellular carcinoma.

Supplementary Materials: The following are available online at http://www.mdpi.com/2077-0383/8/5/720/s1, Figure S1: PLSDA score scatter plot between CG and T2DM-NC, Figure S2: PLSDA score scatter plot between CG and T2DM-C, Figure S3: PLSDA score scatter plot between T2DM-C and T2DM-NC. Table S1: Results from pathway analysis with MetPA for serum samples of CG vs. T2DM-NC patients, Table S2: Results from pathway analysis with MetPA for serum samples of CG vs. T2DM-C patients, Table S3: Results from pathway analysis with MetPA for serum samples of T2DM-C vs. T2DM-NC patients.

Author Contributions: Design of the project, A.M.G., D.V., A.C.-G., M.B., S.D.M., A.R., and F.O. Supervision of the study and contributions to manuscript writing, A.M.G. Analysis and interpretation of the data and writing of the manuscript, L.D.C., D.V. Contributions to clinical data analysis, patient selection, and manuscript revision, E.M., J.S., F.P. (Francesco Prattichizzo), A.R.B., F.O. NMR experiments, metabolomic analyses and contribution to manuscript writing, L.D.C. Interpretation of the data and contribution to manuscript revision, F.P.F., M.M., S.B., F.P. (Francesca Pirini), and G.S.

Acknowledgments: This work has been supported by Pallotti Legacy to M. Bonafè.

Conflicts of Interest: The authors declare no conflict of interest.

References

1. Pandey, A.; Chawla, S.; Guchhait, P. Type-2 diabetes: Current understanding and future perspectives. *IUBMB Life* **2015**, *67*, 506–513. [CrossRef]

2. Bonow, R.O.; Gheorghiade, M. The diabetes epidemic: A national and global crisis. *Am. J. Med.* **2004**, *116*, 2S–10S. [CrossRef] [PubMed]

3. Toniolo, A.; Cassani, G.; Puggioni, A.; Rossi, A.; Colombo, A.; Onodera, T.; Ferrannini, E. The diabetes pandemic and associated infections: Suggestions for clinical microbiology. *Rev. Med. Microbiol.* **2019**, *30*, 1–17. [CrossRef] [PubMed]

4. Kolb, H.; Martin, S. Environmental/lifestyle factors in the pathogenesis and prevention of type 2 diabetes. *BMC Med.* **2017**, *15*, 131. [CrossRef] [PubMed]

5. Chatterjee, S.; Khunti, K.; Davies, M.J. Type 2 diabetes. *Lancet* **2017**, *389*, 2239–2251. [CrossRef]

6. Colayco, D.C.; Niu, F.; McCombs, J.S.; Cheetham, T.C. A1C and cardiovascular outcomes in type 2 diabetes: A nested case-control study. *Diabetes Care* **2011**, *34*, 77–83. [CrossRef]

7. Smooke, S.; Horwich, T.B.; Fonarow, G.C. Insulin-treated diabetes is associated with a marked increase in mortality in patients with advanced heart failure. *Am. Heart J.* **2005**, *149*, 168–174. [CrossRef]

8. Herman, M.E.; O'Keefe, J.H.; Bell, D.S.H.; Schwartz, S.S. Insulin therapy increases cardiovascular risk in Type 2 Diabetes. *Prog. Cardiovasc. Dis.* **2017**, *60*, 422–434. [CrossRef] [PubMed]

9. Basu, S.; Sussman, J.B.; Berkowitz, S.A.; Hayward, R.A.; Yudkin, J.S. Development and validation of risk equations for complications of type 2 diabetes (RECODe) using individual participant data from randomised trial. *Lancet Diabetes Endocrinol.* **2017**, *5*, 788–798. [CrossRef]

10. Wang, T.J.; Larson, M.G.; Vasan, R.S.; Cheng, S.; Rhee, E.P.; McCabe, E.; Lewis, G.D.; Fox, C.S.; Jacques, P.F.; Fernandez, C.; et al. Metabolite profiles and the risk of developing diabetes. *Nat. Med.* **2011**, *17*, 448–453. [CrossRef]

11. Gonzalez-Franquesa, A.; Burkart, A.M.; Isganaitis, E.; Patti, M.E. What have metabolomics approaches taught us about Type 2 Diabetes? *Curr. Diabetes Rep.* **2016**, *16*, 74. [CrossRef] [PubMed]

12. Wang, H.; Zhang, H.; Yao, L.; Cui, L.; Zhang, L.; Gao, B.; Liu, W.; Wu, D.; Chen, M.; Li, X.; et al. Serum metabolic profiling of type 2 diabetes mellitus in Chinese adults using an untargeted GC/TOF-MS. *Clin. Chim. Acta* **2018**, *477*, 39–47. [CrossRef]

13. Papandreou, C.; Bulló, M.; Ruiz-Canela, M.; Dennis, C.; Deik, A.; Wang, D.; Guasch-Ferré, M.; Yu, E.; Razquin, C.; Corella, D.; et al. Plasma metabolites predict both insulin resistance and incident type 2 diabetes: A metabolomics approach within the prevención con dieta mediterránea (PREDIMED) study. *Am. J. Clin. Nutr.* **2019**, *109*, 626–634. [CrossRef] [PubMed]

14. Friedrich, N. Metabolomics in diabetes research. *J. Endocrinol.* **2012**, *215*, 29–42. [CrossRef]

15. Zhou, Y.; Hu, C.; Zhao, X.; Luo, P.; Lu, J.; Li, Q.; Chen, M.; Yan, D.; Lu, X.; Kong, H.; et al. Serum metabolomics study of gliclazide-modified-release-treated type 2 diabetes mellitus patients using a gas chromatography-mass spectrometry method. *Proteome Res.* **2018**, *17*, 1575–1585. [CrossRef]

16. Safai, N.; Suvitaival, T.; Ali, A.; Spégel, P.; Al-Majdoub, M.; Carstensen, B.; Tarnow, L. Effect of metformin on plasma metabolite profile in the copenhagen insulin and metformin therapy (CIMT) trial. *Diabet. Med.* **2018**, *35*, 944–953. [CrossRef] [PubMed]

17. Wu, T.; Qiao, S.; Shi, C.; Wang, S.; Ji, G. Metabolomics window into diabetic complications. *J. Diabetes Investig.* **2018**, *9*, 244–255. [CrossRef]

18. Montesanto, A.; Bonfigli, A.R.; Crocco, P.; Garagnani, P.; De Luca, M.; Boemi, M.; Marasco, E.; Pirazzini, C.; Giuliani, C.; Franceschi, C.; et al. Genes associated with Type 2 diabetes and vascular complications. *Aging* **2018**, *10*, 178–196. [CrossRef] [PubMed]

19. Bonfigli, A.R.; Spazzafumo, L.; Prattichizzo, F.; Bonafè, M.; Mensà, E.; Micolucci, L.; Giuliani, A.; Fabbietti, P.; Testa, R.; Boemi, M.; et al. Leukocyte telomere length and mortality risk in patients with type 2 diabetes. *Oncotarget* **2016**, *7*, 50835–50844. [CrossRef]

20. Olivieri, F.; Spazzafumo, L.; Bonafè, M.; Recchioni, R.; Prattichizzo, F.; Marcheselli, F.; Micolucci, L.; Mensà, E.; Giuliani, A.; Santini, G.; et al. MiR-21-5p and miR-126a-3p levels in plasma and circulating angiogenic cells: Relationship with type 2 diabetes complications. *Oncotarget* **2015**, *6*, 35372–35382. [CrossRef]

21. Testa, R.; Vanhooren, V.; Bonfigli, A.R.; Boemi, M.; Olivieri, F.; Ceriello, A.; Genovese, S.; Spazzafumo, L.; Borelli, V.; Bacalini, M.G.; et al. N-glycomic changes in serum proteins in type 2 diabetes mellitus correlate with complications and with metabolic syndrome parameters. *PLoS ONE* **2015**, *10*, e0119983. [CrossRef]

22. Olivieri, F.; Bonafè, M.; Spazzafumo, L.; Gobbi, M.; Prattichizzo, F.; Recchioni, R.; Marcheselli, F.; La Sala, L.; Galeazzi, R.; Rippo, M.R.; et al. Testa, age- and glycemia-related miR-126-3p levels in plasma and endothelial cells. *Aging* **2014**, *6*, 771–787. [CrossRef]

23. Balistreri, C.R.; Bonfigli, A.R.; Boemi, M.; Olivieri, F.; Ceriello, A.; Genovese, S.; Franceschi, C.; Spazzafumo, L.; Fabietti, P.; Candore, G.; et al. Evidences of +896 A/G TLR4 polymorphism as an indicative of prevalence of complications in T2DM patients. *Mediat. Inflamm.* **2014**, *2014*, 973139. [CrossRef] [PubMed]

24. Testa, R.; Olivieri, F.; Sirolla, C.; Spazzafumo, L.; Rippo, M.R.; Marra, M.; Bonfigli, A.R.; Ceriello, A.; Antonicelli, R.; Franceschi, C.; et al. Leukocyte telomere length is associated with complications of type 2 diabetes mellitus. *Diabet Med* **2011**, *28*, 1388–1394. [CrossRef]

25. Spazzafumo, L.; Mensà, E.; Matacchione, G.; Galeazzi, T.; Zampini, L.; Recchioni, R.; Marcheselli, F.; Prattichizzo, F.; Testa, R.; Antonicelli, R.; et al. Age-related modulation of plasmatic beta-Galactosidase activity in healthy subjects and in patients affected by T2DM. *Oncotarget* **2017**, *8*, 93338–93348. [CrossRef] [PubMed]

26. Beckonert, O.; Keun, H.C.; Ebbels, T.M.; Bundy, J.; Holmes, E.; Lindon, J.C.; Nicholson, J.K. Metabolic profiling, metabolomic and metabonomic procedures for NMR spectroscopy of urine, plasma, serum and tissue extracts. *Nat. Protoc.* **2007**, *2*, 2692. [CrossRef] [PubMed]

27. Nicholson, J.K.; Foxall, P.J.; Spraul, M.; Farrant, R.D.; Lindon, J.C. 750 MHz 1H and 1H-13C NMR spectroscopy of human blood plasma. *Anal. Chem.* **1995**, *67*, 793–811. [CrossRef]

28. Chong, J.; Soufan, O.; Li, C.; Caraus, I.; Li, S.; Bourque, G.; Wishart, D.S.; Xia, J. MetaboAnalyst 4.0: Towards more transparent and integrative metabolomics analysis. *Nucl. Acids Res.* **2018**, *46*, W486–W494. [CrossRef] [PubMed]

29. Liu, X.; Gao, J.; Chen, J.; Wang, Z.; Shi, Q.; Man, H.; Guo, S.; Wang, Y.; Li, Z.; Wang, W. Identification of metabolic biomarkers in patients with type 2 diabetic coronary heart diseases based on metabolomic approach. *Sci. Rep.* **2016**, *6*, 30785. [CrossRef] [PubMed]

30. Trygg, J.; Wold, S. Orthogonal projections to latent structures (O-PLS). *J. Chemom.* **2002**, *16*, 119–128. [CrossRef]

31. Girelli, C.R.; Del Coco, L.; Papadia, P.; De Pascali, S.A.; Fanizzi, F.P. Harvest year effects on Apulian EVOOs evaluated by 1H NMR based metabolomics. *Peer J.* **2016**, *4*, e2740. [CrossRef]

32. De Castro, F.; Benedetti, M.; Antonaci, G.; Del Coco, L.; De Pascali, S.; Muscella, A.; Fanizzi, F.P. Response of cisplatin resistant Skov-3 cells to [Pt (O, O′-Acac)(γ-Acac)(DMS)] treatment revealed by a metabolomic 1H-NMR study. *Molecules* **2018**, *23*, 2301. [CrossRef] [PubMed]

33. Del Coco, L.; Felline, S.; Girelli, C.; Angilè, F.; Magliozzi, L.; Almada, F.; D'Aniello, B.; Mollo, E.; Terlizzi, A.; Fanizzi, F. 1H NMR Spectroscopy and MVA to evaluate the effects of caulerpin-based diet on Diplodus sargus lipid profiles. *Mar. Drugs* **2018**, *16*, 390. [CrossRef] [PubMed]

34. Girelli, C.R.; De Pascali, S.A.; Del Coco, L.; Fanizzi, F.P. Metabolic profile comparison of fruit juice from certified sweet cherry trees (Prunus avium L.) of Ferrovia and Giorgia cultivars: A preliminary study. *Food Res. Int.* **2016**, *90*, 281–287. [CrossRef]

35. Song, B.; Zhang, G.; Zhu, W.; Liang, Z. ROC operating point selection for classification of imbalanced data with application to computer-aided polyp detection in CT colonography. *Int. J. Comput. Assist. Radiol. Surg.* **2014**, *9*, 79–89. [CrossRef] [PubMed]

36. Van den Berg, R.A.; Hoefsloot, H.C.; Westerhuis, J.A.; Smilde, A.K.; van der Werf, M.J. Centering, scaling, and transformations: Improving the biological information content of metabolomics data. *BMC Genom.* **2006**, *7*, 142. [CrossRef]

37. Xia, J.; Wishart, D.S. MetPA: A web-based metabolomics tool for pathway analysis and visualization. *Bioinformatics* **2010**, *26*, 2342–2344. [CrossRef] [PubMed]

38. Piepoli, M.F.; Hoes, A.W.; Agewall, S.; Albus, C.; Brotons, C.; Catapano, A.L.; Cooney, M.T.; Corrà, U.; Cosyns, B.; Deaton, C.; et al. 2016 European guidelines on cardiovascular disease prevention in clinical practice: The sixth joint task force of the European society of cardiology and other societies on cardiovascular disease prevention in clinical practice (constituted by representatives of 10 societies and by invited experts) developed with the special contribution of the European association for cardiovascular prevention & rehabilitation (EACPR). *Eur. Heart J.* **2016**, *37*, 2315–2381. [PubMed]

39. American Diabetes Association. 9. Pharmacologic approaches to glycemic treatment: Standards of medical care in diabetes-2019. *Diabetes Care* **2019**, *42* (Suppl. 1), S90–S102. [CrossRef]

40. Fan, T.W.M. Metabolite profiling by one-and two-dimensional NMR analysis of complex mixtures. *Prog. Nucl. Mag. Res. SP* **1996**, *28*, 161–219. [CrossRef]

41. Schneider, D.J.; Sobel, B.E. PAI-1 and diabetes: A journey from the bench to the bedside. *Diabetes Care* **2012**, *35*, 1961–1967. [CrossRef]

42. Fernández-Real, J.M.; López-Bermejo, A.; Ricart, W. Cross-talk between iron metabolism and diabetes. *Diabetes* **2002**, *51*, 2348–2354. [CrossRef]

43. Ussher, J.R.; Elmariah, S.; Gerszten, R.E.; Dyck, J.R. The emerging role of metabolomics in the diagnosis and prognosis of cardiovascular disease. *J. Am. Coll. Cardiol.* **2016**, *68*, 2850–2870. [CrossRef] [PubMed]

44. Adams, S.H. Emerging perspectives on essential amino acid metabolism in obesity and the insulin-resistant state. *Adv. Nutr.* **2011**, *2*, 445–456. [CrossRef] [PubMed]

45. Xu, F.; Tavintharan, S.; Sum, C.F.; Woon, K.; Lim, S.C.; Ong, C.N. Metabolic signature shift in type 2 diabetes mellitus revealed by mass spectrometry-based metabolomics. *J. Clin. Endocrin. Metab.* **2013**, *98*, E1060–E1065. [CrossRef] [PubMed]

46. Wurtz, P.; Tiainen, M.; Mäkinen, V.P.; Kangas, A.J.; Soininen, P.; Saltevo, J.; Keinänen-Kiukaanniemi, S.; Mäntyselkä, P.; Lehtimäki, T.; et al. Circulating metabolite predictors of glycemia in middle-aged men and women. *Diabetes Care* **2012**, *35*, 1749–1756. [CrossRef]

47. Huffman, K.M.; Shah, S.H.; Stevens, R.D.; Bain, J.R.; Muehlbauer, M.; Slentz, C.A.; Tanner, C.J.; Kuchibhatla, M.; Houmard, J.A.; Newgard, C.B.; et al. Relationships between circulating metabolic intermediates and insulin action in overweight to obese, inactive men and women. *Diabetes Care* **2009**, *32*, 1678–1683. [CrossRef]

48. Qiu, G.; Zheng, Y.; Wang, H.; Sun, J.; Ma, H.; Xiao, Y.; Li, Y.; Yuan, Y.; Yang, H.; Li, X.; et al. Plasma metabolomics identified novel metabolites associated with risk of type 2 diabetes in two prospective cohorts of Chinese adults. *Int. J. Epidemiol.* **2016**, *45*, 1507–1516. [CrossRef]

49. Boden, G.; Chen, X.; Stein, T.P. Gluconeogenesis in moderately and severely hyperglycemic patients with type 2 diabetes mellitus. *Am. J. Physiol. Endocrinol. Metab.* **2001**, *280*, E23–E30. [CrossRef]

50. Magnusson, I.; Rothman, D.L.; Katz, L.D.; Shulman, R.G.; Shulman, G.I. Increased rate of gluconeogenesis in type II diabetes mellitus. *J. Clin. Investig.* **1992**, *90*, 1323–1327. [CrossRef] [PubMed]

51. Gastaldelli, A.; Toschi, E.; Pettiti, M.; Frascerra, S.; Quiñones-Galvan, A.; Sironi, A.M.; Natali, A.; Ferrannini, E. Effect of physiological hyperinsulinemia on gluconeogenesis in nondiabetic subjects and in type 2 diabetic patients. *Diabetes* **2001**, *50*, 1807–1812. [CrossRef] [PubMed]

52. Burrage, L.C.; Nagamani, S.C.S.; Campeau, P.M.; Lee, B.H. Branched-chain amino acid metabolism: From rare Mendelian diseases to more common disorders. *Hum. Mol. Genet.* **2014**, *23*, R1–R8. [CrossRef]

53. Würtz, P.; Soininen, P.; Kangas, A.J.; Rönnemaa, T.; Lehtimäki, T.; Kähönen, M.; Viikari, J.S.; Raitakari, O.T.; Ala-Korpela, M. Branched-chain and aromatic amino acids are predictors of insulin resistance in young adults. *Diabetes Care* **2013**, *36*, 648–655. [CrossRef]

54. Lee, A.; Jang, H.B.; Ra, M.; Choi, Y.; Lee, H.J.; Park, J.Y.; Kang, J.H.; Park, K.H.; Park, S.I.; Song, J. Prediction of future risk of insulin resistance and metabolic syndrome based on Korean boy's metabolite profiling. *Obes. Res. Clin. Pract.* **2015**, *9*, 336–345. [CrossRef] [PubMed]

55. Gar, C.; Rottenkolber, M.; Prehn, C.; Adamski, J.; Seissler, J.; Lechner, A. Serum and plasma amino acids as markers of prediabetes, insulin resistance, and incident diabetes. *Crit. Rev. Clin. Lab. Sci.* **2018**, *55*, 21–32. [CrossRef]

56. Saleem, T.; Dahpy, M.; Ezzat, G.; Abdelrahman, G.; Abdel-Aziz, E.; Farghaly, R. The profile of plasma free amino acids in type 2 diabetes mellitus with insulin resistance: Association with microalbuminuria and macroalbuminuria. *Appl. Biochem. Biotechnol.* **2019**, 1–14. [CrossRef] [PubMed]

57. Shin, A.C.; Fasshauer, M.; Filatova, N.; Grundell, L.A.; Zielinski, E.; Zhou, J.Y.; Scherer, T.; Lindtner, C.; White, P.J.; Lapworth, A.L.; et al. Brain insulin lowers circulating BCAA levels by inducing hepatic BCAA catabolism. *Cell Metab.* **2014**, *20*, 898–909. [CrossRef]

58. Gowda, K.; Zinnanti, W.J.; LaNoue, K. The influence of diabetes on glutamate metabolism in retinas. *J. Neurochem.* **2011**, *117*, 309–320. [CrossRef]

59. Layman, D.K.; Walker, D.A. Potential importance of leucine in treatment of obesity and the metabolic syndrome. *J. Nutr.* **2006**, *136*, 319S–323S. [CrossRef] [PubMed]

60. Mirmiranpour, H.; Khaghani, S.; Bathaie, S.Z.; Nakhjavani, M.; Kebriaeezadeh, A.; Ebadi, M.; Gerayesh-Nejad, S.; Zangooei, M. The preventive effect of L-Lysine on lysozyme glycation in Type 2 Diabetes. *Acta Med. Iran* **2016**, *54*, 24–31.

61. Olio, F.; Vanhooren, V.; Chen, C.C.; Slagboom, P.E.; Wuhrer, M.; Franceschi, C. N-glycomic biomarkers of biological aging and longevity: A link with inflammaging. *Ageing Res. Rev.* **2013**, *12*, 685–698.

62. Labrie, F.; Korner, A. Effect of glucagon, insulin, and thyroxine on tyrosine transaminase and tryptophan pyrrolase of rat liver. *Arch. Biochem. Biophys.* **1969**, *129*, 75–78. [CrossRef]

63. Welsh, P.; Rankin, N.; Li, Q.; Mark, PB.; Würtz, P.; Ala-Korpela, M.; Marre, M.; Poulter, N.; Hamet, P.; Chalmers, J.; et al. Circulating amino acids and the risk of macrovascular, microvascular and mortality outcomes in individuals with type 2 diabetes: Results from the ADVANCE trial. *Diabetologia* **2018**, *61*, 1581–1591. [CrossRef] [PubMed]

64. Kopple, J.D. Phenylalanine and tyrosine metabolism in chronic kidney failure. *J. Nutr.* **2007**, *137*, 1586S–1590S. [CrossRef] [PubMed]

65. Fernstrom, J.D.; Fernstrom, M.H. Tyrosine, phenylalanine, and catecholamine synthesis and function in the brain. *J. Nutr.* **2007**, *137*, 1539S–1547S. [CrossRef] [PubMed]

66. Hellmuth, C.; Kirchberg, F.F.; Lass, N.; Harder, U.; Peissner, W.; Koletzko, B.; Reinehr, T. Tyrosine is associated with insulin resistance in longitudinal metabolomic profiling of obese children. *J. Diabetes Res.* **2016**, *2016*, 2108909. [CrossRef] [PubMed]

67. Kawanaka, M.; Nishino, K.; Oka, T.; Urata, N.; Nakamura, J.; Suehiro, M.; Kawamoto, H.; Chiba, Y.; Yamada, G. Tyrosine levels are associated with insulin resistance in patients with nonalcoholic fatty liver disease. *Hepat. Med.* **2015**, *7*, 29–35. [CrossRef] [PubMed]

68. Jones, L.L.; McDonald, D.A.; Borum, P.R. Acylcarnitines: Role in brain. *Prog. Lipid Res.* **2010**, *49*, 61–75. [CrossRef]
69. Samuel, V.T.; Shulman, G.I. The pathogenesis of insulin resistance: Integrating signaling pathways and substrate flux. *J. Clin. Investig.* **2016**, *126*, 12–22. [CrossRef]

Journal of
Clinical Medicine

Article

Circulating Soluble CD36 is Similar in Type 1 and Type 2 Diabetes Mellitus versus Non-Diabetic Subjects

Esmeralda Castelblanco [1,2], Lucía Sanjurjo [3], Mireia Falguera [4,5], Marta Hernández [6], José-Manuel Fernandez-Real [7,8], Maria-Rosa Sarrias [3,9], Nuria Alonso [2,10,*] and Didac Mauricio [1,2,5,*]

[1]   Department of Endocrinology & Nutrition, University, Hospital de la Santa Creu i Sant Pau & Institut d'Investigació Biomèdica Sant Pau (IIB Sant Pau), 08041 Barcelona, Spain; esmeraldacas@gmail.com

[2]   Centre for Biomedical Research on Diabetes and Associated Metabolic Diseases (CIBERDEM), 08907 Barcelona, Spain

[3]   Innate Immunity Group, Health Sciences Research Institute Germans Trias i Pujol, 08916 Badalona, Spain; lsanjurjo@igtp.cat (L.S.); mrsarrias@igtp.cat (M.-R.S.)

[4]   Primary Health Care Cervera, Gerència d'Atenció Primaria, Institut Català de la Salut, Unitat de Suport a la Recerca, Institut Universitari d'Investigació en Atenció Primària Jordi Gol (IDIAP Jordi Gol), 25200 Cervera, Spain; mireiafalguera@hotmail.com

[5]   Biomedical Research Institute of Lleida (IRBLleida) & University of Lleida, 25198 Lleida, Spain

[6]   Department of Endocrinology & Nutrition, University Hospital Arnau de Vilanova & Biomedical Research Institute of Lleida (IRBLleida), 25198 Lleida, Spain; martahernandezg@gmail.com

[7]   Department of Diabetes, Endocrinology & Nutrition, Hospital Dr Josep Trueta & Biomedical Research Institute of Girona (IDIBGI), 17007 Girona, Spain; jmfreal@idibgi.org

[8]   Centre for Biomedical Research on Physiopathology of Obesity and Nutrition (CIBEROBN), 17007 Girona, Spain

[9]   Centre for Biomedical Research on Liver and Digestive Diseases (CIBEREHD), 28029 Madrid, Spain

[10]   Department of Endocrinology & Nutrition, University Hospital Germans Trias i Pujol & Health Sciences Research Institute, 08916 Badalona, Spain

*   Correspondence: nalonso32416@yahoo.es (N.A.); didacmauricio@gmail.com (D.M.); Tel.: +34-93-497-8860 (N.A.); +34-93-556-5655 (D.M.)

Received: 26 March 2019; Accepted: 16 May 2019; Published: 18 May 2019

Abstract: The aim of this study was to determine whether plasma concentrations of sCD36 (soluble CD36) are associated with the presence of type 1 or type 2 diabetes. Plasma levels of sCD36 were analysed in 1023 subjects (225 type 1 diabetes (T1D) patients, 276 type 2 diabetes (T2D) patients, and 522 non-diabetic control subjects) using an enzyme-linked immunosorbent assay (ELISA). Multinomial and logistic regression models were performed to evaluate associations with sCD36 and its association with diabetes types. There were no significant differences in sCD36 ($p = 0.144$) among study groups, neither in head-to-head comparisons: non-diabetic versus T1D subjects ($p = 0.180$), non-diabetic versus T2D subjects ($p = 0.583$), and T1D versus T2D patients ($p = 0.151$). In the multinomial model, lower sCD36 concentrations were associated with older age ($p < 0.001$), tobacco exposure ($p = 0.006$), T2D ($p = 0.020$), and a higher-platelets count ($p = 0.004$). However, in logistic regression models of diabetes, sCD36 showed only a weak association with T2D. The current findings show a weak association of circulating sCD36 with type 2 diabetes and no association with T1D.

Keywords: sCD36; type 1 diabetes mellitus; type 2 diabetes mellitus

1. Introduction

CD36 belongs to the class B scavenger receptor family, it is a transmembrane glycoprotein of 88-kDa expressed in a wide variety of cell types [1]. CD36 recognises many types of ligands, such as oxidised low-density lipoprotein (oxLDL), advanced glycation products, free fatty acids, or apoptotic cell surfaces [2–5]. This multivariate ligand recognition allows it to exert several functions, depending on the cell type. Among others, CD36 internalise modified lipoproteins (e.g., oxidised LDL), which facilitates cholesterol accumulation in macrophages or the regulation of fatty acid uptake through the plasma membrane and the subsequent metabolism [6,7]. CD36 expression or function are involved in angiogenesis, macrophage and platelet activation, lipid metabolism, and inflammation [8–11], which influences susceptibility to certain metabolic diseases, such as obesity, insulin resistance, impaired glucose tolerance, and fatty liver disease [7,12,13]. CD36 has been reported to be associated with type 2 diabetes (T2D) [14]. In monocytes and macrophages, CD36 is up-regulated by hyperglycaemia, insulin resistance, and oxLDL [8,10,15–17]. Moreover, oxLDL concentrations are elevated in type 2 diabetes [18,19].

The soluble form of CD36 (sCD36) was first described to be associated with metabolic syndrome in a small group of subjects: sCD36 was elevated in T2D patients compared with both lean (5-fold) and obese (2- to 3-fold) non-diabetic subjects [20]. Moreover, it has been described to be associated with insulin resistance, impaired glucose tolerance, and fatty liver in non-diabetic subjects [21–23]. Other studies have found discordant results concerning the circulating concentrations of sCD36 in patients with T2D, resulting in no difference [24,25], or increased concentrations in T2D diabetic subjects with chronic kidney disease [26], compared to non-diabetic subjects. However, a recent study found no association of sCD36 with glucose or additional factors that were described in previous studies to be associated with its concentrations [24]. To our knowledge, no studies have assessed sCD36 concentrations in patients with type 1 diabetes (T1D).

The aim of the current study was to assess whether the circulating concentrations of sCD36 are associated with diabetes, T1D or T2D, compared with a non-diabetic control group, all of them without previous cardiovascular events and free from chronic kidney disease. Taken together, the findings show a weak association of circulating sCD36 with T2D and no association with T1D.

2. Experimental Section

2.1. Subjects

One thousand and twenty-three patients were recruited from the University Hospitals Arnau de Vilanova (Lleida, Spain) and Germans Trias i Pujol (Badalona, Spain) to participate in this study. Subjects with T1D ($n = 225$) and T2D ($n = 276$) were selected from two previous studies [27,28]. The control group ($n = 522$) was selected from a population-based study in our region (subjects without diabetes on the basis of values of HbAc1 <6.5% and glucose <126 mg/dL) [29]. Additionally, 50 subjects, 22 T2D patients and 28 non-diabetic subjects, were recruited at the University Hospital Germans Trias i Pujol to perform the sub-study using flow cytometric analysis. The inclusion criteria for the four groups were as follows: age range 20–85 years, absence of established chronic kidney disease (defined as calculated glomerular filtration rate <60 mL/min and/or urine albumin/creatinine ratio >299 mg/g), and absence of known clinical cardiovascular events or associated revascularization procedures, including coronary heart disease, cerebrovascular disease, or peripheral vascular disease (including the diagnosis of diabetic foot disease).

Blood samples were collected in the fasting state, and blood tests were performed using standard laboratory methods [28]. Urine tests were performed in diabetic patients following standard laboratory methods. Subjects were considered to have hypertension or dyslipidaemia if they were under anti-hypertensive or lipid-lowering agent treatment, respectively. Blood samples for sCD36 measurements were collected in the fasting state with EDTA tubes, processed immediately after extraction, and stored at −80 °C at the biobanks of the participant centres until determination. Peripheral blood samples for flow cytometric and real-time PCR assays were collected on the same condition and processed after one hour on a shaker plate. Local ethics committees of both participating centres approved the study (12/2009 and P11/11) which followed the Declaration of Helsinki. All participants provided written informed consent before inclusion.

2.2. Determination for sCD36 by ELISA

Plasma concentrations of human sCD36 were measured using a commercially available ELISA kit (Nordic BioSite, Täby, Sweden), in accordance with the manufacturer's instructions. Appropriate dilutions of patient samples were measured in duplicates. Briefly, standards and samples were pipetted into the wells and incubated for 2 h on the plate shaker at room temperature (RT). After washing 4 times, a detection antibody was added and incubated for 2 h on the plate shaker at RT. Following washing 4 times, a streptavidin conjugated with Horseradish Peroxidase solution was added to the wells and incubated for 50 min on the plate shaker at RT and protected from light. After washing 4 times, 50 µL substrate solution (TMB-tetramethylbenzidine) was added to the wells and incubated for 20 min at RT protected from light. The colour development was stopped with 50 µL of stop solution. Absorbance was read at 450 nm using SpectraMax 340PC384 (Molecular Devices, LLC, Sunnyvale, California, USA). The results were analysed using a log–log curve fit. The calibration was performed with recombinant human CD36 in the concentration range 1.95–250 ng/mL. For sCD36 concentrations lower than the detection limit, a value of 0.05 ng/mL was assigned (in the non-diabetic control group, $n = 64/522$; T1D, $n = 41/225$; T2D, $n = 24/276$). Intra-assay and inter-assay precision coefficients of the ELISA assay were 4%–6% and 8%–12% respectively, as determined by the manufacturer.

2.3. Consistency Test of sCD36 Determination

Due to the variability found in the concentrations of sCD36 in the published articles, we wanted to rule out that this was a consequence of differences in pre-analytical conditions. We designed an experiment to compare sCD36 assessed by ELISA (Nordic BioSite, Täby, Sweden) that included 96 samples from 14 volunteers. Plasma samples were processed at different centrifugation velocities: $1500\times$ *g* (following our protocol), $1850\times$ g and $3000\times$ g (velocities used in other studies) [20,30]. Moreover, it was assessed whether the number of freeze-thaw cycles affected the sCD36 concentration. Samples without previous thaw, as used in our protocol, and with three freeze-thaw cycles, as requested by the in-house ELISA found in the literature, were evaluated [31].

2.4. Flow Cytometric Analysis

The analysis of CD36 by flow cytometry included 50 subjects, 22 T2D patients and 28 non-diabetic subjects. Peripheral blood samples were collected in the fasting state into vacutainers (BD Bioscience, San Jose, CA, USA) containing EDTA. Erythrocytes were lysed with ammonium chloride (BD Pharm Lyse™, San Jose, CA, USA), then the cells were washed with phosphate buffered saline (PBS), and incubated with monoclonal antibodies against CD36 (Miltenyi Biotec, Bergisch Gladbach, Germany), CD3, and CD14 (BD Biosciences, San Jose, CA, USA). Thereafter, flow cytometric analysis was carried out on a Fortessa SORP flow cytometer (BD Biosciences, San Jose, CA, USA), using sample acquisition and analysis software FACSDiva v6.2 (BD Biosciences, San Jose, CA, USA).

2.5. Real-Time PCR

The analysis of CD36 mRNA included the above-mentioned 50 subjects (22 T2D patients and 28 without diabetes). Erythrocytes were lysed with ammonium chloride (BD Pharm Lyse™), then the cells were washed with PBS and disrupted with QUIzol Lysis Reagent (Quiagen, Hilden, Germany), and total RNA was extracted using the miRNA Mini Kit (Quiagen, Hilden, Germany). Total RNA (1 µg) was reverse-transcribed using the Transcriptor First Strand cDNA Synthesis Kit (Roche, Basel, Switzerland). Each reaction was then amplified in a LightCycler® 480 PCR system using SYBR Green I Master (Roche, Basel, Switzerland). The primer pairs used were: forward primer

5′->3′ (GAGAACTGTTATGGGGCTAT) and reverse primer 5′->3′ (TTCAACTGGAGAGGCAAAGG). Gene expression values were normalized to the expression levels of glyceraldehyde 3-phosphate dehydrogenase (GAPDH).

2.6. Statistical Analyses

Descriptive statistics of mean (standard deviation) or median (interquartile range) were estimated for quantitative variables while, for qualitative variables, absolute and relative frequencies were used. The differences between groups T1D, T2D, and control group, were assessed by Student's *t*-test, ANOVA, the Mann–Whitney test, or the Kruskal–Wallis test. Regarding the differences in qualitative variables, this was assessed by the Chi-squared test or Fisher's exact test. Tukey's correction was used to account for multiple tests. Correlations were assessed by Spearman's rank correlation coefficient. Associations between qualitative variables and sCD36 concentration were tested with the *T* test. Multinomial logistic regression models of sCD36 quartile were performed to study the variables associated with their levels, first quartile was used as reference. In these models, variables of the bivariate analysis with a *p* value <0.2 and clinical relevance were used. Logistic regression models of T1D versus non-diabetic and T2D versus non-diabetic were performed. For the models of diabetes, a non-diabetic age- and sex-matched group from the non-diabetic participants (n = 522) were selected for T1D (n = 314) and T2D (n = 247). In all models, the goodness-of-fit assumption was tested by the Hosmer–Lemeshow test. Logistic regression models were checked by receiver-operating characteristic (ROC) curves and DeLong's test to correlated ROC curves. Estimates of odds ratios was reported with corresponding 95% confidence intervals and statistical significance was established as a *p*-value <0.05. The R statistical software, version 3.3.1, and SPSS software (version 22, IBM, SPSS, Chicago, IL, USA) were used for all of the analyses.

3. Results

A total of 1023 individuals, clinically well-characterised (225 type 1 diabetic patients, 276 type 2 patients, and 522 non-diabetic subjects), were included in the study. In the overall study population, the mean age was 50 years, and up to 45.4% were men. The age in the different groups of participants ranged between 23 and 77 years in patients with T1D, between 40 and 75 years in patients with T2D, and between 27 and 84 years in the non-diabetic control group. There were significant differences among the three study groups for all studied variables, except for tobacco exposure. All the clinical characteristics can be found in Table 1.

3.1. Circulating sCD36 in the Study Groups

There were no significant differences in the median plasma concentrations of sCD36 among the three study groups (p = 0.144). In head-to-head comparisons, concentrations of sCD36 were not different between non-diabetic and T1D subjects (2.84 ng/mL versus 3.66 ng/mL; p = 0.180), non-diabetic and T2D subjects (2.84 ng/mL versus 2.62 ng/mL; p = 0.583), or T1D and T2D (3.66 ng/mL versus 2.62 ng/mL; p = 0.151). In addition, there were no significant differences in sCD36 levels divided in quartiles among the three groups (p = 0.062) (Table 1).

Furthermore, despite median, sCD36 concentrations were similar in women and men with T2D (2.99 ng/mL versus 2.15 ng/mL; p = 0.374), or T1D (3.69 ng/mL versus 3.58 ng/mL; p = 0.812), median sCD36 concentrations were significantly higher in women in the non-diabetic group (3.40 ng/mL versus 2.36 ng/mL; p = 0.047). In addition, sCD36 concentrations tended to be increased in T2D statin users (1.88 ng/mL versus 3.53 ng/mL; p = 0.084) and tended to be decreased in T1D statin users (4.53 ng/mL versus 2.70 ng/mL; p = 0.145).

Table 1. Clinical and anthropometrical characteristics of the study groups.

Variables [1]	Control	T1D	T2D	*p* Value	*p* Control versus T1D	*p* Control versus T2D	*p* T1D versus T2D
	N = 522	*N* = 225	*N* = 276				
Sex, Men	212 (40.6%)	109 (48.4%)	143 (51.8%)	0.006	0.086	0.009	0.509
Age, years	49 (39–59)	44 (38.0–51.0)	59 (51–66)	<0.001	<0.001	<0.001	<0.001
BMI, Kg/m^2	25.5 (23.4–28.3)	25.6 (22.6–28.1)	30.2 (27.9–34.7)	<0.001	0.557	<0.001	<0.001
Waist, cm	94.0 (85.0–102)	88.0 (80.0–97.0)	104 (97.5–112)	<0.001	<0.001	<0.001	<0.001
Alcohol mg/day	2.86 (0–12.3)	2.37 (0–7.19)	1.53 (0–9.92)	0.030	0.095	0.055	0.685
Tobacco	270 (51.7%)	116 (51.6%)	149 (54.0%)	0.805	1.000	0.977	0.977
Hypertension	81 (16%)	55 (24.4%)	153 (55.4%)	<0.001	0.009	<0.001	<0.001
sBP, mmHg	121 (110–132)	128 (116–138)	139 (127–150)	<0.001	<0.001	<0.001	<0.001
dBP, mmHg	76.0 (70–83)	75.0 (68.5–80)	76.0 (70–84)	0.020	0.026	0.889	0.028
Dyslipidaemia	115 (22.6%)	96 (42.7%)	129 (46.7%)	<0.001	<0.001	<0.001	0.412
Antiplatelet	-	63 (28%)	99 (35.9%)	<0.001	-	-	0.076
Statins	114 (22.4%)	96 (42.7%)	116 (42%)	<0.001	<0.001	<0.001	0.958
Glucose, mg/dL	90 (84–96)	159 (107–208)	148 (117–182)	<0.001	<0.001	<0.001	0.619
HbA1c, %	5.5 (5.2–5.8)	7.40 (7.00–7.9)	7.60 (6.80–8.5)	<0.001	<0.001	<0.001	0.230
Creatinine mg/dL	0.79 (0.70–0.93)	0.78 (0.65–0.87)	0.79 (0.68–0.93)	0.011	0.009	0.567	0.046
eGFR	86.2 (77.4–97.6)	95.2 (82.7–108)	86.8 (77.5–100)	<0.001	<0.001	0.383	<0.001
ALT U/L	17 (13–23)	17 (13.8–21)	19 (16–28)	<0.001	0.719	<0.001	<0.001
Triglycerides, mg/dL	87 (67.0–125)	65 (53–86)	118 (84.8–166)	<0.001	<0.001	<0.001	<0.001
Total-C, mg/dL	196 (175–226)	180 (164–201)	184 (163–206)	<0.001	<0.001	<0.001	0.180
HDL, mg/dL	57 (48–68)	62.5 (53–74)	48 (40–58)	<0.001	<0.001	<0.001	<0.001
LDL, mg/dL	119 (100–142)	101 (84.5–116)	108 (88.4–129)	<0.001	<0.001	<0.001	0.003
Haemoglobin, g/dL	14.2 (13.3–15.2)	14 (13.1–14.9)	13.7 (12.9–14.7)	<0.001	0.031	<0.001	0.098
Haematocrit, %	42.4 (3.58)	41.5 (3.51)	41.1 (3.54)	<0.001	0.003	<0.001	0.545
Platelets, 10^9/L	233 (203–273)	223 (197–256)	226 (188–274)	0.019	0.027	0.078	0.759
Diabetes duration, years	-	21 (15–29)	8 (4–13.5)	<0.001	-	-	<0.001
sCD36, ng/mL	2.84 (0.56–9.06)	3.66 (0.69–18.7)	2.62 (0.66–7.57)	0.144	0.180	0.583	0.151
sCD36 Quartiles [2]				0.062	0.259	0.259	0.036
(<−0.562)	134 (25.7%)	53 (23.6%)	64 (23.2%)				
(−0.562,1.05)	127 (24.3%)	49 (21.8%)	80 (29%)				
(1.05,2.24)	133 (25.5%)	51 (22.7%)	77 (27.9%)				
(>2.24)	128 (24.5%)	72 (32.0%)	55 (19.9%)				

[1] Mean (SD) or median (Interquartile range) for continuous variables and frequency (percentage) for categorical variables; [2] Soluble CD36 in quartiles was Ln transformed. BMI, body mass index; sBP, systolic blood pressure; dBP, diastolic blood pressure; ALT, alanine aminotransferase; eGFR; estimated glomerular filtration rate (MDRD4_IDMS equation); HDL, high-density lipoprotein; LDL, low-density lipoprotein.

3.2. Associations of Circulating sCD36 in the Study Groups

When considering all three groups together, age, body mass index (BMI), systolic blood pressure, triglycerides, and haematocrit were negatively correlated with sCD36 (*p*-values between $p < 0.05$ and $p < 0.01$). Moreover, in the control group, age, BMI, systolic blood pressure, triglycerides, glucose, and HbA1c were negatively correlated. The number of platelets was positively correlated with sCD36 (*p*-values between $p < 0.05$ and $p < 0.01$) (Figure S1). Finally, glucose was positively correlated with sCD36 in T1D ($p < 0.05$). In the T2D group, sCD36 was not correlated with any of the analysed variables (Figure 1).

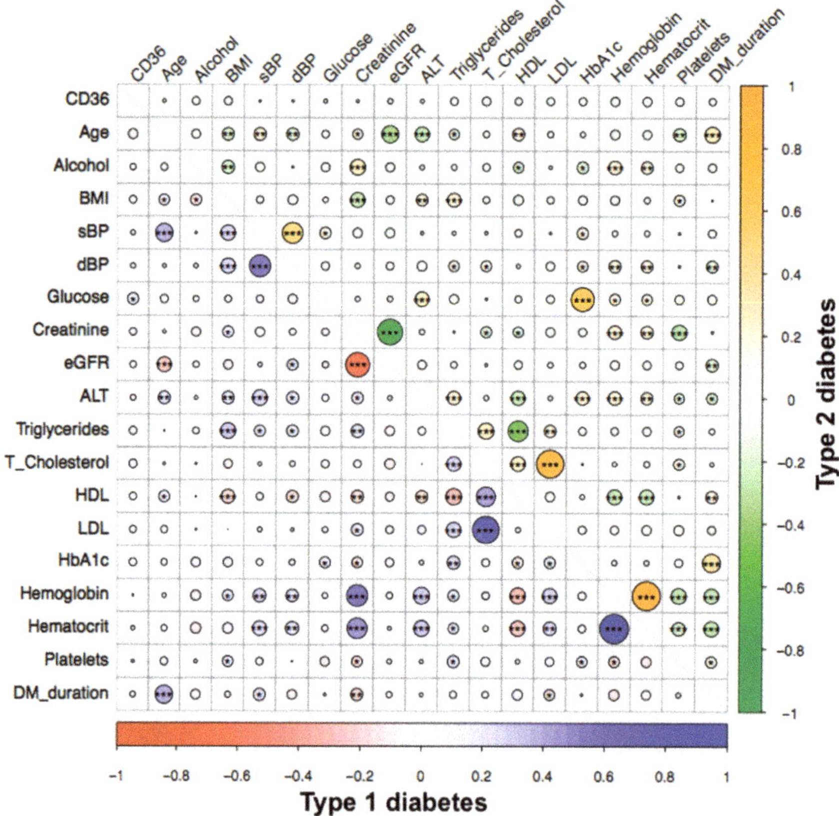

Figure 1. Correlation analysis of circulating sCD36 and clinical variables in type 1 diabetes (T1D) (red–blue) and type 2 diabetes (T2D) (green–orange) groups. Spearman's correlation is significant at the 0.05 *, 0.01 **, and 0.001 *** levels (2-tailed). BMI, Body mass index; sBP, systolic blood pressure; dBP, diastolic blood pressure; eGFR, estimated glomerular filtration rate (MDRD4_IDMS equation); ALT, alanine aminotransferase; HDL, high-density lipoprotein; LDL, low-density lipoprotein; HbA1c, glycated haemoglobin; DM duration, diabetes duration.

On the other hand, in patients with T2D, significantly higher sCD36 concentrations were found in antiplatelet users ($p = 0.039$) and in patients with dyslipidaemia ($p = 0.048$). Significantly lower sCD36 concentrations were found in T1D diabetes patients with tobacco exposure ($p = 0.005$). In non-diabetic subjects, significantly lower sCD36 concentrations were found in subjects with hypertension ($p < 0.001$), dyslipidaemia ($p = 0.044$), and statin users ($p = 0.035$) (Table 2).

Table 2. Association of circulating sCD36 concentration and qualitative parameters in the whole study group, type 1 diabetes, type 2 diabetes, and non-diabetic subjects.

Variables		Whole Group		Control		T1D		T2D	
		Median (IQR)	p-Value	Median (IQR)	p-Value	Median (IQR)	p-Value	Median (IQR)	p-Value
Sex	Men	2.59 (0.48–8.51)	0.089	2.36 (0.36–8.06)	0.047	3.58 (0.69–27.9)	0.812	2.15 (0.65–6.29)	0.374
	Women	3.37 (0.75–9.99)		3.40 (0.74–10.3)		3.69 (0.71–15.1)		2.99 (0.82–8.28)	
Hypertension	No	3.17 (0.69–10.9)	0.024	3.23 (0.73–9.93)	0.001	3.57 (0.70–20.7)	0.751	2.29 (0.64–8.39)	0.875
	Yes	2.65 (0.48–7.17)		1.04 (0.17–5.10)		3.72 (0.68–8.79)		2.83 (0.79–6.14)	
Dyslipidaemia	No	3.09 (0.66–10.3)	0.274	3.15 (0.69–10.0)	0.044	4.53 (0.85–28.5)	0.090	1.80 (0.48–6.95)	0.048
	Yes	2.79 (0.51–8.33)		2.04 (0.26–7.80)		2.75 (0.44–9.89)		3.53 (1.01–7.95)	
Statins	No	3.09 (0.66–10.1)	0.247	3.15 (0.69–9.93)	0.035	4.53 (0.84–23.7)	0.145	1.88 (0.48–6.88)	0.084
	Yes	2.74 (0.48–8.32)		2.01 (0.26–7.64)		2.70 (0.42–13.1)		3.54 (1.02–8.03)	
Obesity	No	3.15 (0.65–10.2)	0.130	3.12 (0.69–9.59)	0.030	3.58 (0.65–24.4)	0.644	2.51 (0.65–7.08)	0.546
	Yes	2.43 (0.53–7.59)		1.70 (0.26–5.91)		3.81 (0.84–10.4)		2.73 (0.81–7.86)	
Tobacco	No	3.68 (0.82–10.5)	0.003	2.92 (0.64–9.58)	0.321	6.38 (1.16–25.2)	0.005	3.24 (0.94–8.58)	0.176
	Yes	2.55 (0.48–7.94)		2.60 (0.48–8.17)		2.37 (0.05–10.9)		2.14 (0.65–6.38)	
Anti-platelet	No	2.86 (0.56–9.55)	0.393	-	-	4.29 (0.68–19.2)	0.410	1.98 (0.48–6.81)	0.039
	Yes	3.27 (0.88–8.42)		-		2.66 (0.73–17.8)		3.59 (1.07–8.06)	
Diabetes	No	2.84 (0.56–9.06)	0.597	-	-	-	-	-	-
	Yes	2.99 (0.67–9.52)		-		-		-	

3.3. Multinomial Logistic Regression Models of sCD36

We generated multinomial logistic regression models for sCD36 quartiles of the whole study group. Additionally, to avoid model meddling because of the group characteristics, we analysed T1D, T2D, and non-diabetic control groups separately. First quartile (Q1) was taken as a referent group and was compared with the other groups, defined as second (Q2), third (Q3), and quarter (Q4) quartile, respectively. In the whole group, the variables independently associated with sCD36 concentrations were: older age with all quartiles of sCD36 (Q2: OR = 0.978, p = 0.014; Q3: OR = 0.980, p = 0.024, and Q4: OR = 0.967, p < 0.001), tobacco exposure with Q4 (OR = 0.581, p = 0.006); T2D with all quartiles of sCD36 (Q2: OR = 2.072, p = 0.006; Q3: OR = 1.852, p = 0.020, and Q4: OR = 1.925, p = 0.020); and Q3 of platelets with the Q2 (OR = 1.858, p = 0.020) and Q4 (OR = 2.204, p = 0.004) (Table S1).

In the multinomial model for T1D, the variables independently associated with sCD36 concentrations were: tobacco exposure with Q3 (OR 0.289, p = 0.009), and Q4 (OR = 0.352, p = 0.015), higher BMI with Q3 (OR = 1.125, p = 0.044), and higher glomerular filtration rate (eGFR) with Q2 (OR = 0.968, p = 0.013) (Table S2).

In the multinomial model of T2D, MPV (mean platelet volume) was associated with Q2 (OR = 0.581, p = 0.027) and Q3 (OR = 0.561, p = 0.023), insulin treatment was associated with Q2 (OR = 5.838, p = 0.033), treatment with oral anti-diabetic agents was associated with Q3 (OR = 4.108, p = 0.029), and antiplatelet treatment was associated with Q3 (OR2.694, p = 0.026) (Table S3).

Finally, in the non-diabetic group, the variables independently associated with sCD36 concentrations were: higher levels of HbA1c with all quartiles (Q2: OR = 0.243, p < 0.001; Q3: OR = 0.260, p < 0.001; and Q4: OR = 0.299, p = 0.002), older age and higher triglycerides with Q4 (OR = 0.975, p = 0.039 and OR = 0.994, p = 0.019 respectively), and Q3 of platelets with Q2 of sCD36 (OR = 2.414, p = 0.029), Q2 of platelets with Q3 of sCD36 (OR = 0.443, p = 0.031) and Q3 and Q4 of platelets with Q4 of sCD36 (OR = 4.318, p < 0.001 and OR = 2.635, p = 0.026, respectively) (Table S4).

3.4. Association of sCD36 with Type 1 and Type 2 Diabetes

The multiple logistic regression model of the non-diabetic control group versus T1D showed that the variables independently associated with the presence of T1D were: older age (OR = 0.969, p = 0.003), the presence of hypertension (OR = 2.020, p = 0.007), presence of dyslipidaemia (OR = 3.127, p < 0.001), higher haematocrit (OR = 0.862, p < 0.001), and the Q4 of platelets (OR = 0.472, p = 0.008) (Figure 2a). The ROC curve showed non-discriminative power when sCD36 was added to the model (Figure S2a).

In the logistic regression model of the non-diabetic controls versus T2D, we observed that the variables independently associated with the presence of T2D were: female gender (OR = 0.294, p < 0.001), hypertension (OR = 2.746, p < 0.001), dyslipidaemia (OR = 1.822, p = 0.011), high BMI (OR = 1.234, p < 0.001), higher haematocrit (OR = 0.763, p < 0.001), Q2 and Q3 of platelets (OR = 0.405, p = 0.004 and OR = 0.483, p = 0.024 respectively), and Q2 and Q3 of sCD36 (OR = 2.479, p = 0.006 and OR = 2.078, p = 0.032, respectively) (Figure 2b). The models with and without sCD36 showed that the addition of sCD36 concentrations did not provide additional discriminative power to the model area under the ROC curve of 0.844, p < 0.001 versus 0.851, p < 0.001 (Figure S2b).

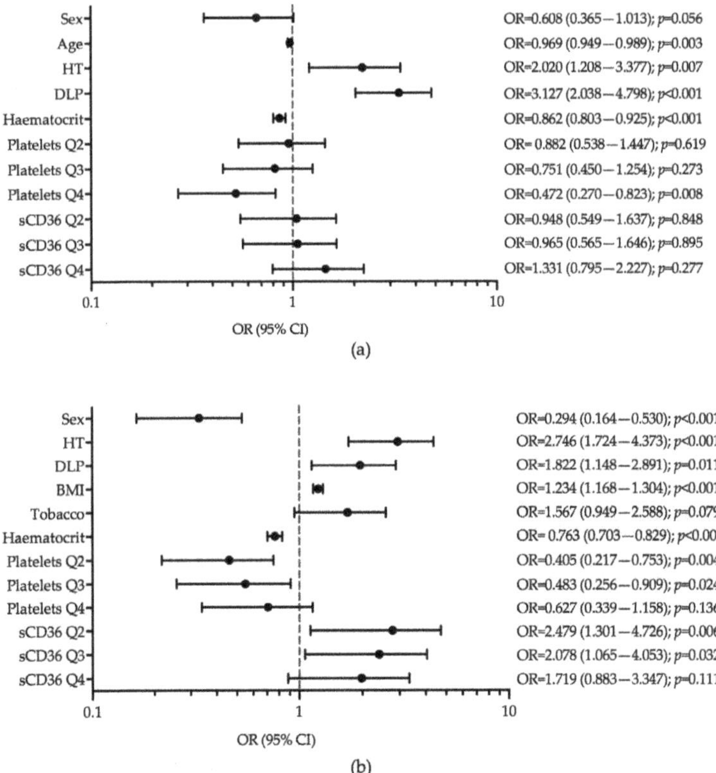

Figure 2. Logistic regression models for presence of diabetes. (**a**) T1D versus non-diabetic group (Hosmer and Lemeshow test *p*-value = 0.789); (**b**) T2D versus non-diabetic (Hosmer and Lemeshow test *p*-value = 0.686). Tobacco, tobacco exposure.

3.5. Flow Cytometric Analysis of CD36 and Real-Time PCR

Flow cytometric analysis was used to detect the expression of CD36 in circulating mononuclear cells from T2D patients and non-diabetic control subjects. In the recruited group, the mean age was 55 years, and up to 46% were men. No significant differences were found in the expression of CD36 scavenger receptor either in monocytes, or in leukocyte population between T2D patients and the non-diabetic group (Figure S3). Regarding mRNA expression, no differences in CD36 expression were found between T2D patients and the non-diabetic control group (Figure S4).

3.6. Consistency Test in sCD36 Assessment

In the assay of sCD36, performed in healthy volunteers, we did not find differences in sCD36 among the different centrifugation forces or among the number of freeze-thaw cycles. Paired plasma samples prepared by 1500× *g*, 1850× *g*, and 3000× *g* gave nearly identical results and a good correlation within each assay. Moreover, in the assays that compared samples with three, or without, freeze-thaw cycles, there were no differences in the results of the three different centrifugation forces, which were highly correlated (Table S5).

4. Discussion

In the present study, we did not observe differences in plasma sCD36 concentrations between patients with T1D, T2D, and non-diabetic control subjects, all of them with normal renal function and free from advanced late diabetic complications, including macrovascular disease. Overall, there was an association between sCD36 plasma concentrations and cardiovascular risk factors such as hypertension, dyslipidaemia, tobacco exposure, age, sex, and BMI. In the current study, we measured most of the variables that have been previously described to be associated with sCD36. These factors included the use of several medications, age, sex, systolic and diastolic blood pressure, tobacco exposure, BMI, HDL, LDL, total cholesterol, triglycerides, haematocrit, haemoglobin, platelets, serum creatinine, blood glucose, and HbA1c. In addition, we have assessed for the first time, the contribution of all of these factors together in a large group of subjects with and without diabetes. Importantly, we found increased sCD36 in patients with T2D that had dyslipidaemia or that were antiplatelet users. As dyslipidaemia and the use of antiplatelet drugs is relatively frequent in T2D, this could have led to spuriously elevated sCD36 in previous studies in patients with T2D.

Regarding the correlations between sCD36 concentrations and clinical variables, such as components of the metabolic syndrome among them insulin resistance, there is a large variability [20,21,23,24,32,33]. In the non-diabetic group, sCD36 concentrations were negatively correlated with glucose and HbA1c, and positively correlated with the number of platelets, as already described in the literature. However, we found a negative correlation of sCD36 with age, BMI, and triglycerides, in contrast to the reported findings of a previous study [21,32,34]. Interestingly, and in accordance with the findings of another study, we found no significant correlations between sCD36 with all variables analysed in T2D [24]. This may explain, at least in part, that our results are discordant with some of the previous studies that found an association of sCD36 with insulin resistance, components of the metabolic syndrome, and the risk of T2D [20,32,35]. An additional reason for these differences may lie in the diverse characteristics of the study populations, which include differences in age, duration of diabetes, metabolic syndrome components, the presence of complications, and the use of antidiabetic medications. The characteristics of our population study faithfully reflects the characteristics of the Spanish population [36]. Regarding medication, and in contrast to the findings reported, we did not find differences in sCD36 concentrations between diabetic patients with or without statin medication [26]. Furthermore, the number of subjects with diabetes in our study was larger than that in previous studies and included patients with T1D for the first time. Previous studies with a lower number of subjects found higher levels of sCD36 in subjects with T2D [20,21,35]. However, in a subsequent study with 200 T2D patients, although these subjects exhibited higher sCD36 concentrations, the only characteristics that predicted sCD36 were diabetes duration and markers of platelet activation [34]. The latter findings are in line with ours.

To our knowledge, this is the first study that addresses the potential association of sCD36 with diabetes as a primary study objective. We did not find a clear association of sCD36 concentrations with T2D, the evaluation of the ROC curve shows that sCD36 does not contribute additional power to the predictive model. In fact, we did find a weak association of sCD36 with T2D, which may partly be due to the clinical and metabolic characteristics, among them waist circumference, which is an indirect measure of insulin resistance. In order to demonstrate that insulin resistance per se may be an associated factor to sCD36, direct measurements of insulin resistance should have been used. Regarding T1D, we found no association with sCD36. Patients with T1D showed younger age and had a high-frequency of hypertension and dyslipidaemia. Since sCD36 concentrations were not related to T1D, this may indicate indirectly that hyperglycaemia, which is a common feature in patients with either T1D or T2D, is not one of the main factors contributing to plasma sCD36 concentrations. Our findings are in line with some recent studies that found no association of sCD36 with diabetes or its associated metabolic disturbances in middle-aged subjects with metabolic syndrome [24], or patients with early-onset coronary artery disease [25]. Another study found that baseline sCD36 did not predict diabetes independently of fasting glucose and insulin in a non-diabetic population [22]. In that study, the addition of sCD36 in a multivariate model of diabetes prediction had no impact on diabetes prediction.

CD36 signalling responds to ox-LDL, and on the one hand, reduces macrophage motility and probably induces the trapping of macrophages in the arterial intima, which promotes atherosclerosis. On the other hand, it induces chronic inflammation and contributes to insulin resistance common in obesity and dyslipidaemia, both associated to metabolic disorders [37,38]. However, this does not seem to be consistently associated with its circulating levels. Regarding the determination of circulating sCD36, there is a lack of well-characterized or standardized methods to assess its concentrations [30]. Despite this scarcity, as reported previously [30], the pre-analytical procedures do not explain the variability of the results reported in different studies. Differences between studies may be explained by several factors: (a) regarding the heterogeneous origin of CD36 particles in the plasma of patients [39,40], our data show no differences in the CD36 surface expression from monocytes and leukocytes of peripheral blood among patients with T2D under antidiabetic treatment and non-diabetic subjects. (b) Although hyperglycaemia was shown to influence CD36 mRNA expression increasing the de novo CD36 synthesis in monocytes from healthy subjects [41], we did not find differences in the CD36 mRNA expression of monocytes and leukocytes from peripheral blood between non-diabetic subjects and patients with T2D under anti-diabetic treatment. (c) In poorly-controlled patients, lower plasma levels of glycated CD36 and higher levels of non-glycated CD36 have been described in comparison with well-controlled diabetic patients [39], and this fact may account also for differences with different assays. (d) Another condition that could influence the concentration of CD36 is the number of microparticles to which it is associated, e.g., CD36 associated with platelet-derived microparticles may be increased in proportion to these cell particles in non-diabetic subjects [40,42]. Actually, our findings showed that increased circulating platelet numbers are independently associated with a moderate and a high-sCD36 concentration in non-diabetic subjects.

Our study has some limitations. First, although we adjusted for risk factors known to be associated with diabetes in our analyses, the possibility that some other confounding factors play a role may have been incompletely accounted for in plasma sCD36. Second, we did not validate our assay with tests from other manufacturers. Third, the cross-sectional design precludes conclusions about causality, and therefore prospective studies are needed to establish the usefulness of the sCD36 plasma concentration as a predictive factor for T1D and T2D. Finally, as we did not measure insulin resistance in the study subjects, this precludes the assessment of the impact of this metabolic variable on plasma sCD36 concentration.

5. Conclusions

Circulating plasma sCD36 concentrations do not appear to be a biomarker of T1D or T2D. However, our study does not completely rule out the possibility that sCD36 may constitute a biomarker of cardiovascular events in diabetes and other populations. The follow-up of the cohorts with and without diabetes subjects may help to clarify this question.

Supplementary Materials: The following are available online at http://www.mdpi.com/2077-0383/8/5/710/s1, Table S1: Multinomial regression model for circulating sCD36 in the whole study group. Table S2. Multinomial regression model for sCD36 in type 1 diabetic group. Table S3. Multinomial regression model for sCD36 in type 2 diabetic group. Table S4. Multinomial regression model for sCD36 in the non-diabetic control group. Table S5. Comparison of pre-analytical conditions of sCD36. Figure S1. Correlation analysis of circulating sCD36 and clinical variables in whole (red–blue) and non-diabetic subject (green–orange) groups. Figure S2. Receiver operating characteristics (ROC) curve showing the relationship between sensitivity and 1-specificity in determining the discriminatory ability of the logistic regression model with and without sCD36 as predictor. Figure S3. Ex vivo flow cytometric analysis of CD36 from type 2 diabetic patients and non-diabetic subjects. Figure S4. CD36 mRNA expression was not different between T2D and non-diabetic controls.

Author Contributions: Conceptualization, M.-R.S., N.A. and D.M.; Data curation, M.F.; Formal analysis, E.C.; Investigation, L.S., M.F. and M.H.; Methodology, E.C. and N.A.; Project administration, D.M.; Writing—original draft, E.C.; Writing—review & editing, J.-M.F.-R., M.-R.S., N.A. and D.M. D.M. was the principal investigator and had the primary responsibility for the final content. All the authors read and approved the final manuscript.

Funding: This research was supported by grants from the European Foundation for the Study of Diabetes (2014-EFSD-00914). CIBER for Diabetes and Associated Metabolic Diseases (CIBERDEM) and CIBER on Liver and Digestive Diseases (CIBEREHD) and CIBER on Physiopathology of Obesity and Nutrition (CIBEROBN) are an initiative from Carlos III National Institute of Health, Spain.

Acknowledgments: We are grateful to Aase Handberg for her highly useful and constructive comments. Also, the authors thank Nuria Villalmanzo (she holds a fellowship from Pla Estratègic i Innovació en Salut PERIS, Departament de Salut de la Generalitat de Catalunya), and Jordi Real for their valuable assistance in conducting the laboratory technics and statistical analysis, respectively. We thank the IGTP Flow Citometry Core Facility and staff (Marco Fernández and Gerard Requena) for technical assistance in flow cytometry experiments and analysis. We want to particularly acknowledge the patients, IGTP-HUGTP, and IRBLleida (B.0000682) Biobanks integrated in the Spanish National Biobanks Network of Instituto de Salud Carlos III (PT17/0015/0045 and PT17/0015/0027 respectively,) and Tumor Bank Network of Catalonia for its collaboration.

Conflicts of Interest: The authors declare no conflict of interest.

References

1. Silverstein, R.L.; Febbraio, M. CD36, a Scavenger receptor involved in immunity, metabolism, angiogenesis, and behavior. *Sci. Signal.* **2009**, *2*, re3. [CrossRef]
2. Asch, A.S.; Barnwell, J.; Silverstein, R.L.; Nachman, R.L. Isolation of the thrombospondin membrane receptor. *J. Clin. Investig.* **1987**, *79*, 1054–1061. [CrossRef]
3. Hoebe, K.; Georgel, P.; Rutschmann, S.; Du, X.; Mudd, S.; Crozat, K.; Sovath, S.; Shamel, L.; Hartung, T.; Zähringer, U.; et al. CD36 is a sensor of diacylglycerides. *Nature* **2005**, *433*, 523–527. [CrossRef]
4. Mikołajczyk, T.P.; Skrzeczyńska-Moncznik, J.E.; Zarębski, M.A.; Marewicz, E.A.; Wiśniewska, A.M.; Dzięba, M.; Dobrucki, J.W.; Pryjma, J.R. Interaction of human peripheral blood monocytes with apoptotic polymorphonuclear cells. *Immunology* **2009**, *128*, 103–113. [CrossRef]
5. Endemann, G.; Stanton, L.W.; Madden, K.S.; Bryant, C.M.; White, R.T.; Protter, A.A. CD36 is a receptor for oxidized low density lipoprotein. *J. Biol. Chem.* **1993**, *268*, 11811–11816. [PubMed]
6. Ibrahimi, A.; Abumrad, N.A. Role of CD36 in membrane transport of long-chain fatty acids. *Curr. Opin. Clin. Nutr. Metab. Care* **2002**, *5*, 139–145. [CrossRef] [PubMed]
7. Coburn, C.T.; Hajri, T.; Ibrahimi, A.; Abumrad, N.A. Role of CD36 in membrane transport and utilization of long-chain fatty acids by different tissues. *J. Mol. Neurosci.* **2001**, *16*, 117–121. [CrossRef]
8. Collotteixeira, S.; Martin, J.; McDermottroe, C.; Poston, R.; McGregor, J. CD36 and macrophages in atherosclerosis. *Cardiovasc. Res.* **2007**, *75*, 468–477. [CrossRef]
9. Febbraio, M.; Hajjar, D.P.; Silverstein, R.L. CD36: A class B scavenger receptor involved in angiogenesis, atherosclerosis, inflammation, and lipid metabolism. *J. Clin. Invest.* **2001**, *108*, 785–791. [CrossRef] [PubMed]
10. Febbraio, M.; Silverstein, R.L. CD36: Implications in cardiovascular disease. *Int. J. Biochem. Cell Biol.* **2007**, *39*, 2012–2030. [CrossRef] [PubMed]
11. Nicholson, A.C.; Han, J.; Febbraio, M.; Silverstein, R.L.; Hajjar, D.P. Role of CD36, the Macrophage Class B Scavenger Receptor, in Atherosclerosis. *Ann. N. Y. Acad. Sci.* **2001**, *947*, 224–228. [CrossRef]
12. He, J.; Lee, J.H.; Febbraio, M.; Xie, W. The emerging roles of fatty acid translocase/CD36 and the aryl hydrocarbon receptor in fatty liver disease. *Exp. Biol. Med. (Maywood)* **2011**, *236*, 1116–1121. [CrossRef]
13. Kennedy, D.J.; Kashyap, S.R. Pathogenic Role of Scavenger Receptor CD36 in the Metabolic Syndrome and Diabetes. *Metab. Syndr. Relat. Disord.* **2011**, *9*, 239–245. [CrossRef]
14. Aitman, T.J.; Glazier, A.M.; Wallace, C.A.; Cooper, L.D.; Norsworthy, P.J.; Wahid, F.N.; Al-Majali, K.M.; Trembling, P.M.; Mann, C.J.; Shoulders, C.C.; et al. Identification of Cd36 (Fat) as an insulin-resistance gene causing defective fatty acid and glucose metabolism in hypertensive rats. *Nat. Genet.* **1999**, *21*, 76–83. [CrossRef]
15. Griffin, E.; Re, A.; Hamel, N.; Fu, C.; Bush, H.; McCaffrey, T.; Asch, A.S. A link between diabetes and atherosclerosis: Glucose regulates expression of CD36 at the level of translation. *Nat. Med.* **2001**, *7*, 840–846. [CrossRef]
16. Liang, C.P.; Han, S.; Okamoto, H.; Carnemolla, R.; Tabas, I.; Accili, D.; Tall, A.R. Increased CD36 protein as a response to defective insulin signaling in macrophages. *J. Clin. Investig.* **2004**, *113*, 764–773. [CrossRef]
17. Sampson, M.J.; Davies, I.R.; Braschi, S.; Ivory, K.; Hughes, D.A. Increased expression of a scavenger receptor (CD36) in monocytes from subjects with Type 2 diabetes. *Atherosclerosis* **2003**, *167*, 129–134. [CrossRef]

18. Kopprasch, S.; Pietzsch, J.; Kuhlisch, E.; Fuecker, K.; Temelkova-Kurktschiev, T.; Hanefeld, M.; Kühne, H.; Julius, U.; Graessler, J. In vivo evidence for increased oxidation of circulating LDL in impaired glucose tolerance. *Diabetes* **2002**, *51*, 3102–3106. [CrossRef]

19. Hoogeveen, R.C.; Ballantyne, C.M.; Bang, H.; Heiss, G.; Duncan, B.B.; Folsom, A.R.; Pankow, J.S. Circulating oxidised low-density lipoprotein and intercellular adhesion molecule-1 and risk of type 2 diabetes mellitus: The Atherosclerosis Risk in Communities Study. *Diabetologia* **2007**, *50*, 36–42. [CrossRef]

20. Handberg, A.; Levin, K.; Levin, K.; Højlund, K.; Beck-Nielsen, H.; Beck-Nielsen, H. Identification of the oxidized low-density lipoprotein scavenger receptor CD36 in plasma: A novel marker of insulin resistance. *Circulation* **2006**, *114*, 1169–1176. [CrossRef]

21. Handberg, A.; Lopez-Bermejo, A.; Bassols, J.; Vendrell, J.; Ricart, W.; Fernandez-Real, J.M. Circulating soluble CD36 is associated with glucose metabolism and interleukin-6 in glucose-intolerant men. *Diabetes Vasc. Dis. Res.* **2009**, *6*, 15–20. [CrossRef]

22. Handberg, A.; Norberg, M.; Stenlund, H.; Hallmans, G.; Attermann, J.; Eriksson, J.W. Soluble CD36 (sCD36) Clusters with Markers of Insulin Resistance, and High sCD36 Is Associated with Increased Type 2 Diabetes Risk. *J. Clin. Endocrinol. Metab.* **2010**, *95*, 1939–1946. [CrossRef]

23. Fernández-Real, J.M.; Handberg, A.; Ortega, F.; Højlund, K.; Vendrell, J.; Ricart, W. Circulating soluble CD36 is a novel marker of liver injury in subjects with altered glucose tolerance. *J. Nutr. Biochem.* **2009**, *20*, 477–484. [CrossRef]

24. Alkhatatbeh, M.; Ayoub, N.; Mhaidat, N.; Saadeh, N.; Lincz, L. Soluble cluster of differentiation 36 concentrations are not associated with cardiovascular risk factors in middle-aged subjects. *Biomed. Rep.* **2016**, *4*, 1–7. [CrossRef]

25. Krzystolik, A.; Dziedziejko, V.; Safranow, K.; Kurzawski, G.; Rać, M.; Sagasz-Tysiewicz, D.; Poncyljusz, W.; Jakubowska, K.; Chlubek, D.; Rać, M.E. Is plasma soluble CD36 associated with cardiovascular risk factors in early onset coronary artery disease patients? *Scand. J. Clin. Lab. Investig.* **2015**, *75*, 1–9. [CrossRef]

26. Chmielewski, M.; Bragfors-Helin, A.C.; Stenvinkel, P.; Lindholm, B.; Anderstam, B. Serum soluble CD36, assessed by a novel monoclonal antibody-based sandwich ELISA, predicts cardiovascular mortality in dialysis patients. *Clin. Chim. Acta* **2010**, *411*, 2079–2082. [CrossRef]

27. Carbonell, M.; Castelblanco, E.; Valldeperas, X.; Betriu, A.; Traveset, A.; Granado-Casas, M.; Hernández, M.; Vázquez, F.; Martín, M.; Rubinat, E.; et al. Diabetic retinopathy is associated with the presence and burden of subclinical carotid atherosclerosis in type 1 diabetes. *Cardiovasc. Diabetol.* **2018**, *17*, 1–10. [CrossRef]

28. Alonso, N.; Traveset, A.; Rubinat, E.; Ortega, E.; Alcubierre, N.; Sanahuja, J.; Hernández, M.; Betriu, A.; Jurjo, C.; Fernández, E.; et al. Type 2 diabetes-associated carotid plaque burden is increased in patients with retinopathy compared to those without retinopathy. *Cardiovasc. Diabetol.* **2015**, *14*, 33–39. [CrossRef]

29. Vilanova, M.B.; Falguera, M.; Marsal, J.; Rubinat, E.; Alcubierre, N.; Castelblanco, E.; Granado-Casas, M.; Miró, N.; Molló, A.; Mata-Cases, M.; et al. Prevalence, clinical features and risk assessment of pre-diabetes in Spain: The prospective Mollerussa cohort study. *BMJ Open* **2017**, *7*, e015158. [CrossRef]

30. Lykkeboe, S.; Larsen, A.L.; Handberg, A. Lack of consistency between two commercial ELISAs and against an in-house ELISA for the detection of CD36 in human plasma. *Clin. Chem. Lab. Med.* **2012**, *50*, 1071–1074. [CrossRef]

31. Petta, S.; Handberg, A.; Marchesini, G.; Cammà, C.; Di Marco, V.; Cabibi, D.; Macaluso, F.S.; Craxì, A. High sCD36 plasma level is associated with steatosis and its severity in patients with genotype 1 chronic hepatitis C. *J. Viral Hepat.* **2012**, *20*, 174–182. [CrossRef]

32. Handberg, A.; Højlund, K.; Gastaldelli, A.; Flyvbjerg, A.; Dekker, J.M.; Petrie, J.; Piatti, P.; Beck-Nielsen, H. RISC Investigators; RISC Investigators Plasma sCD36 is associated with markers of atherosclerosis, insulin resistance and fatty liver in a nondiabetic healthy population. *J. Intern. Med.* **2012**, *271*, 294–304. [CrossRef]

33. Rać, M.; Krzystolik, A.; Rać, M.; Safranow, K.; Dziedziejko, V.; Goschorska, M.; Poncyljusz, W.; Chlubek, D. Is plasma-soluble CD36 associated with density of atheromatous plaque and ankle-brachial index in early-onset coronary artery disease patients? *Kardiol. Pol.* **2016**, *74*, 570–575.

34. Liani, R.; Halvorsen, B.; Sestili, S.; Handberg, A.; Santilli, F.; Vazzana, N.; Formoso, G.; Aukrust, P.; Davì, G. Plasma levels of soluble CD36, platelet activation, inflammation, and oxidative stress are increased in type 2 diabetic patients. *Free Radic. Biol. Med.* **2012**, *52*, 1318–1324. [CrossRef]

35. Handberg, A.; Skjelland, M.; Michelsen, A.E.; Sagen, E.L.; Krohg-Sørensen, K.; Russell, D.; Dahl, A.; Ueland, T.; Oie, E.; Aukrust, P.; et al. Soluble CD36 in plasma is increased in patients with symptomatic atherosclerotic carotid plaques and is related to plaque instability. *Stroke* **2008**, *39*, 3092–3095. [CrossRef]
36. Vinagre, I.; Mata-Cases, M.; Hermosilla, E.; Morros, R.; Fina, F.; Rosell, M.; Castell, C.; Franch-Nadal, J.; Bolíbar, B.; Mauricio, D. Control of Glycemia and Cardiovascular Risk Factors in Patients with Type 2 Diabetes in Primary Care in Catalonia (Spain). *Diabetes Care* **2012**, *35*, 774–779. [CrossRef]
37. Park, Y.M.; Febbraio, M.; Silverstein, R.L. CD36 modulates migration of mouse and human macrophages in response to oxidized LDL and may contribute to macrophage trapping in the arterial intima. *J. Clin. Investig.* **2009**, *119*, 136–145. [CrossRef] [PubMed]
38. Kennedy, D.J.; Kuchibhotla, S.; Westfall, K.M.; Silverstein, R.L.; Morton, R.E.; Febbraio, M. A CD36-dependent pathway enhances macrophage and adipose tissue inflammation and impairs insulin signalling. *Cardiovasc. Res.* **2011**, *89*, 604–613. [CrossRef] [PubMed]
39. Bernal-Lopez, R.M.; Llorente-Cortes, V.; López-Carmona, D.; Mayas, D.M.; Gomez-Huelgas, R.; Tinahones, F.J.; Badimon, L. Modulation of human monocyte CD36 by type 2 diabetes mellitus and other atherosclerotic risk factors. *Eur. J. Clin. Investig.* **2011**, *41*, 854–862. [CrossRef] [PubMed]
40. Alkhatatbeh, M.J.; Enjeti, A.K.; Acharya, S.; Thorne, R.F.; Lincz, L.F. The origin of circulating CD36 in type 2 diabetes. *Nutr. Diabetes* **2013**, *3*, e59. [CrossRef]
41. Lopez-Carmona, M.D.; Plaza-Seron, M.C.; Vargas-Candela, A.; Tinahones, F.J.; Gomez-Huelgas, R.; Bernal-Lopez, M.R. CD36 overexpression: A possible etiopathogenic mechanism of atherosclerosis in patients with prediabetes and diabetes. *Diabetol. Metab. Syndr.* **2017**, *9*, 55. [CrossRef] [PubMed]
42. Alkhatatbeh, M.J.; Mhaidat, N.M.; Enjeti, A.K.; Lincz, L.F.; Thorne, R.F. The putative diabetic plasma marker, soluble CD36, is non-cleaved, non-soluble and entirely associated with microparticles. *J. Thromb. Haemost.* **2011**, *9*, 844–851. [CrossRef] [PubMed]

Journal of
Clinical Medicine

Article

Low-Dose Aspirin for the Primary Prevention of Cardiovascular Disease in Diabetic Individuals: A Meta-Analysis of Randomized Control Trials and Trial Sequential Analysis

Ming-Hsun Lin [1], Chien-Hsing Lee [1], Chin Lin [2,3], Yi-Fen Zou [4], Chieh-Hua Lu [1,5], Chang-Hsun Hsieh [1,*] and Cho-Hao Lee [6,*]

[1]   Division of Endocrinology and Metabolism, Department of Internal Medicine, Tri-Service General Hospital, National Defense Medical Center, Taipe 11490, Taiwan; tim6801@msn.com (M.-H.L.); doc10383@gmail.com (C.-H.L.); undeca2001@gmail.com (C.-H.L.)
[2]   School of Public Health, National Defense Medical Center, Taipe 11490, Taiwan; xup6fup0629@gmail.com
[3]   Department of Research and Development, National Defense Medical Center, Taipe 11490, Taiwan
[4]   Department of Pharmacy, Tri-Service General Hospital, National Defense Medical Center, Taipe 11490, Taiwan; zou.yi.fen@gmail.com
[5]   Department of Medical Research, National Defense Medical Center, Taipe 11490, Taiwan
[6]   Division of Hematology and Oncology Medicine, Department of Internal Medicine, Tri-Service General Hospital, National Defense Medical Center, Taipe 11490, Taiwan
*   Correspondence: 10324@yahoo.com.tw (C.-H.H.); drleechohao@gmail.com (C.-H.L.); Tel.: +886-02-8792-3311 (ext. 12862) (C.-H.L.); Fax: +886-02-8792-7263 (C.-H.L.)

Received: 5 April 2019; Accepted: 3 May 2019; Published: 5 May 2019

Abstract: Background: Evidence of low-dose aspirin as the primary prevention strategy for cardiovascular disease (CVD) in diabetes are unclear. This study was designed to evaluate the effect of low-dose aspirin use for the primary prevention of CVD in diabetes. Methods: We collected randomized controlled trials of low-dose aspirin for the primary prevention of CVD in adults with diabetes lasting at least 12 months from Medline, Embase, and the Cochrane Library up to 10 November 2018. Two reviewers extracted data and appraised the reporting quality according to a predetermined protocol (CRD4201811830). This review was conducted using Cochrane standards, trial sequential analysis, and the Grading of Recommendation. The primary outcomes were major adverse cardiovascular events (MACE, including non-fatal myocardial infarction, ischemia stroke, and cardiovascular death) and an incidence of major hemorrhage (major intracranial hemorrhage and major gastrointestinal bleeding). Results: In this primary prevention (number = 29,814 participants) meta-analysis, low-dose aspirin use reduced the risk of MACE by 9% and increased the risk of major hemorrhage by 24%. The benefits were only observed in subjects of age $\geq$ 60 years while reducing the same risk of MACE. In efficacy, it reduced the risk of stroke but not myocardial infarction. No increase in all-cause mortality or cardiovascular death was observed. Conclusions: We suggested the use of low-dose aspirin as the primary prevention strategy for CVD in diabetes, particularly in an older population. The absolute benefits were largely counterbalanced by the bleeding hazard.

Keywords: aspirin; primary prevention; diabetes mellitus; meta-analysis; trial sequential analysis

---

1. Introduction

The leading cause of mortality and morbidity in diabetes is cardiovascular disease (CVD) [1]. Additionally, diabetes patients have a twofold increased risk of CVD (including coronary heart disease, stroke, and vascular deaths). It has been well established that aspirin shows benefits in reducing the cardiovascular morbidity and mortality as secondary prevention in subjects of diabetes [2]. Although there is no strong evidence for a specific dosage, low-dose aspirin use (81 or 100 mg/day) is common in clinical practice. Higher doses do not increase efficacy but increase the risk of bleeding [3]. However, the role of aspirin in the primary prevention of CVD in subjects of diabetes remains inconclusive [4].

The Antithrombotic Treatment Trialists' Collaboration in 2009 demonstrated that aspirin had no significant effect in diabetic patients. However, this previous consensus was controversial because of an apparent sex and age-related bias [5]. American Diabetes Association 2019 suggested that low-dose aspirin use (75–162 mg/day) was appropriate as the primary prevention strategy in diabetic patients aged more than 50 years old or individual with one additional cardiovascular risk factor and no increased risk of major bleeding [6]. The European Guidelines on CVD prevention do not support this prevention strategy because the evidence for it is weak [7]. Therefore, the appropriate therapy strategy for aspirin is still unclear.

There have been many large trials to detect the true effect of aspirin use; they all included diabetes patients without previous cardiovascular disease. A Study of Cardiovascular Events in Diabetes (ASCEND: 15,480 diabetic participants under 100 mg of aspirin once daily and placebo, median follow-up of 7.4 years), showed that individuals using aspirin had a 12% reduction in risk of major adverse cardiovascular events (MACE) but had a 29% increase in bleeding risk compared with a placebo group [8]. Another large trial, namely, the Aspirin in Reducing Events in the Elderly (ASPREE: 2057 diabetes patients with 100 mg aspirin once daily and placebo, median follow-up of 4.7 years) revealed a non-significant 10% reduction in the risk of MACE and a non-significant 30% increase in bleeding risk [9]. The other large study, namely the Japanese Primary Prevention of Atherosclerosis with Aspirin for Diabetes (JPAD2: 2160 diabetic participants with 81 or 100 mg daily and placebo, median follow-up of 10.3 years), which was a randomized, open-label, standard care-controlled trial, revealed no statistically significant benefit in cardiovascular events [10]. These large randomized control trials indicated that aspirin may have a positive effect on the prevention of CVD, but the increased bleeding risk should be considered. Furthermore, these results were compatible with results from the previous meta-analysis, which also showed inconsistent results about the role of aspirin in primary prevention of CVD and risk for bleeding among subjects with diabetes [11–15]. The possible explanation for this discrepancy may be attributable to several factors, such as dose of aspirin, age, gender, or ethnic factors.

The dosage of aspirin used in previous studies ranged from 100 mg every other day to 650 mg per day, which may have led to a selection bias and affected the results of the meta-analysis. Given the practical problem of aspirin use in diabetes, we aimed to focus on low-dose aspirin use in diabetes for primary prevention in this investigation. Owing to the recent publication of large randomized control trials, including JPAD2, ASCEND, and ASPREE trials, we aimed to effectively summarize the eligible evidence and promote the proper dissemination of this important clinical information. To accomplish this goal in our meta-analysis, we used the GRADE profiler guideline development tool software to assess the overall certainty of the evidence in this topic area. Statistical methods were used to evaluate the benefits and disadvantages of aspirin use in diabetes and to make an objective conclusion regarding its use as the primary prevention method for CVD in diabetic patients.

2. Experimental Section

2.1. Data Sources and Searches

We performed a systematic literature search in electronic datasets (i.e., PubMed, Embase, and Cochrane central) for randomized controlled trials (RCTs) that evaluated low-dose aspirin as the primary prevention strategy in diabetes (additional search details are listed in Table S2).

We also conducted manual screening for references from original articles, conference abstracts, and previous systematic reviews to identify eligible trials. We followed the Preferred Reporting Items for Systematic Reviews and Meta Analyses (PRISMA) guidelines for performing the systematic reviews and meta-analyses of RCTs (Figure 1). The protocol for this systematic review is registered with PROSPERO (no. CRD4201811830).

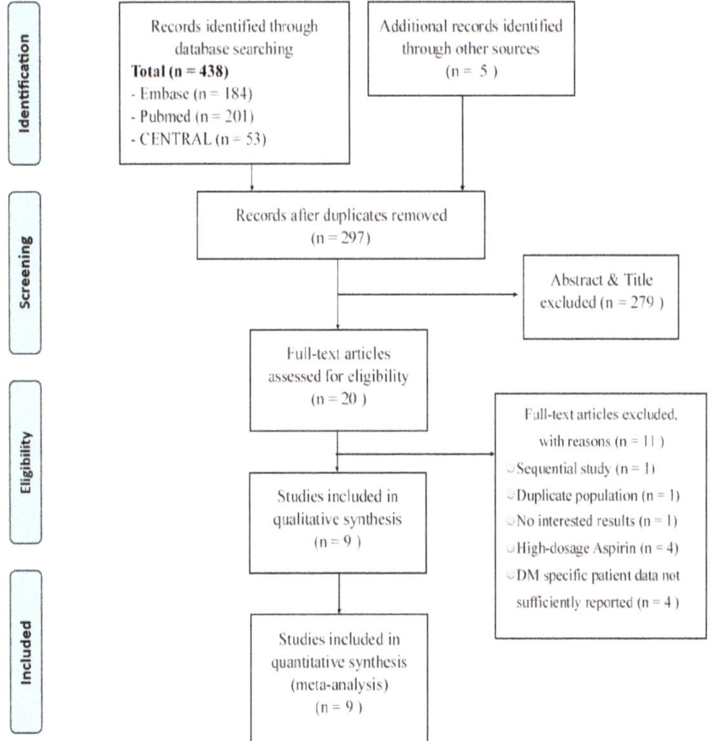

Figure 1. PRISMA flowchart of study selection. Flow diagram of the identification process for eligible studies.

2.2. Study Selection

Low-dose aspirin was defined as a daily aspirin regimen (≤100 mg). Patients with either type 1 or type 2 diabetes without previous major vascular events were included. Furthermore, peer-reviewed RCTs written in various languages were included. Trials examining aspirin use as the primary prevention strategy that had subgroups with diabetes were also included. Language with English or Chinese were included.

Trials identified from the literature search were initially screened for relevance based on titles and abstracts by two independent reviewers (Cho-Hao Lee and Chin Lin). Studies that did not meet these criteria during the title and abstract screen were excluded. Full-text reviews were then performed using the studies included after screening to ensure that they met the eligible criteria. Any disagreement between the two independent reviewers was resolved via group discussions.

2.3. Data Extraction and Quality Assessment

Two independent reviewers (Cho-Hao Lee and Chin Lin) extracted various data, including author names, publication year, geographic regions, trial names, study designs, sample sizes, and participant characteristics (mean age, sex, inclusion criteria, aspirin dosage, follow-up duration, and completion). The outcomes of interest including MACE (a composite of non-fatal myocardial infarction, non-fatal stroke, and death from cardiovascular causes), major hemorrhage (major intracranial hemorrhage and major gastrointestinal bleeding), and all-cause mortality were also extracted. To address the risk of bias with multiple data extractors, standardized "Google Excel" templates were created. Last, the third reviewer (Ming-Hsun Lin) double-checked the data for accuracy.

The quality of trials was appraised using the *Cochrane Handbook for Systematic Reviews of Interventions* [16]. Seven domains—selection bias, attrition bias, performance bias, detection bias, reporting bias, contamination bias, and other risks of bias—are listed in the Figure S1. Any disagreement between the two reviewers was resolved via group discussions.

2.4. Data Synthesis and Analysis

Data analysis was conducted as recommended in the *Cochrane Handbook for Systematic Reviews of Interventions* [17]. We used both random and fixed effects modeling to pool all outcomes and interpreted random-effects meta-analyses with consideration to the complete distribution of effects. We calculated dichotomous outcomes by using the Mantel–Haenszel method and evaluated the risk ratio (RR) with 95% CIs.

Heterogeneity and publication biases were evaluated by I^2 statistic and funnel plots with Egger's test. Statistically significant heterogeneity was defined as an I^2 statistic > 50%. The cause of heterogeneity was investigated for main outcomes by using sensitivity tests and a mixed-effects meta-regression model with variables including follow-up duration, mean age, sample sizes, country, completion, and publication year [17]. All statistical analyses were performed using the 'metafor' and 'meta' [18] packages of R software version 3.3.1 [19]. A two-tailed significance test (p value = 0.05) denoted statistical significance without multiplicity correction in all exploratory analyses.

Furthermore, to assess the certainty of the evidence for each outcome, we used the GRADE profiler guideline development tool software [20] and ranked the quality of the evidence according to the *Cochrane Handbook for Systematic Reviews of Interventions* [16,21]. This method takes issues related to internal validity (e.g., risk of bias, inconsistency, imprecision, and publication bias) and external validity (e.g., directness of results) into account. We downgraded the evidence from "high" certainty by one level for serious concerns and by two levels for very serious concerns.

A trial sequential analysis (TSA), which is similar to interim analyses in a single trial, was also conducted [22,23]. Monitoring boundaries were used to decide if a trial could be terminated early (i.e., when a p value was sufficiently small to show the anticipated effect). We also used monitoring boundaries to determine if a study produced a sufficiently small p value to demonstrate the anticipated power. The chance of random errors may have been increased owing to insufficient comparisons and the repetitive testing of pooled data when the estimated information samples had not been achieved [22,24,25]. TSA version 0.9 beta [26] was used for the quantification of information samples.

3. Results

The systematic search yielded 443 studies, and 297 studies remained after removing duplicates. Then 279 studies were excluded after reviewing the abstract and title. There were 20 eligibility full-text articles; 11 of them were excluded for several reasons after reviewing the full-text articles. The remaining nine randomized control trials involving 29,814 participants were included (Figure 1). Among the total sample, 14,897 patients were randomized to low-dose aspirin use, and 14,917 patients were randomized to a control group. Table 1 shows the characteristics of each study. Four of the nine trials contained subgroups with diabetes, and the other five trials were studies that primarily focused on diabetes. These trials were published from 1998 to 2018 and were conducted in more than five different countries; the mean follow-up period ranged from 3.6 years to 10.3 years. The dosage of aspirin used ranged from 100 mg/every other day to 100 mg/daily.

Meta-regression analyses were performed for MACE, myocardial infarction (MI), stroke, major hemorrhage, and all-cause mortality (see Table S3). There was no significant association of aspirin use with the above outcomes. However, although it was not surprising, all-cause mortality and mean age were related.

Table 1. Characteristics of the included randomized control trials.

Source	Trial Design	Country (Study Name)	Population (DM)	Case (Mean Age)	Males	Aspirin Dosage	Mean Follow up (Completion Rate)
MRC 1998 [27]	PC, RCT	UK (TPT)	Patients at high risk of IHD (type 1 and 2)	68 (57.3 years *)	100.0	75 mg/daily	6.7 years/98.9
Hansson et al. 1998 [28]	DB, RCT	Globally (HOT)	Participants with hypertension (type 1 and 2)	1501 (61.5 years)	NR	75 mg/daily	3.8 years/97.4
Sacco et al. 2003 [29]	MC, OP, RCT, 2 × 2 factorial design	Italy (PPP)	Participants > 50 years with > one cardiovascular risk factor (type 2)	1031 (64.2 years)	48.2	100 mg/daily	3.6 years/99.3
Ridker et al. 2005 [30]	MC, DB, RCT, 2 × 2 factorial design	USA (WHS)	Healthy women (NA)	1027 (54.6 years *)	0	100 mg/every other day	10.1 years/99.4
Belch et al. 2008 [31]	DB, RCT, 2 × 2 factorial design	UK (POPADAD)	Patients aged ≥ 40 years plus ABP ≤ 0.99 (type 1 and 2)	1276 (60 years)	44.1	100 mg/daily	6.7 years/99.5
Ikeda et al. 2014 [32]	OP, RCT	Japan (JPPP)	Elderly with multiple atherosclerotic risk factors (NR)	4903 (70.6 years *)	NR	100 mg/daily	5.0 years/98.7
Saito et al. 2017 [10]	MC, OP, RCT	Japan (JPAD2)	Patients with Type 2 Diabetes (type 2)	2160 (64.4 years)	55.5	81 or 100 mg/daily	10.3 years/63.8
ASCEND group 2018 [8]	DB, RCT	UK (ASCEND)	Patients with diabetes (type 1 and 2)	15,800 (63.3 years)	62.6	100 mg/daily	7.4 years/99.1
McNeil et al. 2018 [9]	DB, RCT	USA/Australia (ASPREE)	Health older (NR)	2057 (74 years *)	44	100 mg/daily	4.7 years/97.2

DM: diabetes mellitus; MC: multi-center; DB: double blind; OP: open label; RCT: randomized control trial; NR: Not reported; ABP: Ankle brachial pressure; TPT: thrombosis prevention trial; HOT: hypertension optimal treatment; PPP: primary prevention project; WHS: women's health study; WHS: women's health study; POPADAD: prevention of progression of arterial disease and diabetes; JPAD2: Japanese primary prevention of atherosclerosis with aspirin for diabetes 2, followed up study from JPAD [33]; JPPP: Japanese primary prevention project; ASCEND: a study of cardiovascular events in diabetes; ASPREE: Aspirin in reducing events in the elderly. *: data extracted from specific diabetes patients.

3.1. Major Adverse Cardiovascular Event and Major Hemorrhage

This study revealed a significant reduction in the risk of MACE (Figure 2A) in the low-dose aspirin group compared with the placebo and no-treatment groups (eight studies, RR 0.91, 95% CI 0.84–0.98). No heterogeneity was found between the studies (I^2 = 0%, p value = 0.99). Furthermore, TSA was used to analyze MACE. After including the three large trials (JPAD2, ASCEND, and ASPREE), the analysis showed a significant reduction in the risk of MACE with low-dose aspirin administration (Figure 2B).

The GRADE system demonstrated a moderate level recommendation for low-dose aspirin administration in diabetic patients (see Table 2).

Major hemorrhaging, including intracranial bleeding and gastrointestinal (GI) bleeding, was the most important adverse effect of aspirin use. The results revealed a significant increase in hemorrhage risk (24%) with aspirin administration (five studies, RR = 1.24, 95% CI =1.03–1.48). The heterogeneity between studies in this analysis was low (I^2 = 23%, p value = 0.27, see Figure 2C).

3.2. Efficacy and Safety End Point

We analyzed several secondary outcomes, including MI, coronary heart disease, and stroke, and summarized the results in Table 2.

There was no reduction in the risk of MI (six studies, RR = 1.01, 95% CI = 0.84–1.22, see Table 2 and Figure S2) or coronary heart disease (six studies, RR = 0.94, 95% CI = 0.84–1.06, see Table 2 and Figure S2). The heterogeneity of evidence was low for MI (I^2 = 21%, p value = 0.27), and no heterogeneity of evidence was found for coronary heart disease (I^2 = 0%, p value = 0.86).

Additionally, there was a reduction in the risk of stroke with aspirin use (seven studies, RR = 0.84, 95% CI = 0.73–0.97, see Table 2 and Figure S2). No heterogeneity of evidence was found in this analysis (I^2 = 0%, p value = 0.65). No significant reduction in the risk of cardiovascular death (five studies, RR = 0.94, 95% CI = 0.77–1.14, see Table 2 and Figure S2) or all-cause mortality (six studies, RR = 0.97, 95% CI = 0.90–1.06, see Table 2 and Figure S2) was found. Furthermore, there was no heterogeneity of evidence in cardiovascular death (I^2 = 0%, p value = 0.54) or all-cause mortality (I^2 = 0%, p value = 0.46).

This present study showed increased hemorrhagic risk by 24% in the aspirin group. The subgroup analyses of major hemorrhage events revealed no statistically significant difference in aspirin use regarding major gastro-intestinal or intra-cranial hemorrhage (Table 2).

3.3. Sensitivity Tests and Publication Bias

We used sensitivity tests, subgroup analyses, and meta-regression analyses to explore potential heterogeneity. Meta-regression analyses were performed for MACE, myocardial infarction (MI), stroke, major hemorrhage, and all-cause mortality (see Figure S2). There was no significant association of aspirin use with the above outcomes. There was no sex difference in our present study. However, although it was not surprising, all-cause mortality and mean age were related. Publication bias was evaluated by visual examination and Egger's regression test for all interested outcomes. Neither visual examinations nor Egger's regression test showed that significant publication bias existed (see Figure S3).

(A) Result of major cardiovascular event (MACE)

Study	Aspirin Event	Aspirin Size	Placebo Event	Placebo Size	Risk Ratio	RR	95%-CI	Weight (fixed)	Weight (random)
HOT, 1998	47	752	54	749		0.87	[0.59; 1.26]	4.1%	3.9%
PPP, 2003	20	519	22	512		0.90	[0.50; 1.62]	1.7%	1.6%
WHS, 2005	58	514	62	513		0.93	[0.67; 1.31]	4.8%	4.9%
POPADAD, 2008	105	638	108	638		0.97	[0.76; 1.24]	8.3%	9.3%
JPPP, 2014	86	2445	98	2458		0.88	[0.66; 1.17]	7.5%	6.9%
JPAD2, 2017	167	1262	171	1277		0.99	[0.81; 1.21]	13.0%	14.2%
ASCEND, 2018	658	7740	743	7740		0.89	[0.80; 0.98]	57.0%	55.8%
ASPREE, 2018	41	1027	47	1030		0.87	[0.58; 1.32]	3.6%	3.3%
Fixed effect model		**14897**		**14917**		**0.91**	**[0.84; 0.98]**	**100.0%**	---
Random effects model						**0.91**	**[0.84; 0.98]**	---	**100.0%**

Heterogeneity: $I^2 = 0\%$, $p = 0.99$

(B) Trial sequential analysis of MACE

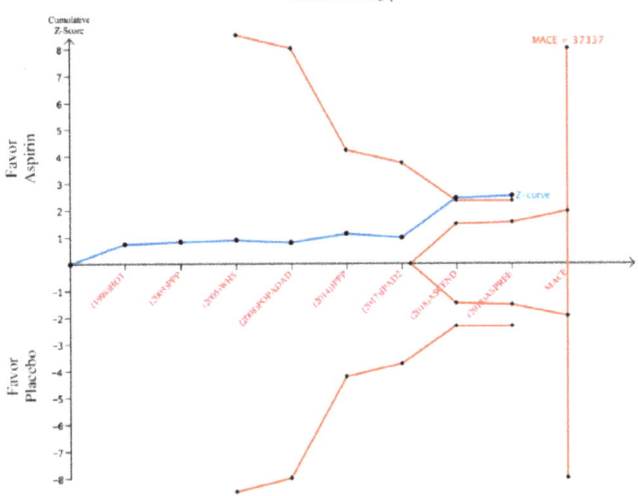

(C) Result of major hemorrhage

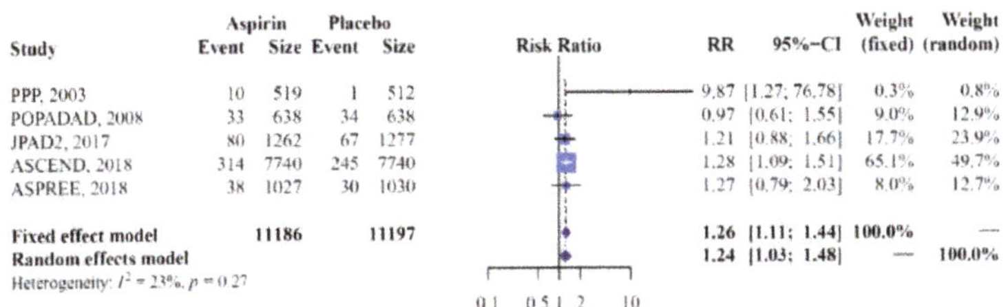

Study	Aspirin Event	Aspirin Size	Placebo Event	Placebo Size	Risk Ratio	RR	95%-CI	Weight (fixed)	Weight (random)
PPP, 2003	10	519	1	512		9.87	[1.27; 76.78]	0.3%	0.8%
POPADAD, 2008	33	638	34	638		0.97	[0.61; 1.55]	9.0%	12.9%
JPAD2, 2017	80	1262	67	1277		1.21	[0.88; 1.66]	17.7%	23.9%
ASCEND, 2018	314	7740	245	7740		1.28	[1.09; 1.51]	65.1%	49.7%
ASPREE, 2018	38	1027	30	1030		1.27	[0.79; 2.03]	8.0%	12.7%
Fixed effect model		**11186**		**11197**		**1.26**	**[1.11; 1.44]**	**100.0%**	---
Random effects model						**1.24**	**[1.03; 1.48]**	---	**100.0%**

Heterogeneity: $I^2 = 23\%$, $p = 0.27$

Figure 2. Meta-analysis result, (**A**) Forest plot of major adverse cardiovascular events (MACE) in aspirin compared with control, (**B**) Trial sequential analysis (TSA) of MACE, heterogeneity adjustment required an information size of 37,337 participants calculated on the basis of the proportion of MACE incidence of 8.7% in the placebo group, as well as $\alpha = 5\%$, $\beta = 20\%$, power = 0.80, and $I^2 = 0.00\%$. Cumulative Z-curve (solid blue line) crosses the trial sequential monitoring boundary, which shows sufficient evidence of a statistically significant reduction in MACE risk with low-dose aspirin administration after involving large trials such as ASCEND [8] and ASPREE [9]. Horizontal dark red lines illustrate the traditional level of statistical significance ($p = 0.05$), (**C**) forest plot of major hemorrhage in aspirin compared with control. Main outcomes of low-dose aspirin in the primary prevention of cardiovascular disease (CVD). Risk ratio and 95% CI were used as a measure of effect for dichotomous variables. (Trial abbreviations are listed in the footnote of Table 1).

Table 2. Effects of low-dose aspirin use as the primary prevention strategy on clinical outcomes in diabetic patients.

Outcome Assessment	No. of Trials (Patients)	Anticipated Absolute Effects (95% CI)		Risk Ratio (95% CI) Random-Effect Estimate	p-Value Random-Effect Estimate	Heterogeneity I^2(%) Cochrane Q p-Value	Certainty of Evidence (GRADE) [†]
		Risk with Placebo	Risk with Aspirin				
MACE	8 (29,814)	87 per 1000	80 per 1000 (73 to 86)	**0.91 (0.84–0.98)**	**0.018** *	0.0% $p = 1.00$	⊕⊕⊕◯ MODERATE
MACE with age ≥ 60 years	5 (18,664)	111 per 1000	101 per 1000 (92 to 108)	**0.91 (0.83–0.98)**	**0.023** *	0.0% $p = 1.00$	⊕⊕◯◯ LOW
MACE with age < 60 years	3 (7212)	90 per 1000	91 per 1000 (65 to 124)	1.00 (0.73–1.39)	0.973	67.2% $p = 0.0476$	⊕◯◯◯ Very LOW
Myocardial infarction	6 (22,854)	31 per 1000	32 per 1000 (26 to 38)	1.01 (0.84–1.22)	0.891	21.1% $p = 0.275$	⊕⊕◯◯ LOW
Fatal myocardial infarction	3 (19,295)	15 per 1000	14 per 1000 (9 to 23)	0.98 (0.60–1.60)	0.942	44.1% $p = 0.166$	⊕◯◯◯ Very LOW
Stroke	7 (22,922)	36 per 1000	31 per 1000 (27 to 35)	**0.84 (0.73–0.97)**	**0.017** *	0.0% $p = 0.645$	⊕⊕⊕◯ MODERATE
Fatal stroke	3 (19,025)	5 per 1000	6 per 1000 (4 to 9)	1.20 (0.82–1.77)	0.349	0.0% $p = 0.773$	⊕◯◯◯ Very LOW
Coronary heart disease	6 (22,923)	48 per 1000	45 per 1000 (40 to 50)	0.94 (0.84–0.1.06)	0.323	0.0% $p = 0.864$	⊕⊕◯◯ LOW
Major hemorrhage	5 (22,383)	34 per 1000	42 per 1000 (35 to 50)	**1.24 (1.03–1.48)**	**0.0221** *	22.9% $p = 0.2687$	⊕⊕◯◯ LOW
Major intracranial hemorrhage	5 (21,353)	6 per 1000	7 per 1000 (5 to 9)	1.08 (0.77–1.50)	0.654	0.0% $p = 0.666$	⊕⊕◯◯ LOW
Major GI bleeding	4 (20,326)	14 per 1000	20 per 1000 (13 to 32)	1.43 (0.92–2.22)	0.117	56.7% $p = 0.074$	⊕⊕◯◯ LOW
All-cause death	6 (23,884)	87 per 1000	84 per 1000 (78 to 92)	0.97 (0.90–1.06)	0.537	0.0% $p = 0.457$	⊕⊕◯◯ LOW
Cardiovascular death	5 (21,827)	18 per 1000	17 per 1000 (14 to 21)	0.94 (0.77–1.14)	0.517	0.0% $p = 0.545$	⊕⊕◯◯ LOW

MACE: major adverse cardiovascular event; GI: gastrointestinal; CI: confidence interval; I^2: index for assessing heterogeneity; value > 50% indicates a moderate to high heterogeneity. *: The significance level in the classical model was set as <0.05; [†]: Certainty of effect estimates: ⊕⊕⊕⊕, high; ⊕⊕⊕, moderate; ⊕⊕, low; ⊕, very low.

4. Discussion

This present study demonstrated that use of low dose aspirin in diabetic patients as primary prevention of cardiovascular disease is warranted. Our sophisticated analysis showed a moderate level recommendation to use aspirin for reducing the risk of MACE. Although aspirin use increased the risk of major hemorrhage (i.e., major GI bleeding and major intracranial bleeding), it could be prescribed among subjects of diabetes without obvious bleeding risk, or otherwise other research has suggested that the use of proton pump inhibitors (PPIs) can decrease the risk of GI bleeding from long-term aspirin use [34]. Long-term aspirin therapy achieves the benefit of MACE prevention with a number needed to treat per year of 122 and carries a risk of GI hemorrhage with a number needed to harm per year of 115. It is possible to estimate the balance between the benefit and harm for aspirin use in diabetic patients with different cardiovascular risk levels by comparing the pooled estimate of the number needed to harm against the number needed to treat from individual trials.

The results confirmed the current guideline of ADA that low dose aspirin may be considered as a primary prevention strategy in those with diabetes mellitus (DM) who are at increased cardiovascular risk and are not at increased risk of bleeding [6].

Age is an important factor to determine the usage of low dose aspirin in the primary prevention of CVD in subjects of DM. A subgroup analysis was performed for those aged 60 years and older. For this subgroup, aspirin use also showed a protective effect in major adverse cardiovascular events compared with the control group (RR = 0.91, CI = 0.83–0.98, Table 2). However, GRADE analysis showed that the recommendation level was low. For those younger than 60 years old, there was no statistically significant effect in MACE (RR = 1.00, CI = 0.73–1.93, Table 2). This could be attributable to the high heterogeneity between studies (I^2 = 67.2%) and the low power. In conclusion, the benefit only existed among subjects aged greater than 60 years old instead of below 60 years old in our subgroup analysis. Elderly diabetic populations may have more cardiovascular risk factors, which may explain this difference.

4.1. Comparison of Previous Published Meta-Analysis

This study only included low-dosage aspirin (less than 100 mg daily) and used trial sequential analysis to assess the true results compared with previous published studies. Our finding of a 9% reduction in the risk of MACE is similar to previously published meta-analyses published between 2009 and 2016 with a non-significant 8 to 10% reduction in MACE [5,14,15,35,36]. The present meta-analysis enrolled a large sample size and was adequate to detect the benefit of aspirin in the prevention of CVD among subjects of diabetes. Furthermore, the present study included primary and secondary end points that revealed little to no heterogeneity of evidence, further supporting the important role of aspirin.

Compared with previous work, this analysis showed no statistically significance reduction of MI in the aspirin group, which was similar to previous published meta-analysis that revealed a non-statistically significance 14% to 17% reduction in MI. Although previous studies revealed the trend of decreased risk, our study showed a neutral effect after including a large trial of ASCEND, which enrolled the most diabetic population. In contrast, the individual participant meta-analysis reported that aspirin may significantly reduce the incidence of MI [2]. Further research is warranted.

Previous published meta-analyses reported a trend for the reduction in stroke risk, and the mean RR reduction ranged from 14% to 30% [4,5,11–14,35]. Our analysis demonstrated a moderate level of evidence for a reduction in stroke risk with low-dose aspirin administration. The adequacy of the present investigation may explain the difference between studies. For major bleeding risk, all previous results showed increased bleeding risk ranging from 54% to 69%, but no statistical significance. Our present study also showed similar results that statistically significantly increased bleeding risk by 24%. Furthermore, the bleeding risk in our study was lower than previous results because of three large trials included (JPAD2, ASCEND, and ASPREE).

4.2. Strengths and Limitations

The major strength of the present analysis is the focus on the low dose of aspirin in the primary prevention of CVD among subjects with diabetes. Besides, owing to the well-designed protocol and detailed inclusion criteria, we conducted the largest meta-analysis of low-dose aspirin in diabetic patients to date (nearly 30,000 people). This large sample allowed the examination of primary points, secondary points, efficacy, and safety. To further clarify the effect of aspirin, we also performed evaluations of internal validity (risk of bias, inconsistency, imprecision, and publication bias) and external validity (directness of results). Even though the risk of hemorrhage increased, our analysis is the first analysis to show a significant reduction of MACE with aspirin use for diabetic patients.

One limitation of the current investigation is the exclusion of observational studies, which may provide more real-world data and longer follow-up times. Compared with RCT, the outcomes of real-world evidence (RWE) continue to be assigned lower credibility. It must be emphasized that RWE research is a real-world practice that does not need to be executed as RCT research for it to be reliable. RCT research, characterized as having the highest reliability, and RWE research, which reflects the actual clinical aspects, can have a mutually supplementary relationship. We believe that our present meta-analysis of randomized control trial data proved the efficacy and safety of low dose aspirin; however, we still need large real-world evidence to inspect the effect on clinical practice [37].

Furthermore, the diversity of the sample is low because five of the trials were conducted in the United States, two were in Europe (United Kingdom and Italy), three were in Japan, and only one study utilized a global design. Therefore, the generalizability of these results to other countries, races, and ethnicities is limited.

An unpublished large trial still in progress, namely the Aspirin and Simvastatin Combination for Cardiovascular Events Prevention Trial in Diabetes (ISRCTN48110081), enrolled approximately 50,000 participants and followed participants for approximately 5 years [38]. This trial will help clarify the benefits of aspirin use. Although previous randomized control trials exist, they primarily include Caucasian and Japanese participants. Additional large-scale trials with a global design are necessary to better understand the benefits and disadvantages of aspirin use in diabetic individuals around the world.

5. Conclusions

Previous published data do not recommend aspirin use as the primary prevention strategy for diabetic individuals. The current investigation has shown a statistically significant protective effect and a moderate level of confidence for the administration of low-dose aspirin for the primary prevention in diabetes, particularly in older individuals (age $\geq$ 60 years). However, clinicians should assess the effect of aspirin use in each patient on a case-by-case basis and use evidence-based medicine to guide clinical decision making, especially considered the absolute benefits were largely counterbalanced by the bleeding hazard.

Supplementary Materials: The following are available online at http://www.mdpi.com/2077-0383/8/5/609/s1, Table S1. PRISMA checklist, Table S2. Search details, Table S3. Meta-regression results, Figure S1. Assessment of risk of bias, Figure S2. Forest plots of secondary outcomes, Figure S3. Funnel plots and the results of Egger's test of major outcomes in participants with diabetes for aspirin intervention trials.

Author Contributions: Conceptualization, M.-H.L. and C.-H.L. (Cho-Hao Lee); methodology, C.-H.L. (Cho-Hao Lee) and C.L.; software, C.-H.L. (Cho-Hao Lee); validation, M.-H.L. and C.-H.L. (Cho-Hao Lee); formal analysis, C.-H.L. (Chieh-Hua Lu); investigation, M.-H.L. and C.-H.L. (Cho-Hao Lee); resources data, C.-H.L. (Cho-Hao Lee) and C.L.; curation Y.-F.Z.; writing—original draft preparation, M.-H.L.; writing—review and editing, M.-H.L., C.-H.L. (Cho-Hao Lee), and C.-H.H.; supervision C.-H.L. (Chien-Hsing Lee) and C.-H.H.; project administration, C.-H.H.

Funding: The current study was funded by the Tri-Service General Hospital under grant number TSGH-C106-006-S01.

Conflicts of Interest: The funders had no role in the design of the study; in the collection, analyses, or interpretation of data; in the writing of the manuscript, or in the decision to publish the results.

References

1. Emerging Risk Factors Collaboration; Sarwar, N.; Gao, P.; Seshasai, S.R.; Gobin, R.; Kaptoge, S.; Di Angelantonio, E.; Ingelsson, E.; Lawlor, D.A.; Selvin, E.; et al. Diabetes mellitus, fasting blood glucose concentration, and risk of vascular disease: A collaborative meta-analysis of 102 prospective studies. *Lancet* **2010**, *375*, 2215–2222. [PubMed]
2. Antithrombotic Trialists' Collaboration. Collaborative meta-analysis of randomised trials of antiplatelet therapy for prevention of death, myocardial infarction, and stroke in high risk patients. *BMJ* **2002**, *324*, 71–86. [CrossRef]
3. Campbell, C.L.; Smyth, S.; Montalescot, G.; Steinhubl, S.R. Aspirin dose for the prevention of cardiovascular disease: A systematic review. *JAMA* **2007**, *297*, 2018–2024. [CrossRef]
4. Capodanno, D.; Angiolillo, D.J. Aspirin for primary cardiovascular risk prevention and beyond in diabetes mellitus. *Circulation* **2016**, *134*, 1579–1594. [CrossRef]
5. Pignone, M.; Alberts, M.J.; Colwell, J.A.; Cushman, M.; Inzucchi, S.E.; Mukherjee, D.; Rosenson, R.S.; Williams, C.D.; Wilson, P.W.; Kirkman, M.S. Aspirin for primary prevention of cardiovascular events in people with diabetes: A position statement of the American Diabetes Association, a scientific statement of the American Heart Association, and an expert consensus document of the American College of Cardiology Foundation. *Diabetes Care* **2010**, *33*, 1395–1402. [PubMed]
6. American Diabetes Association. 10. Cardiovascular Disease and Risk Management: *Standards of Medical Care in Diabetes—2019*. *Diabetes Care* **2019**, *42*, S103–S123. [CrossRef]
7. Piepoli, M.F.; Hoes, A.W.; Agewall, S.; Albus, C.; Brotons, C.; Catapano, A.L.; Cooney, M.-T.; Corrà, U.; Cosyns, B.; Deaton, C.; et al. 2016 European Guidelines on cardiovascular disease prevention in clinical practice: The Sixth Joint Task Force of the European Society of Cardiology and Other Societies on Cardiovascular Disease Prevention in Clinical Practice (constituted by representatives of 10 societies and by invited experts) Developed with the special contribution of the European Association for Cardiovascular Prevention & Rehabilitation (EACPR). *Eur. Heart J.* **2016**, *37*, 2315–2381. [PubMed]
8. The ASCEND Study Collaborative Group. Effects of aspirin for primary prevention in persons with diabetes mellitus. *N. Engl. J. Med.* **2018**, *379*, 1529–1539.
9. McNeil, J.J.; Wolfe, R.; Woods, R.L.; Tonkin, A.M.; Donnan, G.A.; Nelson, M.R.; Reid, C.M.; Lockery, J.E.; Kirpach, B.; Storey, E. Effect of aspirin on cardiovascular events and bleeding in the healthy elderly. *N. Engl. J. Med.* **2018**, *379*, 1509–1518. [CrossRef]
10. Saito, Y.; Okada, S.; Ogawa, H.; Soejima, H.; Sakuma, M.; Nakayama, M.; Doi, N.; Jinnouchi, H.; Waki, M.; Masuda, I.; et al. Low-Dose Aspirin for Primary Prevention of Cardiovascular Events in Patients with Type 2 Diabetes Mellitus: 10-Year Follow-Up of a Randomized Controlled Trial. *Circulation* **2017**, *135*, 659–670. [CrossRef]
11. De Berardis, G.; Sacco, M.; Strippoli, G.F.; Pellegrini, F.; Graziano, G.; Tognoni, G.; Nicolucci, A. Aspirin for primary prevention of cardiovascular events in people with diabetes: Meta-analysis of randomised controlled trials. *BMJ* **2009**, *339*, b4531. [CrossRef] [PubMed]
12. Butalia, S.; Leung, A.A.; Ghali, W.A.; Rabi, D.M. Aspirin effect on the incidence of major adverse cardiovascular events in patients with diabetes mellitus: A systematic review and meta-analysis. *Cardiovasc. Diabetol.* **2011**, *10*, 25. [CrossRef] [PubMed]
13. Kokoska, L.A.; Wilhelm, S.M.; Garwood, C.L.; Berlie, H.D. Aspirin for primary prevention of cardiovascular disease in patients with diabetes: A meta-analysis. *Diabetes Res. Clin. Pract.* **2016**, *120*, 31–39. [CrossRef] [PubMed]
14. Zhang, C.; Sun, A.; Zhang, P.; Wu, C.; Zhang, S.; Fu, M.; Wang, K.; Zou, Y.; Ge, J. Aspirin for primary prevention of cardiovascular events in patients with diabetes: A meta-analysis. *Diabetes Res. Clin. Pract.* **2010**, *87*, 211–218. [CrossRef]
15. Kunutsor, S.; Seidu, S.; Khunti, K. Aspirin for primary prevention of cardiovascular and all-cause mortality events in diabetes: Updated meta-analysis of randomized controlled trials. *Diabet. Med.* **2017**, *34*, 316–327. [CrossRef]
16. Higgins, J.; Green, S. Cochrane Handbook for Systematic Reviews of Interventions Version 5.1.0. The Cochrane Collaboration. Available online: https://handbook-5-1.cochrane.org/ (accessed on 1 April 2019).

17. Deeks, J.J.; Higgins, J.P.T.; Altman, D.G. Analysing data and undertaking meta-analyses. In *Cochrane Handbook for Systematic Reviews of Interventions: Cochrane Book Series*; Higgins, J.P.T., Green, S., Eds.; John Wiley & Sons: Chichester, UK, 2008; pp. 243–296.

18. Viechtbauer, W. Conducting meta-analyses in R with the metafor package. *J. Stat. Softw.* **2010**, *36*, 1–48. [CrossRef]

19. Ntali, G.; Wass, J.A. Epidemiology, clinical presentation and diagnosis of non-functioning pituitary adenomas. *Pituitary* **2018**, *21*, 111–118. [CrossRef] [PubMed]

20. GRADEpro Guideline Development Tool (Software). Available online: https://gradepro.org/ (accessed on 1 April 2019).

21. Schünemann, H.J.; Oxman, A.D.; Higgins, J.P.T.; Vist, G.E.; Glasziou, P.; Guyatt, G.H. Chapter 11: Presenting resultsand 'Summary of findings' tables. In *Cochrane Handbook for Systematic Reviewsof Interventions Version 5.1.0 (Updated March 2011)*; Higgins, J.P., Green, S., Eds.; The Cochrane Collaboration: London, UK, 2011.

22. Wetterslev, J.; Thorlund, K.; Brok, J.; Gluud, C. Trial sequential analysis may establish when firm evidence is reached in cumulative meta-analysis. *J. Clin. Epidemiol.* **2008**, *61*, 64–75. [CrossRef] [PubMed]

23. Thorlund, K.; Imberger, G.; Walsh, M.; Chu, R.; Gluud, C.; Wetterslev, J.; Guyatt, G.; Devereaux, P.J.; Thabane, L. The number of patients and events required to limit the risk of overestimation of intervention effects in meta-analysis—A simulation study. *PLoS ONE* **2011**, *6*, e25491. [CrossRef] [PubMed]

24. Brok, J.; Thorlund, K.; Gluud, C.; Wetterslev, J. Trial sequential analysis reveals insufficient information size and potentially false positive results in many meta-analyses. *J. Clin. Epidemiol.* **2008**, *61*, 763–769. [CrossRef] [PubMed]

25. Higgins, J.P.; Whitehead, A.; Simmonds, M. Sequential methods for random-effects meta-analysis. *Stat. Med.* **2011**, *30*, 903–921. [CrossRef] [PubMed]

26. Wetterslev, J.; Engstrøm, J.; Gluud, C.; Thorlund, K. Trial sequential analysis: Methods and software for cumulative meta-analyses. *Cochrane Methods Cochrane Database Syst. Rev. Suppl.* **2012**, *1*, 29–31.

27. Medical Research Council's General Practice Research Framework. Thrombosis prevention trial: Randomised trial of low-intensity oral anticoagulation with warfarin and low-dose aspirin in the primary prevention of ischaemic heart disease in men at increased risk. *Lancet* **1998**, *351*, 233–241. [CrossRef]

28. Hansson, L.; Zanchetti, A.; Carruthers, S.G.; Dahlöf, B.; Elmfeldt, D.; Julius, S.; Ménard, J.; Rahn, K.H.; Wedel, H.; Westerling, S. Effects of intensive blood-pressure lowering and low-dose aspirin in patients with hypertension: Principal results of the Hypertension Optimal Treatment (HOT) randomised trial. *Lancet* **1998**, *351*, 1755–1762. [CrossRef]

29. Sacco, M.; Pellegrini, F.; Roncaglioni, M.C.; Avanzini, F.; Tognoni, G.; Nicolucci, A. Primary prevention of cardiovascular events with low-dose aspirin and vitamin E in type 2 diabetic patients: Results of the Primary Prevention Project (PPP) trial. *Diabetes Care* **2003**, *26*, 3264–3272. [CrossRef]

30. Ridker, P.M.; Cook, N.R.; Lee, I.-M.; Gordon, D.; Gaziano, J.M.; Manson, J.E.; Hennekens, C.H.; Buring, J.E. A randomized trial of low-dose aspirin in the primary prevention of cardiovascular disease in women. *N. Engl. J. Med.* **2005**, *352*, 1293–1304. [CrossRef] [PubMed]

31. Belch, J.; MacCuish, A.; Campbell, I.; Cobbe, S.; Taylor, R.; Prescott, R.; Lee, R.; Bancroft, J.; MacEwan, S.; Shepherd, J. The prevention of progression of arterial disease and diabetes (POPADAD) trial: Factorial randomised placebo controlled trial of aspirin and antioxidants in patients with diabetes and asymptomatic peripheral arterial disease. *BMJ* **2008**, *337*, a1840. [CrossRef]

32. Ikeda, Y.; Shimada, K.; Teramoto, T.; Uchiyama, S.; Yamazaki, T.; Oikawa, S.; Sugawara, M.; Ando, K.; Murata, M.; Yokoyama, K. Low-dose aspirin for primary prevention of cardiovascular events in Japanese patients 60 years or older with atherosclerotic risk factors: A randomized clinical trial. *JAMA* **2014**, *312*, 2510–2520. [CrossRef]

33. Ogawa, H.; Nakayama, M.; Morimoto, T.; Uemura, S.; Kanauchi, M.; Doi, N.; Jinnouchi, H.; Sugiyama, S.; Saito, Y.; Japanese Primary Prevention of Atherosclerosis with Aspirin for Diabetes (JPAD) Trial Investigators. Low-dose aspirin for primary prevention of atherosclerotic events in patients with type 2 diabetes: A randomized controlled trial. *JAMA* **2008**, *300*, 2134–2141. [CrossRef] [PubMed]

34. Lai, K.C.; Lam, S.K.; Chu, K.M.; Wong, B.C.; Hui, W.M.; Hu, W.H.; Lau, G.K.; Wong, W.M.; Yuen, M.F.; Chan, A.O. Lansoprazole for the prevention of recurrences of ulcer complications from long-term low-dose aspirin use. *N. Engl. J. Med.* **2002**, *346*, 2033–2038.

35. Stavrakis, S.; Stoner, J.A.; Azar, M.; Wayangankar, S.; Thadani, U. Low-dose aspirin for primary prevention of cardiovascular events in patients with diabetes: A meta-analysis. *Am. J. Med. Sci.* **2011**, *341*, 1–9. [CrossRef] [PubMed]
36. Younis, N.; Williams, S.; Ammori, B.; Soran, H. Role of aspirin in the primary prevention of cardiovascular disease in diabetes mellitus: A meta-analysis. *Expert Opin. Pharmacother.* **2010**, *11*, 1459–1466. [CrossRef] [PubMed]
37. Kim, H.S.; Lee, S.; Kim, J.H. Real-world Evidence versus Randomized Controlled Trial: Clinical Research Based on Electronic Medical Records. *J. Korean Med. Sci.* **2018**, *33*, e213. [CrossRef] [PubMed]
38. De Berardis, G.; Sacco, M.; Evangelista, V.; Filippi, A.; Giorda, C.B.; Tognoni, G.; Valentini, U.; Nicolucci, A. Aspirin and Simvastatin Combination for Cardiovascular Events Prevention Trial in Diabetes (ACCEPT-D): Design of a randomized study of the efficacy of low-dose aspirin in the prevention of cardiovascular events in subjects with diabetes mellitus treated with statins. *Trials* **2007**, *8*, 21. [PubMed]

Journal of
Clinical Medicine

Article

Increased Arterial Stiffness in Prediabetic Subjects Recognized by Hemoglobin A1c with Postprandial Glucose but Not Fasting Glucose Levels

Chung-Hao Li [1,2], Feng-Hwa Lu [2,3], Yi-Ching Yang [2,3], Jin-Shang Wu [1,2,3,*] and Chih-Jen Chang [2,3,*]

1   Department of Health Management Center, National Cheng Kung University Hospital, College of Medicine, National Cheng Kung University, Tainan 70403, Taiwan; smallhear@gmail.com
2   Department of Family Medicine, National Cheng Kung University Hospital, College of Medicine, National Cheng Kung University, Tainan 70403, Taiwan; fhlu@mail.ncku.edu.tw (F.-H.L.); yichings@gmail.com (Y.-C.Y.)
3   Department of Family Medicine, College of Medicine, National Cheng Kung University, Tainan 70101, Taiwan
*   Correspondence: jins@mail.ncku.edu.tw (J.-S.W.); changcj.ncku@gmail.com (C.-J.C.); Tel.: +886-6-2353535 (ext. 5191) (J.-S.W.); +886-6-2353535 (ext. 5210) (C.-J.C.); Fax: +886-6-2386650 (J.-S.W. & C.-J.C.)

Received: 9 March 2019; Accepted: 30 April 2019; Published: 1 May 2019

Abstract: Previous studies exploring the association between arterial stiffness and prediabetes remain controversial. This study aimed to investigate the association of the different domains of prediabetes categorized by glycated hemoglobin A1c (A1c) 5.7–6.4%, impaired fasting glucose (IFG), fasting plasma glucose of 5.6–6.9 mmol/L, and impaired glucose tolerance (IGT), two-hour post-load glucose of 7.8–11.0 mmol/L, on arterial stiffness. These were measured by brachial–ankle pulse-wave velocity (baPWV). We enrolled 4938 eligible subjects and divided them into the following nine groups: (1) normoglycemic; (2) isolated A1c 5.7–6.4%; (3) isolated IFG; (4) IFG with A1c 5.7–6.4%; (5) isolated IGT; (6) combined IGT and IFG with A1c <5.7%; (7) IGT with A1c 5.7–6.4%; (8) combined IGT and IFG with A1c 5.7–6.4%; and (9) newly diagnosed diabetes (NDD). The baPWV values were significantly high in subjects with NDD (β = 47.69, 95% confidence interval (CI) = 29.02–66.37, $p < 0.001$), those with IGT with A1c 5.7–6.4% (β = 36.02, 95% CI = 19.08–52.95, $p < 0.001$), and those with combined IGT and IFG with A1c 5.7–6.4% (β = 27.72, 95% CI = 0.68–54.76, $p = 0.044$), but not in the other subgroups. These findings suggest that increased arterial stiffness was found in prediabetes individuals having an A1c 5.7–6.4% with IGT, but not IFG. Isolated A1c 5.7–6.4% and isolated IGT were not associated with elevated arterial stiffness.

Keywords: prediabetes; arterial stiffness; baPWV

---

1. Introduction

Type 2 diabetes mellitus is one of the leading causes of premature morbidity and mortality worldwide, and is an increasing public health concern [1]. Subjects with a condition lying between normoglycemia and dysglycemia are considered to be prediabetic, and these clinical conditions have been associated with the development of microvascular and macrovascular complications [2]. Since 2009, the American Diabetes Association (ADA) has revised the criteria for the detection of prediabetes and diabetes to include glycated hemoglobin A1c (A1c) levels as diagnostically relevant. This is in addition to pre-existing diagnostics establishing impaired fasting glucose (IFG) and impaired glucose tolerance (IGT) [3,4]. Accordingly, these ADA recommendations are based mainly on associations between A1c levels and microvascular complications [3].

Diabetes both precedes and contributes to the development of macrovascular disease [1], and the related diagnosis of prediabetes has become an important public issue [5]. One underlying mechanism of diabetes and macrovascular disease is the associated elevated serum advanced glycation end products (AGEs) that may cause the crosslinking of collagen molecules and the loss of collagen elasticity. The glycosylation of the vessel walls has also been related to the thinning of elastin fibers, subsequently resulting in arterial stiffness [6–8], which is an independent predictor of future cardiovascular events and mortality [9,10]. Several non-invasive methods are currently available to evaluate the severity of arterial stiffness, including one widely used method that measures pulse-wave velocity (PWV). Although carotid-femoral PWV (cfPWV) is considered the gold standard indicator of arterial stiffness, brachial–ankle PWV (baPWV) has been validated and is used in most Asian countries due to ease of execution, reproducibility, and the strong correlations with cfPWV [11,12].

Previous studies exploring the association between arterial stiffness and prediabetes have lacked consistency [13–17]; one explanation for the discrepancies might be related to the different ways of defining prediabetes, including determining the fasting plasma glucose (FPG), two-hour post-load glucose (2hPG), and A1c levels. For example, Ohnishi et al. [13] found significant differences in the baPWV values of the normal versus the IFG groups based on FPG level alone. In contrast, using both FPG and 2hPG level measurements, our previous study demonstrated the association of IGT with greater arterial stiffness and isolated IFG with insignificant differences in baPWV, compared with normal glucose tolerance [14]. Because 2hPG levels were not determined, the characteristics of the subjects in the IFG group of the study by Ohnishi et al. may have had similar characteristics to the subjects in the IGT group. Hence, the effects of IGT may have been misunderstood to be that of IFG. In addition, in a large study of Caucasian adults [18], the concordance of prediabetes, diagnosed via IFG, IGT, or A1c 5.7–6.4%, was limited, and, thus, the agreement among the three diagnostic criteria was only 10.4%. The relationship between arterial stiffness and prediabetes requires consensus among these three blood glucose indices. The aim of this study was to investigate the different domains of prediabetes categorized by A1c 5.7–6.4%, IFG, and IGT, as per the ADA criteria parameters of arterial stiffness shown by the baPWV values of a large Chinese population without known hypertension, diabetes, or cardiovascular disease.

2. Experimental Section

2.1. Study Population

Subjects who received self-motivated physical check-ups were included in the study, which covered the period between October 2006 and August 2009. The subjects' data were extracted retrospectively for secondary data analysis, without personal identification information, from the Health Management Center of the National Cheng Kung University Hospital. None of the female participants were pregnant. To avoid confounding effects, individuals were excluded if they: (1) had a history of hypertension, diabetes, coronary heart disease, or stroke; (2) were using medications that are known to influence blood pressure, plasma glucose, and lipid profiles; (3) had either of their lower limbs amputated; (4) claimed alcohol consumption levels of more than 30 gm per week [19]; (5) had estimated glomerular filtration rates (eGFR) of <30 mL/min/1.73 m^2; (6) were anemic (hemoglobin levels either <8.38 mmol/L in men or <7.45 mmol/L in women) [20]; or (7) had brachial–ankle indexes of <0.95 [21]. A total of 4938 subjects (3076 men, 62.3%; 1862 women, 37.7%) met the inclusion criteria and were enrolled. The study protocol was approved by the Ethics Committee for Human Research at the National Cheng Kung University Hospital, Taiwan (Approval No. ER-100-164).

2.2. Clinical Parameter Assessment

All participants completed a structured questionnaire containing questions regarding medical history, medication use, and lifestyle habits including smoking, alcohol consumption, and regular exercise. Smoking habits were categorized as former, current, or never; alcohol consumption was categorized as either current or non-user. Habitual exercise was defined as vigorous exercise at least three times per week [22]. The subjects' height and weight were measured to the nearest 0.1 cm and 0.1 kg, respectively using a certified anthropometry instrument, and body mass indexes (BMI; kilograms per meter squared) were calculated for all subjects. Systolic and diastolic blood pressure (SBP and DBP, respectively) were determined using a DINAMAP vital sign monitor (model 1846SX, Critikon Inc., Tampa, FL, USA). The mean levels, from right and left, of brachial blood pressure were measured while the subjects were in the supine position and wrapped in appropriately-sized pneumatic pressure cuffs, after at least 15 min of rest.

All blood samples were collected after overnight fasting of at least 10 h, and creatinine, FPG, cholesterol, triglyceride, high-density lipoprotein cholesterol (HDL-C), A1c, and hemoglobin levels were determined. Following a 75 g oral glucose tolerance test, 2hPG levels were measured in subjects without a history of diabetes. FPG and 2hPG levels were determined using a hexokinase method (Roche Diagnostic GmbH, Mannheim, Germany). A1c levels were measured using ion-exchange HPLC (HbA1c, BIO-RAD V-II TURBO Hemoglobin A1c program, Bio-Rad Laboratories, Inc., Kent, England). Initially, subjects were divided into five groups, according to the ADA diagnostic criteria [4]: (1) normoglycemic—FPG <5.6 mmol/L, 2hPG <7.8 mmol/L, and A1c <5.7%; (2) isolated A1c 5.7–6.4%—FPG <5.6 mmol/L, 2hPG <7.8 mmol/L, and A1c 5.7–6.4%; (3) IFG without IGT—FPG <7.0 mmol/L, 2hPG <7.8 mmol/L, and A1c <6.5%; (4) IGT—2hPG of 7.8–11.0 mmol/L, FPG <7.0 mmol/L, and A1c <6.5%; and (5) with newly diagnosed diabetes (NDD)—FPG ≥7.0 mmol/L, 2hPG ≥11.1 mmol/L, or A1c ≥6.5% (Figure 1). Groups 3 and 4 were further categorized into two and four groups, respectively: (3.1) isolated IFG—FPG of 5.6–6.9 mmol/L, 2hPG <7.8 mmol/L, and A1c <5.7%; (3.2) IFG with A1c 5.7–6.4%—FPG of 5.6–6.9 mmol/L, 2hPG <7.8 mmol/L, and A1c 5.7–6.4%; (4.1) isolated IGT—FPG <5.6 mmol/L, 2hPG of 7.8–11.0 mmol/L, and A1c <5.7%; (4.2) combined IGT and IFG with A1c <5.7%—FPG of 5.6–6.9 mmol/L, 2hPG of 7.8–11.0 mmol/L, and A1c <5.7%; (4.3) IGT with A1c 5.7–6.4%—FPG <5.6 mmol/L, 2hPG of 7.8–11.0 mmol/L, and A1c 5.7–6.4%; and (4.4) combined IGT and IFG with A1c 5.7–6.4%—FPG of 5.6–6.9 mmol/L, 2hPG of 7.8–11.0 mmol/L, and A1c 5.7–6.4% (Figure 2).

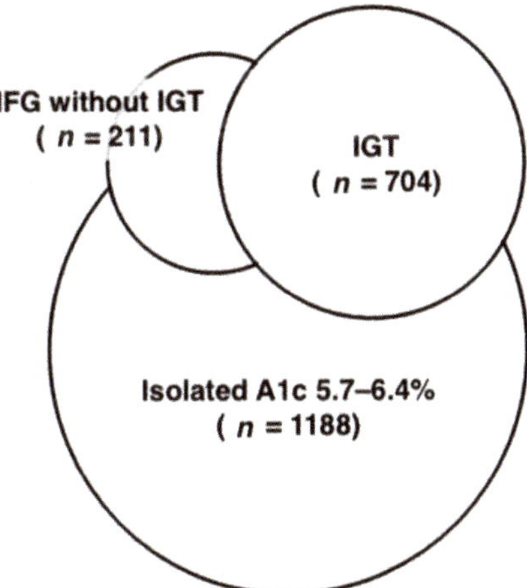

Figure 1. Venn diagram representing different domains of prediabetes categorized by glycated hemoglobin A1c (A1c) of 5.7–6.4%, impaired fasting glucose (IFG), and impaired glucose tolerance (IGT).

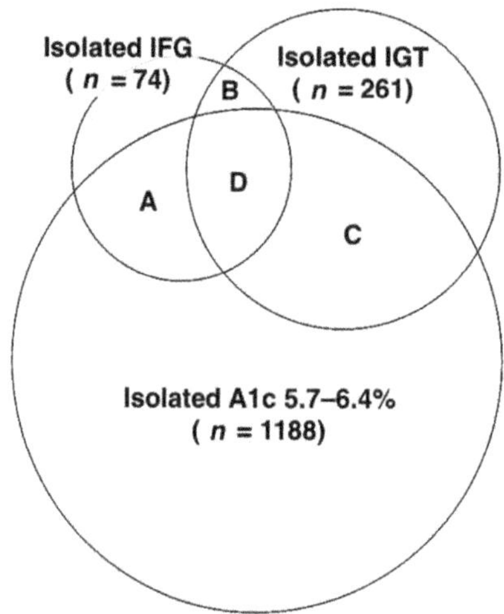

Figure 2. Venn diagram representing different domains and agreements of prediabetes categorized by glycated hemoglobin A1c (A1c) 5.7–6.4%, impaired fasting glucose (IFG), and impaired glucose tolerance (IGT). A) IFG with A1c 5.7–6.4% (n = 137); B) combined IGT and IFG with A1c <5.7% (n = 29); C) IGT with A1c 5.7–6.4% (n = 305); and D) combined IGT and IFG with A1c 5.7–6.4% (n = 109).

2.3. Vascular Assessment

After 5 min of supine rest, the subjects' baPWV values were measured using a non-invasive vascular screening device (BP-203RPE II; Colin Medical Technology, Komaki, Japan). This device allows pulse-wave measurements from sensors in four wrapped pneumatic pressure cuffs that simultaneously measure the blood pressure and pulse waves in the brachial arteries of both arms and the tibial arteries of both legs. Then, we automatically computed baPWV (cm/s) by dividing the brachial–ankle distances (L = 0.5934 × body height (cm) + 14.4014) by the time interval between rising waveforms of the brachial region and the ankle (ΔT). Because the left and right baPWV values were significantly positively correlated (r = 0.968, p < 0.001), the mean baPWV values were used in the final analyses.

2.4. Statistical Methods

Statistical analyses were performed using SPSS software for Windows (version 17.0; SPSS, Inc., Chicago, IL, USA). Continuous variables are expressed as mean ± standard deviation, and categorical variables are expressed as percentages. Differences between the unadjusted group means of continuous variables were identified using ANOVA with Scheffé's post hoc test. Categorical variables were compared using Chi-square tests among groups. Multiple linear regression analyses were performed to test the relationship between baPWV values (dependent variables) and different glycemic states, with adjustment for other confounding variables, including age, sex, BMI, SBP, total cholesterol, triglyceride, HDL, former smoking vs. never, current smoking vs. never, current alcohol consumption, and habitual exercise. In Models 1 and 2 of the multiple linear regression models, glycemic states were divided into five and nine subgroups, respectively. Unstandardized regression coefficients and 95% confidence intervals (CI) were derived from each regression model. Through the statistical tests, the differences and associations were considered significant when p < 0.05.

3. Results

Initially, 4938 subjects were classified into groups, normoglycemic ($n = 2583$), isolated A1c 5.7–6.4% ($n = 1188$), IFG without IGT ($n = 211$), IGT ($n = 704$), and NDD ($n = 252$). The Venn diagram representing different domains of prediabetes, including isolated A1c 5.7–6.4%, IFG without IGT, and IGT, is shown in Figure 1. The baseline characteristics of the subjects in these five groups are presented in Table 1. Significant differences were identified in age, sex, BMI, SBP, DBP, FPG, 2hPG, cholesterol, triglyceride, HDL-C, and prevalence of current alcohol drinking. Within these five groups, the mean baPWV values were 1253.1 ± 187.6, 1353.7 ± 223.9, 1376.9 ± 227.8, 1406.3 ± 251.7, and 1491.6 ± 276.6 cm/s (ANOVA; $p < 0.001$), respectively. The Scheffé's post hoc test also revealed that baPWV values increased more in the isolated A1c 5.7–6.4%, IFG without IGT, IGT and NDD groups than in the normoglycemic group ($p < 0.001$). Following the discordance in the categorization resulting from the three tests (A1c, FPG, 2hPG), the prediabetic groups were further clarified, as shown in Figure 2, and all subjects were then classified into nine groups, including normoglycemic ($n = 2583$), isolated A1c 5.7–6.4% ($n = 1188$), isolated IFG ($n = 74$), IFG with A1c 5.7–6.4% ($n = 137$), isolated IGT ($n = 261$), combined IGT and IFG with A1c <5.7% ($n = 29$), IGT with A1c 5.7–6.4% ($n = 305$), combined IGT and IFG with A1c 5.7–6.4% ($n = 109$), and NDD ($n = 252$).

In the multiple linear regression analyses of the clinical variables and the baPWV values of Table 2, Model 1 shows independently higher baPWV values only in the prediabetic group of IGT ($\beta = 16.59$, 95% CI = 4.41–28.76, $p = 0.008$) and the NDD group ($\beta = 46.35$, 95% CI = 27.18–65.04, $p < 0.001$), but not in the isolated A1c 5.7–6.4% or IFG without IGT groups, compared with normoglycemic subjects. In addition, age, current smoking, and SBP were positively associated with the baPWV values, whereas BMI and habitual exercise were inversely associated with the baPWV values. These above factors accounted for 63.2% of the total variance (adjusted $R^2 = 0.632$). In Model 2, when the agreement domains were stratified further using the three tests, the baPWV remained significantly higher in the NDD group ($\beta = 47.69$, 95% CI = 29.02–66.37, $p < 0.001$) and the prediabetic groups of IGT with A1c 5.7–6.4% ($\beta = 36.02$, 95% CI = 19.08–52.95, $p < 0.001$) and combined IGT and IFG with A1c 5.7–6.4% ($\beta = 27.72$, 95% CI = 0.68–54.76, $p = 0.044$). However, compared to the normoglycemic subgroup, the elevated baPWV values were not significant in the two subgroups—isolated and combined IGT and IFG with A1c <5.7%—originally included from the IGT group. The above factors accounted for 63.3% of the total variance (adjusted $R^2 = 0.633$).

Table 1. Demographic and clinical characteristics of study subjects with differing glycemic statuses according to A1c, FPG, and 2hPG.

	Normoglycemic [1]	Prediabetes			NDD [5]	p *	Post Hoc Test **
		Isolated A1c 5.7–6.4% [2]	IFG without IGT [3]	IGT [4]			
Variables	(n = 2583)	(n = 1188)	(n = 211)	(n = 704)	(n = 252)		
Age (years)	42.3 ± 10.7	49.6 ± 10.5	50.0 ± 11.0	50.8 ± 11.2	53.6 ± 11.1	<0.001	a, b, c, d, g, i, j
Gender, male (%)	59.5	62.4	65.9	69.7	66.3	<0.001	-
Body mass index (kg/m^2)	23.3 ± 3.3	24.4 ± 3.3	25.5 ± 2.9	25.0 ± 3.5	25.7 ± 3.6	<0.001	a, b, c, d, e, f, g
Former/Current smoking (%)	5.2/9.6	5.8/9.8	7.1/10.9	8.0/9.8	9.1/9.1	0.115	-
Current alcohol drinking (%)	10.5	9.6	12.3	14.1	12.3	0.029	-
Habitual exercise (%)	7.0	8.0	6.2	6.1	4.4	0.230	-
SBP (mmHg)	113.5 ± 12.8	118.1 ± 13.6	122.4 ± 13.8	121.9 ± 15.2	125.8 ± 14.8	<0.001	a, b, c, d, e, f, g, j
DBP (mmHg)	67.4 ± 9.8	70.9 ± 9.8	73.8 ± 9.9	73.4 ± 10.3	76.1 ± 10.1	<0.001	a, b, c, d, e, f, g, j
FPG (mmol/L)	4.67 ± 0.37	4.84 ± 0.37	5.77 ± 0.22	4.96 ± 0.63	6.92 ± 2.99	<0.001	a, b, c, d, e, f, g, h, i, j
2hPG (mmol/L)	5.29 ± 1.13	5.64 ± 1.16	5.98 ± 1.13	8.93 ± 0.87	13.78 ± 4.97	<0.001	a, b, c, d, e, f, g, h, i, j
Cholesterol (mmol/L)	4.96 ± 0.89	5.27 ± 0.96	5.34 ± 0.92	5.28 ± 0.90	5.58 ± 1.06	<0.001	a, b, c, d, g, j
Triglyceride (mmol/L)	1.24 ± 0.74	1.43 ± 0.86	1.75 ± 1.22	1.69 ± 1.11	1.90 ± 1.19	<0.001	a, b, c, d, e, f, g, j
HDL-C (mmol/L)	1.39 ± 0.38	1.31 ± 0.36	1.28 ± 0.34	1.23 ± 0.33	1.22 ± 0.34	<0.001	a, b, c, d, f, g
baPWV (cm/sec)	1253.1 ± 187.6	1353.7 ± 223.9	1376.9 ± 227.8	1406.3 ± 251.7	1491.6 ± 276.6	<0.001	a, b, c, d, f, g, i, j

Data presented as means ± standard deviation or percentage (%); [1] Group 1: normoglycemic (A1c <5.7%, FPG<100 mg/dL, and 2hPG <140 mg/dL); [2] Group 2: isolated A1c 5.7–6.4% (A1c 5.7–6.4%, FPG<100 mg/dL, and 2hPG < 140 mg/dL); [3] Group 3: IFG without IGT (FPG 100–125 mg/dL, 2hPG <140 mg/dL, and A1c <6.5%); [4] Group 4: IGT (FPG <126 mg/dL, 2hPG 140–199 mg/dL, and A1c <6.5%); [5] Group 5: NDD (FPG ≥126 mg/dL, or 2hPG ≥200 mg/dL, or A1c ≥6.5%); IFG and IGT was not included in isolated A1c 5.7–6.4%; * p for the difference, between the five groups, by ANOVA; ** Scheffé's tests: a—1 vs. 2; b—1 vs. 3; c—1 vs. 4; d—1 vs. 5; e—2 vs. 3; f—2 vs. 4; g—2 vs. 5; h—3 vs. 4; i—3 vs. 5; j—4 vs. 5. Abbreviations: A1c—glycated hemoglobin A1c; IFG—impaired fasting glucose; IGT—impaired glucose tolerance; NDD—newly diagnosed diabetes; SBP—systolic blood pressure; DBP—diastolic blood pressure; FPG—fasting plasma glucose; 2hPG—two-hour post-load glucose; HDL-C—high-density lipoprotein cholesterol; baPWV—brachial–ankle pulse-wave velocity.

Table 2. Coefficients (β) and 95% confidence intervals (CI) for the independent effects of clinical variables on baPWV level.

Independent Variables	Model 1		Model 2	
	β (95% CI)	p Value	β (95% CI)	p Value
Age (years)	–	<0.001	8.27 (7.89~8.65)	<0.001
Gender (female vs. male)	5.39 (−4.00~14.77)	0.260	5.02 (−4.36~14.39)	0.294
Body mass index (kg/m^2)	−8.09 (−9.41~−6.77)	<0.001	−8.20 (−9.52~−6.87)	<0.001
Smoking (former vs. never)	8.16 (−8.77~25.08)	0.345	8.60 (−8.32~25.52)	0.319
Smoking (current vs. never)	20.56 (6.42~34.71)	0.004	19.73 (5.59~33.87)	0.006
Current alcohol drinking (yes vs. no)	-8.51 (−22.03~5.01)	0.217	−8.13 (−21.65~5.39)	0.238
Habitual exercise (≥3 times/week vs. <3 times/week)	−22.12 (−37.26~−6.98)	0.004	−22.47 (−37.61~−7.34)	0.004
Cholesterol (mmol/L)	1.81 (−2.84~6.45)	0.445	1.59 (−3.04~6.24)	0.500
Triglyceride (mmol/L)	3.98 (−1.23~9.19)	0.134	3.81 (−1.41~9.02)	0.152
HDL-C (mmol/L)	−9.96 (−23.79~3.87)	0.158	−9.47 (−23.92~4.36)	0.180
SBP (mmHg)	9.06 (8.74~9.37)	<0.001	9.04 (8.72~9.36)	<0.001
Isolated A1c 5.7–6.4% * vs. normoglycemic	4.84 (−5.08~14.75)	0.339	5.56 (−4.36~15.47)	0.272
IFG without IGT vs. normoglycemic	−7.52 (−27.21~12.17)	0.454	–	–
Isolated IFG vs. normoglycemic	–	–	−14.36 (−46.13~17.40)	0.375
IFG with A1c 5.7–6.4% vs. normoglycemic	–	–	−2.32 (−26.41~21.78)	0.851
IGT vs. normoglycemic	16.59 (4.41~28.76)	0.008	–	–
Isolated IGT vs. normoglycemic	–	–	−6.90 (−24.55~10.74)	0.443
Combined IGT and IFG with A1c <5.7% vs. normoglycemic	–	–	4.51 (−45.88~54.91)	0.861
IGT with A1c 5.7–6.4% vs. normoglycemic	–	–	36.02 (19.08~52.95)	<0.001
Combined IGT and IFG with A1c 5.7–6.4% vs. normoglycemic	–	–	27.72 (0.68~54.76)	0.044
NDD vs. normoglycemic	46.35 (27.18~65.04)	<0.001	47.69 (29.02~66.37)	<0.001
Adjusted R^2 (%)	63.2		63.3	

* IFG and IGT were not included in isolated A1C 5.7–6.4%. Abbreviations: baPWV—brachial–ankle pulse-wave velocity; HDL-C—high-density lipoprotein cholesterol; SBP—systolic blood pressure; A1c—glycated hemoglobin A1c; IFG—impaired fasting glucose; IGT—impaired glucose tolerance; NDD—newly diagnosed diabetes.

4. Discussion

The present study is the first to show the different effects of prediabetes, categorized concomitantly by A1c, FPG, and 2hPG levels, on arterial stiffness. The results showed that increased arterial stiffness was found in individuals having either IGT with A1c 5.7–6.4% or combined IGT and IFG with A1c 5.7–6.4% but not in those having either IGT with A1c <5.7% or combined IGT and IFG with A1c <5.7%, independent of other cardiovascular risk factors. The elevated arterial stiffness was insignificant in subjects having either IFG with A1c <5.7% or IFG with A1c 5.7–6.4%. The association between prediabetes and arterial stiffness is still not consistent [13–17]. The MARK study [17] showed that A1c, FPG, and postprandial glucose were positively related to arterial stiffness in diabetes subjects, but not in prediabetes subjects. Ohnishi et al. [13] found that baPWV values were significantly higher than normal fasting glucose levels in IFG subjects; however, they only had fasting glucose data and not 2hPG data. Our previous study highlighted the importance of IGT, categorized by FPG and 2hPG values, and the associated risk of increased arterial stiffness based on FPG and 2hPG levels [14]. Di Pino et al.'s study [16] revealed that arterial stiffness, as shown by an augmentation index, was higher in subjects with normal glucose tolerance with A1c 5.7–6.4% compared to that in subjects with normal glucose tolerance with A1c <5.7%; however, these values were similar to those of the IGT and type 2 diabetes patients. Liang et al. [15] found that prediabetic subjects with abnormal A1c, FPG, and 2hPG had higher cfPWV values, but they did not test the concomitant influence of A1c, FPG, and 2hPG on cfPWV. The discrepancy between the results of this study and those of Liang et al. may be related to the different methodologies. Subjects were excluded from our study if they had history of hypertension, diabetes, coronary heart disease, stroke, or peripheral vascular disease and if they were taking medications known to influence blood pressure, plasma glucose, and lipid profiles, whereas such subjects were not excluded from Liang et al.'s study. In Liang et al.'s study, subjects with cardiovascular disease or those taking medications were included among subjects with A1c 5.7–6.4%, IFG, or IGT, resulting in elevated arterial stiffness in subjects with abnormal A1c, FPG, and 2hPG.

Discrepancies in the criteria used in the three different tests for diagnosing either prediabetes or diabetes have been reported and suggest the importance of categorizing the different characteristics of the subjects. IFG indicates impaired first-phase insulin secretion and reduced hepatic insulin sensitivity, whereas IGT suggests markedly peripheral insulin resistance and defective second-phase insulin secretion, contributing to prolonged defects [23]. A1c is a marker representing both basal and postprandial hyperglycemia, and it provides a "picture" of the average blood glucose level over the past 2–3 months [24]. In this study, increased arterial stiffness was found in prediabetes individuals having an A1c 5.7–6.4% with IGT, but not IFG. This result highlighted the importance of 2hPG, as possibly having a stronger associated risk of arterial stiffness than that of FPG. The main mechanism of hyperglycemia affecting arterial stiffness is the generation and formation of AGEs, whose biochemical process may involve the glycosylation of vessel walls, subsequent crosslinking of collagen molecules, loss of collagen elasticity, and thinning of elastin fibers. Otherwise, AGEs can interact with certain receptors, inducing intracellular signaling with enhanced oxidative stress, triggering the inflammatory process, and, finally, increasing arterial stiffness [6–8,25,26]. Furthermore, the serum levels of AGEs are positive correlates of insulin resistance [27], which is related to the reduced production of nitric oxide and its related vasodilation [28], post-load hyperinsulinemia-induced oxidative stress [29], decreased endothelial progenitor cells with impaired vascular repair qualities [30], and related metabolic alterations, such as either dyslipidemia or elevated blood pressure [31]. Importantly, IGT showed a more pronounced degree of insulin resistance than did IFG [32]. Therefore, the above findings may partially explain elevated arterial stiffness in prediabetes individuals having an A1c 5.7–6.4% with IGT, but not IFG.

As shown previously, we have identified the independent effects of age, blood pressure, and smoking status on arterial stiffness [33–35]. Aging is widely associated with compromised arterial stiffness; the effects of aging on baPWV might be mediated by the intermediate parameters of cardiovascular risk factors, such as blood pressure [36]. The positive relationship between smoking and increased arterial stiffness may reflect increased inflammation, thrombosis, oxidation of low-density lipoprotein cholesterol, and oxidative stress [34]. Although the association between obesity and arterial stiffness remained inconclusive, in agreement with Tomiyama et al. [37], BMI was negatively associated with baPWV in this study. In the work of Ben et al. [38], arterial stiffness was reduced until middle age, and it was speculated that vascular adaptation to obesity was lost with advanced age, meaning that the adverse association between obesity and arterial stiffness becomes apparent only later in life. Moreover, the effects of obesity on arterial stiffness might be mediated by intermediate cardiovascular risk factors, such as blood pressure, and the highly associated variables included might lead to further collinearity [39].

Despite the large study cohort, our study was limited to a cross-sectional design, which cannot be used to establish causal relationships. Our study was also confined to the Chinese population, and the data complemented the differences between previous studies of different ethnic groups. All subjects were extracted from a health management center and had received regular health examinations. Therefore, the present results should be used carefully because they are not indicative of the general population. In addition, since the arterial stiffness in our study was measured by baPWV, a peripheral stiffness parameter but not a direct measure of central stiffness, our results are limited and would be more relevant if the subjects were then used for a longitudinal study, either in terms of future cardiovascular events or mortality. Additionally, we did not measure either insulin or high sensitivity C-reactive protein to provide more information about insulin resistance or vascular inflammation status.

5. Conclusions

In conclusion, increased arterial stiffness was found in prediabetes individuals having an A1c 5.7–6.4% with IGT, but not IFG. Isolated A1c 5.7–6.4%, isolated IGT, and IGT with A1c <5.7% were not associated with elevated arterial stiffness. Despite the practicality and convenience of FPG and A1c, it is still important to consider postprandial glucose values to facilitate early recognition of the significant proportion of the at-risk population.

Author Contributions: No potential conflicts of interest relevant to this article were reported. C.-H.L., J.-S.W., and C.-J.C. contributed to the study design, statistical analyses, research data interpretation, discussion, and review of the manuscript. F.-H.L. and Y.-C.Y. contributed to the analyses by means of suggestions and advice. C.-H.L. drafted the article, and all authors were involved in the acquisition of data.

Funding: This study was supported by grants from the Department of Family Medicine, National Cheng Kung University Hospital (NCKUHFM-100-001 and NCKUH-10804038).

Acknowledgments: The authors thank Enago (https://www.enago.tw/) for the assistance editing the manuscript.

Conflicts of Interest: The authors declare no conflicts of interest.

References

1. Tancredi, M.; Rosengren, A.; Svensson, A.M.; Kosiborod, M.; Pivodic, A.; Gudbjornsdottir, S.; Wedel, H.; Clements, M.; Dahlqvist, S.; Lind, M. Excess mortality among persons with Type 2 diabetes. *N. Engl. J. Med.* **2015**, *373*, 1720–1732. [CrossRef] [PubMed]

2. Brannick, B.; Wynn, A.; Dagogo-Jack, S. Prediabetes as a toxic environment for the initiation of microvascular and macrovascular complications. *Exp. Biol. Med.* **2016**, *241*, 1323–1331. [CrossRef]

3. International expert committee report on the role of the A1c assay in the diagnosis of diabetes. *Diabetes Care.* **2009**, *32*, 1327–1334. [CrossRef] [PubMed]

4. American Diabetes Association. Standards of medical care in diabetes—2011. *Diabetes Care* **2011**, *34*, S11–S61. [CrossRef] [PubMed]

5. Wilson, M.L. Prediabetes: Beyond the borderline. *Nurs. Clin. N. Am.* **2017**, *52*, 665–677. [CrossRef] [PubMed]

6. Aronson, D. Cross-linking of glycated collagen in the pathogenesis of arterial and myocardial stiffening of aging and diabetes. *J. Hypertens.* **2003**, *21*, 3–12. [CrossRef] [PubMed]

7. Schram, M.T.; Henry, R.M.; van Dijk, R.A.; Kostense, P.J.; Dekker, J.M.; Nijpels, G.; Heine, R.J.; Bouter, L.M.; Westerhof, N.; Stehouwer, C.D. Increased central artery stiffness in impaired glucose metabolism and type 2 diabetes: The hoorn study. *Hypertension* **2004**, *43*, 176–181. [CrossRef]

8.   Stephen, E.; Venkatasubramaniam, A.; Good, T.; Topoleski, L.D.T. The effect of glycation on arterial microstructure and mechanical response. *J. Biomed. Mater. Res. A* **2014**, *102*, 2565–2572. [CrossRef]

9.   Vlachopoulos, C.; Aznaouridis, K.; Stefanadis, C. Prediction of cardiovascular events and all-cause mortality with arterial stiffness: A systematic review and meta-analysis. *J. Am. Coll. Cardiol.* **2010**, *55*, 1318–1327. [CrossRef] [PubMed]

10. Jain, S.; Khera, R.; Corrales-Medina, V.F.; Townsend, R.R.; Chirinos, J.A. Inflammation and arterial stiffness in humans. *Atherosclerosis* **2014**, *237*, 381–390. [CrossRef] [PubMed]

11. Tanaka, H.; Munakata, M.; Kawano, Y.; Ohishi, M.; Shoji, T.; Sugawara, J.; Tomiyama, H.; Yamashina, A.; Yasuda, H.; Sawayama, T.; et al. Comparison between carotid-femoral and brachial-ankle pulse wave velocity as measures of arterial stiffness. *J. Hypertens.* **2009**, *27*, 2022–2027. [CrossRef] [PubMed]

12. Cavalcante, J.L.; Lima, J.A.; Redheuil, A.; Al-Mallah, M.H. Aortic stiffness current understanding and future directions. *J. Am. Coll. Cardiol.* **2011**, *57*, 1511–1522. [CrossRef] [PubMed]

13. Ohnishi, H.; Saitoh, S.; Takagi, S.; Ohata, J.; Isobe, T.; Kikuchi, Y.; Takeuchi, H.; Shimamoto, K. Pulse wave velocity as an indicator of atherosclerosis in impaired fasting glucose: The Tanno and Sobetsu study. *Diabetes Care* **2003**, *26*, 437–440. [CrossRef]

14. Li, C.H.; Wu, J.S.; Yang, Y.C.; Shih, C.C.; Lu, F.H.; Chang, C.J. Increased arterial stiffness in subjects with impaired glucose tolerance and newly diagnosed diabetes but not isolated impaired fasting glucose. *J. Clin. Endocrinol. Metab.* **2012**, *97*, E658–E662. [CrossRef]

15. Liang, J.; Zhou, N.; Teng, F.; Zou, C.; Xue, Y.; Yang, M.; Song, H.; Qi, L. Hemoglobin A1c levels and aortic arterial stiffness: The cardiometabolic risk in Chinese (CRC) study. *PLoS ONE* **2012**, *7*, e38485. [CrossRef] [PubMed]

16. Di Pino, A.; Scicali, R.; Calanna, S.; Urbano, F.; Mantegna, C.; Rabuazzo, A.M.; Purrello, F.; Piro, S. Cardiovascular risk profile in subjects with prediabetes and new-onset type 2 diabetes identified by HbA(1c) according to American Diabetes Association criteria. *Diabetes Care* **2014**, *37*, 1447–1453. [CrossRef] [PubMed]

17. Gomez-Sanchez, L.; Garcia-Ortiz, L.; Patino-Alonso, M.C.; Recio-Rodriguez, J.I.; Feuerbach, N.; Marti, R.; Agudo-Conde, C.; Rodriguez-Sanchez, E.; Maderuelo-Fernandez, J.A.; Ramos, R.; et al. Glycemic markers and relation with arterial stiffness in Caucasian subjects of the MARK study. *PLoS ONE* **2017**, *12*, e0175982. [CrossRef] [PubMed]

18. Marini, M.A.; Succurro, E.; Castaldo, E.; Cufone, S.; Arturi, F.; Sciacqua, A.; Lauro, R.; Hribal, M.L.; Perticone, F.; Sesti, G. Cardiometabolic risk profiles and carotid atherosclerosis in individuals with prediabetes identified by fasting glucose, postchallenge glucose, and hemoglobin A1c criteria. *Diabetes Care* **2012**, *35*, 1144–1149. [CrossRef]

19. Ji, A.; Lou, P.; Dong, Z.; Xu, C.; Zhang, P.; Chang, G.; Li, T. The prevalence of alcohol dependence and its association with hypertension: A population-based cross-sectional study4 in Xuzhou city, China. *BMC Public Health* **2018**, *18*, 364. [CrossRef]

20. Lippi, G.; Targher, G. Glycated hemoglobin (HbA1c): Old dogmas, a new perspective? *Clin. Chem. Lab. Med.* **2010**, *48*, 609–614. [CrossRef]

21. Motobe, K.; Tomiyama, H.; Koji, Y.; Yambe, M.; Gulinisa, Z.; Arai, T.; Ichihashi, H.; Nagae, T.; Ishimaru, S.; Yamashina, A. Cut-off value of the ankle-brachial pressure index at which the accuracy of brachial-ankle pulse wave velocity measurement is diminished. *Circ. J.* **2005**, *69*, 55–60. [CrossRef] [PubMed]

22. Whaley, M.H.; Brubaker, P.H.; Otto, R.M.; Armstrong, L.E.; American College of Sports Medicine. *ACSM's Guidelines for Exercise Testing and Prescription*, 7th ed.; Lippincott Williams & Wilkins: Philadelphia, PA, USA, 2006. [CrossRef]

23. Nathan, D.M.; Davidson, M.B.; DeFronzo, R.A.; Heine, R.J.; Henry, R.R.; Pratley, R.; Zinman, B. Impaired fasting glucose and impaired glucose tolerance: Implications for care. *Diabetes Care* **2007**, *30*, 753–759. [CrossRef] [PubMed]

24. Barr, R.G.; Nathan, D.M.; Meigs, J.B.; Singer, D.E. Tests of glycemia for the diagnosis of type 2 diabetes mellitus. *Ann. Int. Med.* **2002**, *137*, 263–272. [CrossRef] [PubMed]

25. Basta, G.; Sironi, A.M.; Lazzerini, G.; Del Turco, S.; Buzzigoli, E.; Casolaro, A.; Natali, A.; Ferrannini, E.; Gastaldelli, A. Circulating soluble receptor for advanced glycation end products is inversely associated with glycemic control and S100A12 protein. *J. Clin. Endocrinol. Metab.* **2006**, *91*, 4628–4634. [CrossRef] [PubMed]

26. Di Pino, A.; Urbano, F.; Zagami, R.M.; Filippello, A.; Di Mauro, S.; Piro, S.; Purrello, F.; Rabuazzo, A.M. Low endogenous secretory receptor for advanced glycation end-products levels are associated with inflammation and carotid atherosclerosis in prediabetes. *J. Clin. Endocrinol. Metab.* **2016**, *101*, 1701–1709. [CrossRef] [PubMed]

27. Tahara, N.; Yamagishi, S.; Matsui, T.; Takeuchi, M.; Nitta, Y.; Kodama, N.; Mizoguchi, M.; Imaizumi, T. Serum levels of advanced glycation end products (AGEs) are independent correlates of insulin resistance in nondiabetic subjects. *Cardiovasc. Ther.* **2012**, *30*, 42–48. [CrossRef] [PubMed]

28. Kim, J.A.; Montagnani, M.; Koh, K.K.; Quon, M.J. Reciprocal relationships between insulin resistance and endothelial dysfunction: Molecular and pathophysiological mechanisms. *Circulation* **2006**, *113*, 1888–1904. [CrossRef] [PubMed]

29. Arcaro, G.; Cretti, A.; Balzano, S.; Lechi, A.; Muggeo, M.; Bonora, E.; Bonadonna, R.C. Insulin causes endothelial dysfunction in humans: Sites and mechanisms. *Circulation* **2002**, *105*, 576–582. [CrossRef]

30. Cubbon, R.M.; Kahn, M.B.; Wheatcroft, S.B. Effects of insulin resistance on endothelial progenitor cells and vascular repair. *Clin. Sci.* **2009**, *117*, 173–190. [CrossRef]

31. Libman, I.M.; Barinas-Mitchell, E.; Bartucci, A.; Chaves-Gnecco, D.; Robertson, R.; Arslanian, S. Fasting and 2-hour plasma glucose and insulin: Relationship with risk factors for cardiovascular disease in overweight nondiabetic children. *Diabetes Care* **2010**, *33*, 2674–2676. [CrossRef]

32. Festa, A.; D'Agostino, R., Jr.; Hanley, A.J.; Karter, A.J.; Saad, M.F.; Haffner, S.M. Differences in insulin resistance in nondiabetic subjects with isolated impaired glucose tolerance or isolated impaired fasting glucose. *Diabetes* **2004**, *53*, 1549–1555. [CrossRef] [PubMed]

33. Cecelja, M.; Chowienczyk, P. Dissociation of aortic pulse wave velocity with risk factors for cardiovascular disease other than hypertension: A systematic review. *Hypertension* **2009**, *54*, 1328–1336. [CrossRef] [PubMed]

34. Tomiyama, H.; Hashimoto, H.; Tanaka, H.; Matsumoto, C.; Odaira, M.; Yamada, J.; Yoshida, M.; Shiina, K.; Nagata, M.; Yamashina, A. Continuous smoking and progression of arterial stiffening: A prospective study. *J. Am. Coll. Cardiol.* **2010**, *55*, 1979–1987. [CrossRef] [PubMed]

35. Lee, H.Y.; Oh, B.H. Aging and arterial stiffness. *Circ. J.* **2010**, *74*, 2257–2262. [CrossRef] [PubMed]

36. Yiming, G.; Zhou, X.; Lv, W.; Peng, Y.; Zhang, W.; Cheng, X.; Li, Y.; Xing, Q.; Zhang, J.; Zhou, Q.; et al. Reference values of brachial-ankle pulse wave velocity according to age and blood pressure in a central Asia population. *PLoS ONE* **2017**, *12*, e0171737. [CrossRef] [PubMed]

37. Tomiyama, H.; Yamashina, A.; Arai, T.; Hirose, K.; Koji, Y.; Chikamori, T.; Hori, S.; Yamamoto, Y.; Doba, N.; Hinohara, S. Influences of age and gender on results of noninvasive brachial-ankle pulse wave velocity measurement—A survey of 12517 subjects. *Atherosclerosis* **2003**, *166*, 303–309. [CrossRef]

38. Corden, B.; Keenan, N.G.; de Marvao, A.S.; Dawes, T.J.; Decesare, A.; Diamond, T.; Durighel, G.; Hughes, A.D.; Cook, S.A.; O'Regan, D.P. Body fat is associated with reduced aortic stiffness until middle age. *Hypertension* **2013**, *61*, 1322–1327. [CrossRef] [PubMed]

39. Thomas, G.N.; Lao, X.Q.; Jiang, C.Q.; McGhee, S.M.; Zhang, W.S.; Adab, P.; Lam, T.H.; Cheng, K.K. Implications of increased weight and waist circumference on vascular risk in an older Chinese population: The guangzhou biobank cohort study. *Atherosclerosis* **2008**, *196*, 682–688. [CrossRef] [PubMed]

Journal of
Clinical Medicine

Article

Feasibility and Effectiveness of Electrochemical Dermal Conductance Measurement for the Screening of Diabetic Neuropathy in Primary Care. Decoding Study (Dermal Electrochemical Conductance in Diabetic Neuropathy)

Juan J. Cabré [1,*], Teresa Mur [2,*], Bernardo Costa [1], Francisco Barrio [1], Charo López-Moya [2], Ramon Sagarra [1], Montserrat García-Barco [1], Jesús Vizcaíno [1], Immaculada Bonaventura [3], Nicolau Ortiz [4], Gemma Flores-Mateo [1], Oriol Solà-Morales [5] and the Catalan Diabetes Prevention Research Group [†]

[1] Catalan Diabetes Prevention Research Group, IDIAP Jordi Gol., 43202 Reus, Barcelona, Spain; costaber@gmail.com (B.C.); ciscobarrio@gmail.com (F.B.); rsagarra.tarte.ics@gencat.cat (R.S.); mgarcia.tarte.ics@gencat.cat (M.G.-B.); jvizcaino.tgn.ics@gencat.cat (J.V.); gemmaflores@gmail.com (G.F.-M.)
[2] Primary Care, CAP Rubí, Mutua Terrassa, 08191 Rubí, Spain; clopez@mutuaterrassa.cat
[3] Neurology Department, Hospital Mutua Terrassa, 08221 Terrassa, Spain; ibonaventura@mutuaterrassa.cat
[4] Neurology Department, Hospital Universitari Sant Joan de Reus, 43202 Reus, Spain; nortiz@grupsagessa.com
[5] HITT (Health Innovation Technology Transer), 08006 Barcelona, Spain; osola@hittbcn.com
[*] Correspondence: juanjocabre@gmail.com (J.J.C.); tmm387@gmail.com (T.M.); Tel.: +34-977778512 (J.J.C.); +34-636066185 (T.M.)
[†] Membership of the Catalan Diabetes Prevention Research Group is provided in the acknowledgments.

Received: 31 March 2019; Accepted: 29 April 2019; Published: 1 May 2019

Abstract: Diabetes mellitus (DM) is the leading cause of polyneuropathy in the Western world. Diabetic neuropathy (DNP) is the most common complication of diabetes and is of great clinical significance mainly due to the pain and the possibility of ulceration in the lower limbs. Early detection of neuropathy is essential in the medical management of this complication. Early unmyelinated C-fiber dysfunction is one of the typical findings of diabetic neuropathy and the first clinical manifestation of dysfunction indicating sudomotor eccrine gland impairment. In order to assess newly developed technology for the measurement of dermal electrochemical conductance (DEC), we analyzed the feasibility and effectiveness of DEC (quantitative expression of sudomotor reflex) as a screening test of DNP in primary health care centers. The study included 197 people (with type 2 diabetes, prediabetes and normal tolerance) who underwent all the protocol tests and electromyography (EMG). On comparing DEC with EMG as the gold standard, the area under the receiver operating characteristic (ROC) curve (AUC, area under the curve) was 0.58 in the whole sample, AUC = 0.65 in the diabetes population and AUC = 0.72 in prediabetes, being irrelevant in subjects without glucose disturbances (AUC = 0.47). Conclusions: In usual clinical practice, DEC is feasible, with moderate sensitivity but high specificity. It is also easy to use and interpret and requires little training, thereby making it a good screening test in populations with diabetes and prediabetes. It may also be useful in screening general populations at risk of neuropathy.

Keywords: dermal electrochemical conductance; diabetes mellitus; neuropathy; primary care; screening; sudomotor reflex

1. Introduction

The prevalence of diabetes mellitus (DM) is very high in Spain, being approximately 14% according to oral glucose tolerance test (OGTT) results [1]. The management of DM requires a significant consumption of health care resources, mainly in relation to the care of vascular complications. Among the late microvascular events which may develop in patients with DM, polyneuropathy (PN) is the most common and disabling, and is the leading cause of morbidity and mortality in these patients [2]. Indeed, in Spain, the leading cause of neuropathy is DM, with its prevalence increasing with the presence of DM and other risk factors such as obesity [3].

PN is defined as the presence of symptoms and/or signs of peripheral nerve dysfunction in people with DM, after ruling out other possible causes [4]. The Toronto Panel Consensus on PN defined this disorder as a symmetrical, length-dependent sensorimotor PN attributed to metabolic and microvessel alterations due to chronic exposure to hyperglycemia and other risk factors. In patients with PN, thin fibers (autonomic system—sweating) and thermal and tactile sensitivity are first affected, followed by the involvement of large fibers, presenting an altered vibrating sensation which eventually alters electromyography (EMG) patterns. Therefore, dysfunction of the sweat reflex in small distal fibers is one of the earliest changes detected in these patients [5].

The most common clinical presentation of PN is distal symmetric polyneuropathy (DSPN), being predominantly sensory in 80% of cases [3]. Pain is the most important symptom and is described as burning or flashing, lancinating, deep, and with frequent exacerbations during rest [4]. Pain often affects the quality of life of these patients, and it is a frequent cause of depression and/or anxiety [6]. Moreover, some patients may develop hypoesthesia, which may lead to severe foot lesions [7]. The prevalence of DSPN varies greatly according to the population, definition, and detection method.

In the Rochester study including >64,000 patients, the prevalence of PN was between 66% and 59% for type1 DM and type 2 DM, respectively [8]. The 3rd report of the Technical Study Group of Diabetes of the World Health Organization (WHO) described a prevalence of 40% [7], and 50% in patients with >25 years of DM evolution. Pirart et al. [9] reported a prevalence ranging from 25% to 48% [7,10–17], whereas in Spain, Cabezas-Cerrato et al. published a figure of 24.1% [10].

DSPN-related factors are age, DM duration, metabolic control, male gender, acute myocardial infarction, hyperlipidemia (especially hypertriglyceridemia), smoking, and general cardiovascular risk factors [2,15,16,18]. Puig et al. [11] also included urinary albumin excretion as a risk factor of presenting DSPN. The diagnosis of DSPN is commonly made based on signs and symptoms and usually includes the use of several scores such as the Neuropathy Disability Score (NDS), the Neuropathy Symptoms Score (NSS), the Michigan Neuropathy Instrument (MNI) and the Douleur Neuropathique 4 Questions (DN4). These methods are easy to perform and are reproducible, sensitive, and adequate for use in a screening program [14].

There are many confirmatory tests, including measurements of nerve conduction velocity (EMG) and biothesiometry or skin biopsy. However, the most commonly used are the measurement of altered sensations using a vibrating tuning fork with 128 Hz and/or pressure with Semmes-Weinstein 5:07 monofilament [15]. Monofilament testing (MFT) is widely accepted and recommended by all scientific societies because of its validity, predictive risk, efficiency, and simplicity. Feng et al. [16] reported that monofilament (MFT) has a sensitivity of 57% to 93%, a specificity of 75% to 100%, a positive predictive value (PPV) of 36% to 94%, and a negative predictive value (NPV) of 84% to 100% compared with the measurement of nerve velocity by EMG.

Although electrophysiological measures are more objective and reproducible, they are limited in that they only detect dysfunction based on the presence of thicker and faster (myelinated) fibers and show their involvement later. Consequently, EMG is a specific, albeit very insensitive, test. Recently developed noninvasive techniques are more reproducible and reliable for the detection of early dysfunction of small fibers. One of these new techniques involves the measurement of dermal electrochemical conductance (DEC) or the sudomotor dysfunction index and has been evaluated by well-designed studies which support its use as a screening test [17–21].

Ramachandran et al. [22] studied the use of DEC to detect diabetes and other disorders of glucose metabolism. In a study on the use of DEC, Casellini et al. [5] applied a PN test which showed a low sensitivity of 78% and a specificity of 92% in diabetic patients without neuropathy compared to other subjects with neuropathy and a control group. In this latter study, correlation with clinical parameters showed adequate reproducibility of the results, particularly in regard to the measurements of the feet [5]. Several other studies [23] also obtained significantly lower DEC values on comparing diabetic patients and controls. In a study of patients following a 12-month program of intense physical activity, Raisanen et al. [18] observed a greater improvement in DEC compared to weight, waist circumference, or maximum oxygen volume (VO2 max).

Although we will not discuss the best algorithm to screen diabetic neuropathy, certain tests have shown promising results in predicting future diabetes-related complications. Some of these tests include related factors such as obesity, macroangiopathy (i.e., kidney disease) or retinopathy [24–28]. Therefore, taking into account the large number of methods used and the learning curve required to correctly implement these techniques, as well as the absence of consensus as to which method is the most adequate to diagnose DSPN, the aim of this study is to validate the usefulness of DEC measurement in the early diagnosis of DSPN compared with traditional techniques in the primary care setting.

2. Experimental Section

2.1. Hypothesis and Objectives

The hypothesis of our study was that the measurement of DEC is feasible, sensitive, and specific and more or equally effective to other techniques commonly used in the initial screening of diabetic neuropathy in primary care.

2.1.1. Main Objective

To evaluate the feasibility, effectiveness and performance of a new technique which measures DEC (sudomotor reflex) in the screening of diabetic neuropathy in primary care.

2.1.2. Specific Objectives

Determine the performance of DEC (quantitative assessment of sudomotor reflex) as a tool for the screening of diabetic neuropathy in primary care compared with the Semmes-Weinstein 5:07 MFT when an EMG is used to confirm the presence of diabetic PN in patients with prediabetes and type 2 diabetes.

Determine the performance of DEC (quantitative assessment of sudomotor reflex) as a tool for screening diabetic neuropathy in primary care compared with the Semmes-Weinstein 5:07 MFT when the NDS is used to confirm the presence of diabetic PN in patients with prediabetes and type 2 diabetes.

2.2. Material and Methods

The study design has been extensively described elsewhere [29]. In brief, this was a prospective study aimed at evaluating (a) the feasibility of the protocol in primary care and (b) the capacity of achieving early diagnosis of PN. A prospective and blinded comparison was made between DEC (sudomotor reflex) (Sudoscan®, Impeto Medical, Paris, France) and EMG; with MFT 5.07 (10 g) Semmes-Weinstein; with Rydel-Seiffer 128 Hz tuning fork (vibrating sensibility); and the NDS score, in a consecutive sample of individuals attended in the primary care setting in the Vallès Occidental and Baix Camp counties (Catalonia, Spain), and their reference hospitals.

2.2.1. Design

We performed a blind, prospective study comparing DEC (sudomotor reflex) (Sudoscan™, Impeto Medical, France), EMG, the Semmes-Weinstein 5:07 MFT (10 g), the sensitivity of a vibrating tuning fork 128 Hz, NDS score and DN4 score in a consecutive series of patients treated in primary care.

2.2.2. Sites

The primary care teams of Terrassa-Sud and other partners belonging to the Mútua Terrassa reference hospital and those of the primary care teams of Reus (CAP Sant Pere) and the University Hospital Sant Joan de Reus reference hospital participated in the study.

2.2.3. Study Subjects

We consecutively included patients with type 2 DM over 40 years of age, with or without symptoms of neuropathy, attended in primary care. We also included the following 2 groups of patients matched by age and sex: one including patients with prediabetes (intermediate alterations of glucose metabolism defined as impaired fasting glucose (IFG)) and/or impaired glucose tolerance (IGT) determined by OGTT after 2-h 75 g oral glucose administration and another including patients without glucose alterations (normal glucose tolerance) (control group).

Three main diagnostic categories (normal, prediabetes, and diabetes) were defined using the WHO criteria based on 2-h postload glucose (<7.8 (140 mg/dL), 7.8–11.0 mmol/L (140–200 mg/dL)) and/or fasting plasma glucose (6.1–6.9 mmol/L; 110–126 mg/dL) and >11.1 mmol/L (>200 mg/dL), respectively.

The exclusion criteria were type 1 DM, upper or lower limb amputation (except phalanges), diagnosis of neuropathy not related to diabetes, neuropathy by entrapment, use of psychoactive substances, chronic alcoholism, malnutrition; treatment with beta-blockers, presence of terminal disease, or life expectancy <3 years. Pregnancy was ruled out in women (negative pregnancy test) and a history of gestational diabetes was also taken into account.

The study period was from 1 January 2017 to 31 January 2019.

2.2.4. Sample Size

It was estimated that the study should include a total of 160 participants in order to evaluate the validity and performance of a screening test showing a sensitivity of 82%, a precision of 9%, and a confidence interval of 95% (95% CI: 128–192), and considering a drop-out rate of 20%. The proportion of diabetes/prediabetes/normal glucose tolerance was 2:1:1.

The contribution of patients by centers was 66% from the Mutua de Terrassa (minimum 106 cases) and 34% from Reus (minimum 54 cases), achieving a final sample size of 197 cases.

2.2.5. Variables and Data Collection

This has been described previously elsewhere [29]. In brief, a three-step procedure was carried out to obtain the study data:

First Visit: In Primary Care

After verifying the inclusion criteria and receiving written informed consent to participate, the medical history of the patients was obtained and a physical examination was performed using the MFT, and the NDS score was given to screen for PN. The patients also underwent DEC quantification using the Sudoscan device. A score of 8 out of 8 with the Semmes-Weinstein 5:07 MFT (10 g) was considered as sensitive [16]. An NDS score ≥6 points was considered as the presence of moderate or severe PN [30]. The determination of vibration sensitivity was performed using a 128 Hz Rydel-Seiffer tuning fork.

Second Visit: In Hospital

Done at the reference hospital, where a neurologist blinded to previous test results, performed a neurographic test, including a sensory conduction study of the median, ulnar and sural nerves, and motor conduction study of the deep peroneal nerve. The variables studied included the amplitude of compound muscle action potential and distal latency of the motor nerves, and amplitude and distal latency of sensory nerves. DEC determination and the other neuropathy screening and electrophysiological tests took no longer than 1 month.

Third Visit: In Primary Care

The results of the previous visit were recorded and the patients were informed of the results and the diagnosis.

2.2.6. Statistical Analysis

The χ^2 test was used to analyze qualitative variables and the Student's t test was performed for quantitative variables. Logistic regression was used to identify predictors of diabetic neuropathy. Dependent variables (response) included the presence of diabetic neuropathy diagnosed by EMG or the NDS questionnaire. The performance of DEC was compared with the MFT as screening tests of PN. To determine the validity and reliability of DEC, the sensitivity, specificity, the PPV and NPV, and the positive and negative likelihood ratios were calculated. A receiver operating characteristic (ROC) curve was used and the area under the curve was calculated. A $p < 0.05$ was considered as statistically significant. The analyses were performed with the statistical packages STATA/SE 12.0 (StataCorp, College Station, TX, USA) and R for Windows (R Foundation, Vienna, Austria).

2.2.7. Ethics Approval and Consent to Participate

The research Ethics Committee board at the Jordi Gol Research Institute (Barcelona) (www.idiapjordigol.org) approved the protocol (January 2015, reference number P14/147) and each participant signed written informed consent.

2.2.8. Registry

The study is registered in https://clinicaltrials.gov/ with the reference number NCT03495089.

3. Results

A total of 197 participants were evaluated: 100 with T2D, 50 with prediabetes and 47 controls free of glucose disturbances.

Table 1 shows the socio-demographical and clinical characteristics of the whole sample. The three groups showed significant differences in variables usually associated with diabetes, including family history, body mass index (BMI) and waist circumference, mean systolic blood pressure and comorbidities (i.e., high blood pressure or dyslipidemia, linked to their pharmacological treatment). Differences were also observed in triglyceride values and the glomerular filtration rate. Normal Douleur Neuropathique 4 Questions (DN4) questionnaire results were more frequently observed in the control group without DM.

Table 1. Sociodemographic and clinical data of the study participants (type 2 diabetes (T2D), prediabetes and normal glucose tolerance).

Variable	Participants with T2D ($n = 100$)	Participants with Prediabetes ($n = 50$)	Participants with Normal Glucose Tolerance ($n = 47$)	p
Age (mean ± SD)	64.65 ± 7.5	62.7 ± 6.8	63.47 ± 7.4	0.30
Gender (% men)	50	38	38.3	0.33
Origin country (% Spain)	93	100	93.6	0.70
Family history of diabetes (%)	83 (83)	26 (50.2)	33 (70.2)	<0.001
Current smoker; n (%)	8 (8)	2 (4)	3 (6.4)	0.99
Alcohol consumption; n (%)	59 (59)	34 (68)	30 (63.8)	0.70
T2D evolution (mean ± SD, year)	10.6 ± 7.52	-	-	-
T2D complications (%)				
Cardiopathy	4	1	0	-
retinopathy	6	0	0	-
peripheral vascular disease	7	0	0	-
stroke	3	0	1	-
nephropathy	6	0	0	-
NIAD treatment (n)	152	-	-	-
Insulin treatment (n)	29	-	-	-
Antihypertensive treatment; n (%)	74 (74)	28 (56)	25 (53.1)	0.02
Cholesterol-lowering agents; n (%)	62 (62)	18 (36)	15 (31.9)	0.001
Antiaggregant treatment; n (%)	21 (21)	8 (16)	6 (12.7)	0.33
Height (mean ± SD, cm)	162.4 ± 9.7	162.7 ± 8.8	160.8 ± 7.5	0.54
Weight (mean ±SD, kg)	79.6 ± 13.6	85.3 ± 18.3	74.9 ± 14.9	0.004
BMI (mean ±SD, kg/m^2)	30.3 ± 4.5	33.28 ± 10.8	28.6 ± 4.6	0.003
Waist circumference (mean ± SD, cm)	105.0 ± 11.2	108,0 ± 15.3	109.5 ± 11.4	0.016
SBP (mean ± SD, mmHg)	136.5 ± 14.1	132.3 ± 13.6	125.8 ± 15.2	<0.001
DBP (mean ± SD, mmHg)	78.2 ± 10.4	79.8 ± 7.9	75.3 ± 9.5	0.06
HbA1c (mean ± SD, %)	7.2 ± 1.2	5.9 ± 0,3	5.3 ± 0.4	0.02
Glomerular Filtration Rate (mean ± SD, mL/min/1.73 m^2)	67.3 ± 14.4	87.0 ± 7.3	72.9 ± 19.3	<0.001
Total cholesterol (mean ± SD, mg/mL)	191.9 ± 39.4	196.8 ± 33.9	190.9 ± 30.9	0.67
HDL-chol (mean ± SD, mg/dL)	48.7 ± 11.0	58.9 ± 14.5	54.9 ± 12.2	<0.001
LDL-chol (mean ± SD, mg/dL)	111.7 ± 33.3	114.4 ± 32.0	116.5 ± 28.0	0.69
Triglycerides (mean ± SD, mg/dL)	159.6 ± 90.0	120.7 ± 49.2	117.0 ± 44.7	<0.001
DN4 questionnaire (% score = 0)	61	70	74.5	<0.001

SD: standard deviation; DM: diabetes mellitus; NIAD: non-insulin antidiabetic drugs; BMI: body mass index; SBP: systolic blood pressure; DBP; diastolic blood pressure; HbA1c: glycohemoglobin A1c; HDL-chol: high density lipoprotein cholesterol; LDL-chol: low density lipoprotein cholesterol.

The prevalence of PN was 14.4% by DEC and 21% by EMG. Considering EMG as the gold standard, on evaluating the DEC in the whole sample the results were: sensitivity 21%, specificity 95%, PPV 47% and NPV 81%, with a likelihood ratio of 4.13 and an area under the receiver operating characteristic (OC) curve (AUC) of 0.58.

The high specificity found indicates that DEC is a reliable test for the detection of individuals without PN while the low sensitivity suggests that this test is not very useful for the detection of disease among people with PN.

On comparing DEC with other tests such as the MFT, NDS or DN4 questionnaires, the sensitivity, specificity, PPV and NPV were similar, in the whole population and in individuals without DM and in those with prediabetes (Tables 2 and 3).

Table 2. Sensitivity (Sen), specificity (Spe), positive predictive value (PPV), negative predictive value (NPV) of dermal electrochemical conductance (DEC) compared to different tests as the gold standard.

		Gold Standard		
		EMG	MFT	NDS	DN4
DEC		Sen = 21.2%	Sen = 15.3%	Sen = 33.3%	Sen = 21.4%
		Spe = 94.8%	Spe = 93.5%	Spe = 94%	Spe = 93.4%
		PPV = 46.6%	PPV = 26.6%	PPV = 26.6%	PPV = 20%
		NPV = 81.3%	NPV = 87.9%	NPV = 95.6%	NPV = 93.4%
		LR+ = 4.13	LR+ = 2.35	LR+ = 5.55	LR+ = 3.24
		LR− = 0.83	LR− = 0.90	LR− = 0.71	LR− = 0.77

DEC: dermal electrochemical conductance; EMG: electromyography; MFT: monofilament; NDS: Neuropathy Disability Score; DN4: Douleur Neuropathique-4 Questions; Sen: sensitivity; Spe: specificity; PPV: positive predictive value; NPV: negative predictive value. LR+: positive likelihood ratio; LR−: negative likelihood ratio.

Table 3. Comparison between dermal electrochemical conductance (DEC) results and other tests such as the gold standard based on the glycemic state of the participants.

		Gold Standard										
		MFT			EMG			NDS			DN4		
		DM	No-DM	Pre-DM	DM	No-DM	Pre-DM	DM	No-DM	Pre-DM	DM	No-DM	Pre-DM
DEC	Sen	25%	0%	14%	20%	0%	5%	37%	0%	1%	30%	0	0
	Spe	93%	95%	93%	96%	95%	93%	93%	95%	93%	93%	95.5%	91%
	PPV	33%	0%	25%	67%	0%	25%	33%	0%	25%	33%	0	0
	NPV	90%	84%	87%	74%	98%	96%	94.5%	93%	1%	92%	95.5%	93%
	LR+	3.57	0	2	5	0	0.71	5.28	0	0.14	4.28	0	0
	LR−	0.80	20	0.92	0.83	20	1.02	0.67	20	1.06	0.75	22.2	11.1

DEC: dermal electrochemical conductance; EMG: electromyography; MFT: monofilament; NDS: Neuropathy Disability Score; DN4: Douleur Neuropathique-4 Questions; Sen: sensitivity; Spe: specificity; PPV: positive predictive value; NPV: negative predictive value; DM: type 2 diabetes; No-DM: non-diabetic; pre-DM: prediabetic. LR+: positive likelihood ratio; LR−: negative likelihood ratio.

Table 4 reports conductance (hands and feet) for the 3 different groups. All the prediabetic and normotolerant patients underwent oral glucose tolerance test since this was the method used to rule out ignored diabetes. It was therefore considered mandatory in the 97 cases included without known diabetes.

Table 4. Comparison between DEC results in the three groups of participants.

		Conductance (µS)		
		Normal Tolerance	Pre-DM	DM
DEC	H-DEC (hands)	70.3 ± 12.7	63.9 ± 19.2	51.8 ± 17.2
	F-DEC (feet)	76.8 ± 15.8	73.3 ± 22.5	71.3 ± 18.0

Painful neuropathy (DN4 >4 points) was observed in a total of 12 patients: 11 in the diabetes group (11%) and 1 in the prediabetics (0.02%) and none in the group with normal tolerance.

Figures 1–3 show the ROC curves comparing DEC versus the other tests evaluated.

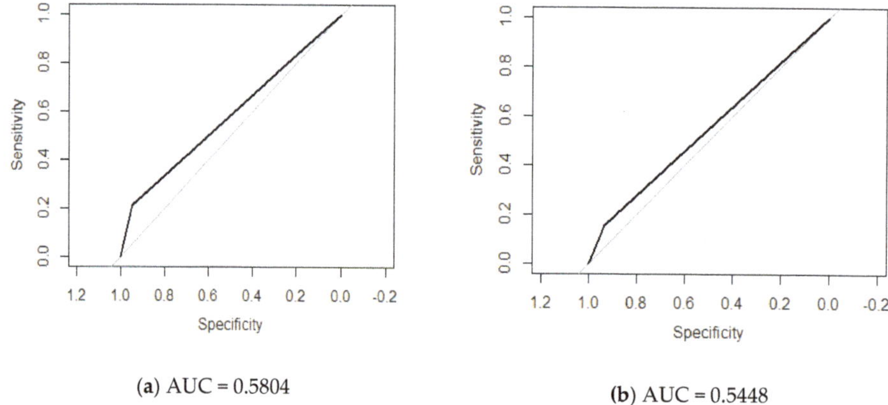

(**a**) AUC = 0.5804

(**b**) AUC = 0.5448

Figure 1. (**a**) Receiver operating characteristic (ROC) curve of dermal electrochemical conductance (DEC) versus electromyography (EMG) as the gold standard in the whole study population. (**b**) ROC curve of DEC versus monofilament (MFT) as the gold standard in the whole study population. AUC: Area under the ROC Curve.

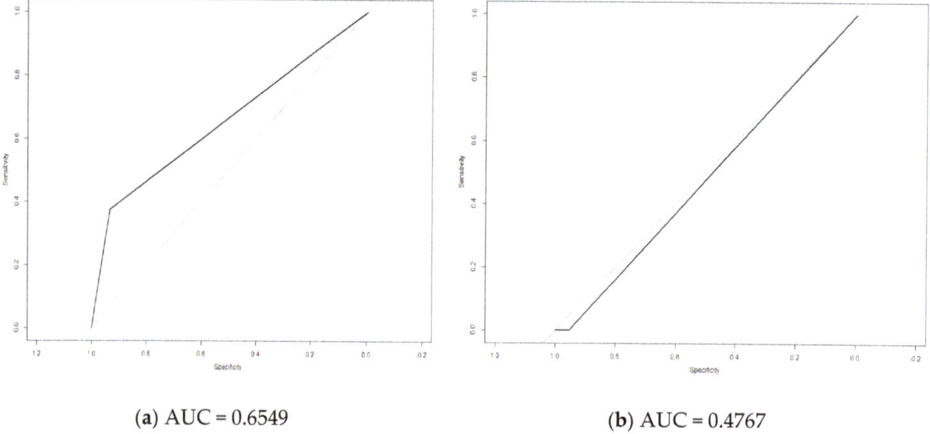

(**a**) AUC = 0.6549

(**b**) AUC = 0.4767

Figure 2. (**a**) Receiver operating characteristic (ROC) curve of dermal electrochemical conductance (DEC) versus electromyography (EMG) as the gold standard in the population with diabetes mellitus. (**b**) ROC curve of DEC versus EMG as the gold standard in the population without glucose disturbances. AUC: Area under the ROC curve.

In the whole sample, the ROC curve comparing DEC with MTF (AUC = 0.65) showed greater effectiveness than that comparing DEC vs. EMG (AUC = 0.54). On the other hand, in the population with type 2 diabetes, the comparison of DEC with EMG showed the greatest significance (AUC = 0.66). Although the results are not very high, DEC may be a good screening test for populations with PN compared to the gold standard (EMG).

It was of note that in the screening of prediabetic subjects the AUC was 0.72, which is reasonable for a screening test.

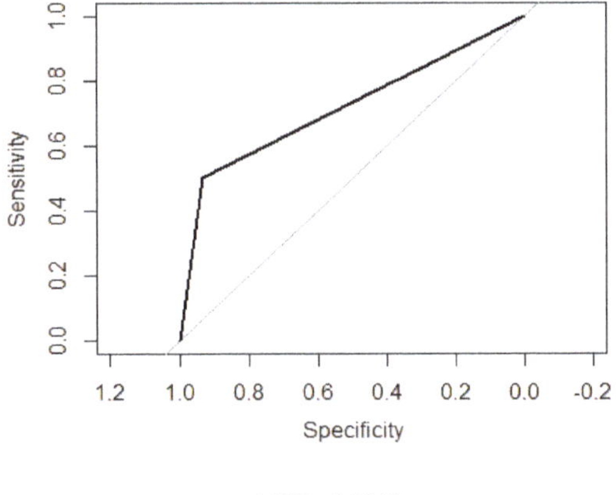

$$\text{AUC} = 0.7181$$

Figure 3. Receiver operating characteristic (ROC) curve of DEC versus electromyography (EMG) as the gold standard in the population with prediabetes.

4. Discussion

As mentioned in the Introduction, state-of-the-art methods demonstrate a high variability in PN diagnostic methods, being directly related to the variability in prevalence rates. For example, with the use of EMG, Puig and Portillo obtained prevalences of 75 and 86%, respectively [11,25]. However, most authors use clinical methods based on exploratory symptoms and signs, obtaining diverse prevalences ranging from 13% up to 81% [10,12,14,26–28]. It is clear that the tests used to determine the diagnosis of PN and the type of patients included have an impact on the prevalence, as does the cut-off to consider pathology. Gordon Smith [17] found poor DEC results in feet with PN, reporting a sensitivity higher than in our study (77%) but with a lower specificity (67%) and with a higher cut-off (70 µS vs. 60 µS, respectively). Casellini [5] compared three groups (diabetic individuals with and without PN, and healthy volunteers) and also concluded that DEC is less effective in diabetic subjects with PN, showing similar results to our study (sensitivity 78%, specificity 92%). In a study of 75 individuals, but including 45 patients with type 1 diabetes, Selvarajah [23] also described similar results with DEC (sensitivity 87%, specificity 76% and AUC 0.85). Nonetheless, unlike our study, these three studies included subjects with known confirmed (or discarded) PN. The aim of the present study was to validate DEC as a method to diagnose PN, and the presence of PN or causes other than diabetes were an exclusion criterion. Eranki [20] used DEC for the detection of cardiovascular complications in 308 patients with diabetes, reporting complications in 120 cases, 79% due to neuropathy. They reported a sensitivity of 83%, a specificity of 55% and an AUC = 0.73 for PN. The results of the study by Erank et al. differed greatly from ours, which showed a very low sensitivity but a very high specificity due to the low number of cases.

One important difference of our study is that we also evaluated a group with prediabetes. With the addition of this group, the definitive results differ from the preliminary results, with the NDS questionnaire showing the best AUC compared to MFT before the addition of prediabetes. There are few studies in this regard. However, some have suggested the utility of DEC for detecting diabetes in subjects with glucose intolerance, being even better than basal glycemia. With the use of DEC in 212 subjects, Ramachandran [22] diagnosed 24 type 2 diabetes, 30 cases with glucose intolerance and 57 subjects with normal glucose tolerance but with metabolic syndrome.

In a study evaluating 47 patients with type 2 diabetes and 16 controls, Kneger et al. [31] reported very similar results to those of the present study. DEC was effective in detecting subjects with PN and correlated well with clinical signs and symptoms of neuropathy.

Bins-Hall et al. [32] studied a one-step approach for the diagnosis of PN in 236 patients with diabetes undergoing funduscopy. These authors evaluated a battery of tests including: the Toronto scale, MFT, and two devices: DPN-Check™ and Sudoscan™. The results showed high performance (with the Toronto scale as the gold standard (30.9%)) with DPN-check (sensitivity 84.3%, specificity 68.3%) and Sudoscan (sensitivity 77.4%, specificity 68.3%) compared to MFT (underestimation, only 14.4% detection).

In a Chinese cohort, Sudoscan™ was a useful method to detect PN in a screening of 394 asymptomatic persons with diabetes [33].

In a systematic review and meta-analysis on MFT, Wang et al. [34] reported that the sensitivity of this test was very low.

It is of note that in present study we prioritized the NDS and DN4 scores; the first provides clinical data (questionnaire) and exploration data (vibration, temperature, reflexes) while the second is an abbreviation of the first, and provides a greater amount of clinical data.

Limitations and Strengths of the Study

The main limitation of the present study is accurate diagnosis of diabetic neuropathy since some studies have shown that some cases of diabetic neuropathy present no alterations in the EMG.

Indeed, both EMG and MFT examine large nerve fibers. However, since they are the usual tests available, they represent a limitation of the study. Therefore, several considerations should be taken into account. First, the EMG test is more specific, albeit not very sensitive, showing positive results in advanced stages of PN. The fingerboard and the NDS questionnaire are commonly used for the diagnosis of diabetic neuropathy, probably because the NDS is carried out before EMG and is actually often used to avoid the need for EMG.

Another inherent limitation of this study may be the low number of participants, although this number fulfilled that obtained in the calculation of the sample size.

Therefore, both the EMG and the NDS score, which mainly assess the dysfunction of myelinated fibers, provide a good profile for diagnostic confirmation. On the contrary, both the DEC and the MFT are able to diagnose and stage diabetic neuropathy earlier than the previous 2 tests by the detection of unmyelinated fiber dysfunction. Nonetheless, another limitation is that MFT is a good test for the prediction of foot ulcers but is certainly insensitive for the detection of early neuropathy. Therefore, for purposes of simplification and taking into account the possible limitations, we compared the effectiveness of the measurement of DEC and the use of MFT as diagnostic tools in primary care according to whether the true diagnosis is achieved by the EMG or the score of the NDS questionnaire.

Although other tests are available to detect early neuropathy, such as the Utah Early Neuropathy Scale (UENS) and the Norfolk Quality of Life Questionnaire-Diabetic Neuropathy (QOL-DN) scale, these questionnaires are not practical in the clinical scenario. However, the latter test is a good tool to detect neuropathy unawareness in patients with diabetes [35]. Indeed, in the study by Veresiu et al., the Norfolk QOL-DN scale was used in 25,000 patients with diabetes and it was found that neither 6600 patients nor their physicians were aware of the presence of neuropathy.

In regard to technical aspects, the use of DEC in clinical examinations has a few limitations as described in previous studies (the use of drugs such as tricyclic antidepressants, non-cardioselective beta-blockers, or extensive dermatitis on the palms of the hands or soles of the feet). However, we do not believe that these aspects apply to the present study since they were exclusion criteria and were not affected by age, sex or previous physical exercise or changes in temperature or acute alterations in blood pressure. According to these studies, the results of DEC do not depend on the rate of perspiration [36], having a high correlation (correlation coefficient 0.814; coefficient of variation 1.15%) [5].

One of the strengths of this study was that DEC and physical examinations were performed by only a few investigators in order to avoid biases in sample selection or erroneous categorization of the diagnosis of PN. This is one of the first studies to compare DEC measurement with the gold standard (EMG). Other studies, even those by the manufacturer, are based on other devices or clinical questionnaires, thereby making our study innovative.

Considering the limitation of resources in primary care and the health care system as a whole, we are currently performing a cost-effectiveness analysis of the use of DEC in the primary care setting.

5. Conclusions

In conclusion, in the usual clinical practice, DEC is feasible, with a moderate sensitivity but a high specificity. It is also easy to use and interpret and requires little training, thereby making it a good screening test in populations with diabetes and prediabetes. It may also be useful in screening general populations at risk of neuropathy.

Nonetheless, the results of cost-effectiveness analyses must first be analyzed before recommending the implementation of DEC in primary care in Catalonia.

Author Contributions: Conceptualization, B.C., T.M. and J.J.C.; methodology, B.C., J.J.C., T.M., G.F.-M.; formal analysis, J.J.C., B.C., T.M.; investigation, T.M., M.G.-B., C.L.-M., F.B., R.S., J.V.; resources, B.C., J.J.C., O.S.-M., F.B.; data curation, G.F.-M.; writing—original draft preparation, B.C., J.J.C., T.M., O.S.-M., I.B., N.O.; writing—review and editing, J.J.C., B.C., T.M., I.B., N.O.; visualization, J.J.C., B.C.; supervision, B.C.; project administration, B.C., J.J.C.; funding acquisition, B.C., J.J.C. The Catalan Diabetes Prevention Research Group contributed and assessed globally to this paper in order to prioritize topics and funding acquisition, and innovate technology advances in the usual clinical scenario.

Funding: The DECODING project was funded by (1) First prize of the Research Grants of the CAMFiC (Catalonian Society of Family and Community Medicine), grant agreement no. XV/2014), (2) Grant from the Laboratorios del Esteve, S.A.U., (3) Department of Health, Generalitat de Catalunya and the PERIS (Pla Estratègic de Recerca i Innovació en Salut) 2016–2020, grant agreement SLT002/16/00093, and (4) *La Marató de TV3* Foundation (2015 grant agreement 73201609.10). As DECODING is part of diabetes prevention projects (Catalan Diabetes Prevention Research Group), its team also participated in the grant agreements SLT002/16/00045, SLT002/16/00154 within this national call for research.

Acknowledgments: The authors would like to acknowledge the contribution of the Catalan Diabetes Prevention Research Group. In addition to the authors, the study group consists of the following members: Bernardo Costa (IDIAP Jordi Gol); Conxa Castell (Departamento de Salud, Generalitat de Catalunya); Joan Josep Cabré, Francesc Barrio, Montserrat García, Susanna Dalmau, Ramon Sagarra, Santiago Mestre, Cristina Jardí, Josep Basora, (Reus—Tarragona); Xavier Cos, Marta Canela, Dolors López (Sant Martí—Barcelona); Ana Martínez, Pilar Martí (El Carmel—Barcelona); Teresa Mur, Charo López, Núria Porta (Mútua Terrassa—Barcelona); Antoni Boquet, Eva María Cabello (Hospitalet-Cornellà—Barcelona); Juan-Luis Bueno, Marc Olivart (Tàrrega-Cervera—Lleida); Mercè Bonfill, Montserrat Nadal (Amposta-Ulldecona—Terres de l'Ebre); Albert Alum, Concepción García (Blanes-Tordera—Girona); Marta Roura, Anna Llens (Figueres-Olot—Girona Nord); María Boldú, Natàlia Vall (Lleida—Ciutat).

Conflicts of Interest: The authors declare no conflict of interest.

References

1. Soriguer, F.; Goday, A.; Bosch-Comas, A.; Bordiú, E.; Calle-Pascual, A.; Carmena, R.; Casamitjana, R.; Castaño, L.; Castell, C.; Catalá, M.; et al. Prevalence of diabetes mellitus and impaired glucose regulation in Spain: The Di@bet.es Study. *Diabetologia* **2012**, *55*, 88–93. [CrossRef] [PubMed]

2. Tesfaye, S.; Selvarajah, D. Advances in the epidemiology, pathogenesis and management of diabetic peripheral neuropathy. *Diabetes Metab. Res. Rev.* **2012**, *28*, 8–14. [CrossRef] [PubMed]

3. Said, G. Diabetic neuropathy—A review. *Nat. Clin Pract. Neurol.* **2007**, *3*, 331–340. [CrossRef] [PubMed]

4. Bernal, D.S.; Tabasco, M.M.; Riera, M.H.; Pedrola, M.S. Etiología y manejo de la neuropatía diabética dolorosa. *Rev. Soc. Esp. Dolor.* **2010**, *17*, 286–296. (In Spanish) [CrossRef]

5. Casellini, C.M.; Parson, H.K.; Richardson, M.S.; Nevoret, M.L.; Vinik, A.I. Sudoscan, a Noninvasive Tool for Detecting Diabetic Small Fiber Neuropathy and Autonomic Dysfunction. *Diabetes Technol. Ther.* **2013**, *15*, 948–953. [CrossRef] [PubMed]

6. Gore, M.; Brandenburg, N.A.; Dukes, E.; Hoffman, D.L.; Tai, K.S.; Stacey, B. Pain severity in diabetic peripheral neuropathy is associated with patient functioning, symptom levels of anxiety and depression, and sleep. *J. Pain Symptom Manag.* **2005**, *30*, 374–385. [CrossRef] [PubMed]

7. Tesfaye, S.; Stevens, L.K.; Stephenson, J.M.; Fuller, J.H.; Plater, M.; Ionescu-Tirgoviste, C.; Nuber, A.; Pozza, G.; Ward, J.D. Prevalence of diabetic peripheral neuropathy and its relation to glycaemic control and potential risk factors: The EURODIAB IDDM Complications Study. *Diabetologia* **1996**, *39*, 1377–1384. [CrossRef]

8. Dyck, P.J.; Kratz, K.M.; Lehman, K.A.; Karnes, J.L.; Melton, L.J.; O'Brien, P.C.; Litchy, W.J.; Windebank, A.J.; Smith, B.E.; Low, P.A.; et al. The Rochester Diabetic Neuropathy Study: Design, criteria for types of neuropathy, selection bias, and reproducibility of neuropathic tests. *Neurology* **1991**, *41*, 799–807. [CrossRef]

9. Pirart, J. Diabetes mellitus and its degenerative complications: A prospective study of 4400 patients observed between 1947 and 1973. *Diabetes Care* **1978**, *1*, 168–188. [CrossRef]

10. Cabezas-Cerrato, J.; Neuropathy Spanish Study Group of the Spanish Diabetes Society. The prevalence of clinical diabetic polyneuropathy in Spain: A study in primary care and hospital clinic groups. *Diabetologia* **1998**, *41*, 1263–1269. [CrossRef] [PubMed]

11. Puig, M.L.; Aguirre, D.R.; Rodríguez, M.C. Neuropatía periférica de los miembros inferiores en diabéticos tipo 2 de diagnóstico reciente. *Av Endocrinol* **2006**, *22*, 149. (In Spanish)

12. Samur, J.A.A.; Rodríguez, M.Z.C.; Olmos, A.I.; Bárcena, D.G. Prevalencia de neuropatía periférica en diabetes mellitus. *Acta Médica Grupo Ángeles* **2006**, *4*, 13. (In Spanish)

13. Franklin, G.M.; Shetterly, S.M.; Cohen, J.A.; Baxter, J.; Hamman, R.F. Risk factors for distal symmetric neuropathy in NIDDM: The San Luis Valley Diabetes Study. *Diabetes Care* **1994**, *17*, 1172–1177. [CrossRef] [PubMed]

14. Kanji, J.N.; Anglin, R.E.; Hunt, D.L.; Panju, A. Does this patient with diabetes have large-fiber peripheral neuropathy? *JAMA* **2010**, *303*, 1526–1532. [CrossRef]

15. González, C.P. Monofilamento de Semmes-Weinstein. *Diabetes Práctica Actual Habilidades En Aten Primaria* **2010**, *1*, 8–19. (In Spanish)

16. Feng, Y.; Schlösser, F.J.; Sumpio, B.E. The Semmes Weinstein monofilament examination as a screening tool for diabetic peripheral neuropathy. *J. Vasc. Surg.* **2009**, *50*, 675–682. [CrossRef]

17. Gordon Smith, A.; Lessard, M.; Reyna, S.; Doudova, M.; Robinson Singleton, J. The Diagnostic Utility of Sudoscan for Distal Symmetric Peripheral Neuropathy. *J. Diabetes Complicat.* **2014**. Available online: http://www.sciencedirect.com/science/article/pii/S1056872714000555 (accessed on 7 February 2018). [CrossRef]

18. Raisanen, A.; Eklund, J.; Calvet, J.H.; Tuomilehto, J. Sudomotor Function as a Tool for Cardiorespiratory Fitness Level Evaluation: Comparison with Maximal Exercise Capacity. *Int. J. Environ. Res. Public Health* **2014**, *11*, 5839–5848. [CrossRef] [PubMed]

19. Yajnik, C.S.; Kantikar, V.V.; Pande, A.J.; Deslypere, J.P. Quick and Simple Evaluation of Sudomotor Function for Screening of Diabetic Neuropathy. *ISRN Endocrinol.* **2012**, *2012*, 103714. Available online: http://downloads.hindawi.com/journals/isrn.endocrinology/2012/103714.pdf (accessed on 25 January 2018). [CrossRef] [PubMed]

20. Eranki, V.; Santosh, R.; Rajitha, K.; Pillai, A.; Sowmya, P.; Dupin, J.; Calvet, J. Sudomotor function assessment as a screening tool for microvascular complications in type 2 diabetes. *Diabetes Res. Clin. Pract.* **2013**, *101*, e11–e13. [CrossRef]

21. Luk, A.O.Y.; Fu, W.-C.; Li, X.; Ozaki, R.; Chung, H.H.Y.; Wong, R.Y.M.; So, W.-Y.; Chow, F.C.C.; Chan, J.C.N. The Clinical Utility of SUDOSCAN in Chronic Kidney Disease in Chinese Patients with Type 2 Diabetes. *PLoS ONE* **2015**, *10*, e0134981. [CrossRef]

22. Ramachandran, A.; Moses, A.; Shetty, S.; Thirupurasundari, C.J.; Seeli, A.C.; Snehalatha, C.; Singvi, S.; Deslypere, J.-P. A new non-invasive technology to screen for dysglycaemia including diabetes. *Diabetes Res. Clin. Pract.* **2010**, *88*, 302–306. [CrossRef]

23. Selvarajah, D.; Cash, T.; Davies, J.; Sankar, A.; Rao, G.; Grieg, M.; Pallai, S.; Gandhi, R.; Wilkinson, I.D.; Tesfaye, S. SUDOSCAN: A Simple, Rapid, and Objective Method with Potential for Screening for Diabetic Peripheral Neuropathy. *PLoS ONE* **2015**, *10*, 0138224. [CrossRef]

24. Pop-Busui, R.; Boulton, A.J.; Feldman, E.L.; Bril, V.; Freeman, R.; Malik, R.A.; Sosenko, J.M.; Ziegler, D. Diabetic neuropathy: A position statement by the American Diabetes Association. *Diabetes Care* **2017**, *40*, 136–154. [CrossRef]

25. Portillo, R.; Lira, D.; Quiñónez, M. Evaluación Neurofisiológica y Clínica en Pacientes con Diabetes Mellitus. 2005, pp. 11–18. Available online: http://www.scielo.org.pe/scielo.php?script=sci_arttext&pid=S1025-55832005000100003 (accessed on 27 February 2018).

26. Lu, B.; Hu, J.; Wen, J.; Zhang, Z.; Zhou, L.; Li, Y.; Hu, R. Determination of peripheral neuropathy prevalence and associated factors in Chinese subjects with diabetes and pre-diabetes–ShangHai Diabetic neuRopathy Epidemiology and Molecular Genetics Study (SH-DREAMS). *PLoS ONE* **2013**, *8*, e61053. [CrossRef] [PubMed]
27. Ziegler, D.; Rathmann, W.; Dickhaus, T.; Meisinger, C.; Mielck, A. Prevalence of polyneuropathy in pre-diabetes and diabetes is associated with abdominal obesity and macroangiopathy. *Diabetes Care* **2008**, *31*, 464–469. [CrossRef] [PubMed]
28. Teruel, C.M.; Fernández-Real, J.M.; Ricart, W.; Valent, R.F.; Vallés, M.P. Prevalence of diabetic retinopathy in the region of Girona. Study of related factors. *Arch. Soc. Espanola Oftalmol.* **2005**, *80*, 85–91.
29. Cabré, J.J.; Mur, T.; Costa, B.; Barrio, F.; López-Moya, C.; Sagarra, R.; García-Barco, M.; Vizcaíno, J.; Bonaventura, I.; Ortiz, N.; et al. Feasibility and effectiveness of electrochemical dermal conductance measurement for the screening of diabetic neuropathy in primary care. DECODING Study (Dermal Electrochemical Conductance in Diabetic Neuropathy). Rationale and design. *Medicine* **2018**, *97*, e10750. [CrossRef]
30. Meijer, J.W.; Van Sonderen, E.; Blaauwwiekel, E.E.; Smit, A.J.; Groothoff, J.W.; Eisma, W.H.; Links, T.P. Diabetic neuropathy examination: A hierarchical scoring system to diagnose distal polyneuropathy in diabetes. *Diabetes Care* **2000**, *23*, 750–753. [CrossRef]
31. Krieger, S.M.; Reimann, M.; Haase, R.; Henkel, E.; Hanefeld, M.; Ziemssen, T. Sudomotor Testing of Diabetes Polyneuropathy. *Front. Neurol.* **2018**, *9*, 803. [CrossRef]
32. Binns-Hall, O.; Selvarajah, D.; Sanger, D.; Walker, J.; Scott, A.; Tesfaye, S. One-stop microvascular screening service: An effective model for the early detection of diabetic peripheral neuropathy and the high-risk foot. *Diabet. Med.* **2018**, *35*, 887–894. [CrossRef]
33. Mao, F.; Liu, S.; Qiao, X.; Zheng, H.; Xiong, Q.; Wen, J.; Zhang, S.; Zhang, Z.; Ye, H.; Shi, H.; et al. SUDOSCAN, an effective tool for screening chronic kidney disease in patients with type 2 diabetes. *Exp. Ther. Med.* **2017**, *14*, 1343–1350. [CrossRef]
34. Wang, F.; Zhang, J.; Yu, J.; Liu, S.; Zhang, R.; Ma, X.; Yang, Y.; Wu, P. Diagnostic Accuracy of Monofilament Tests for Detecting Diabetic Peripheral Neuropathy: A Systematic Review and Meta-Analysis. *J. Diabetes Res.* **2017**, *2017*, 8787261. [CrossRef]
35. Veresiu, A.I.; Bondor, C.I.; Florea, B.; Vinik, E.J.; Vinik, A.I.; Gavan, N.A. Detection of undisclosed neuropathy and assessment of its impact on quality of life: A survey in 25,000 Romanian patients with diabetes. *J. Diabetes Its Complicat.* **2015**, *29*, 644–649. [CrossRef] [PubMed]
36. Sudoscan Use Instructions. Available online: https://www.sudoscan.com/for-physicians/research-articles/ (accessed on 18 March 2019).

Journal of
Clinical Medicine

Article

FGF23 and Klotho Levels are Independently Associated with Diabetic Foot Syndrome in Type 2 Diabetes Mellitus

Javier Donate-Correa [1,*,†], Ernesto Martín-Núñez [1,2,†], Carla Ferri [1,2],
Carolina Hernández-Carballo [1,2], Víctor G. Tagua [1], Alejandro Delgado-Molinos [3],
Ángel López-Castillo [3], Sergio Rodríguez-Ramos [4], Purificación Cerro-López [4],
Victoria Castro López-Tarruella [5], Miguel Angel Arévalo-González [6], Nayra Pérez-Delgado [7],
Carmen Mora-Fernández [1,‡] and Juan F. Navarro-González [1,8,9,*,‡]

[1] Research Unit, University Hospital Nuestra Señora de Candelaria (UHNSC), 38010 Santa Cruz de Tenerife,
    Spain; emarnu87@gmail.com (E.M.-N.); carlamferri@gmail.com (C.F.);
    carolinahdezcarballo@gmail.com (C.H.-C.); vtagua@funcanis.es (V.G.T.);
    carmenmora.fdez@gmail.com (C.M.-F.)
[2] Doctoral and Graduate School, University of La Laguna, 38200 San Cristóbal de La Laguna, Spain
[3] Vascular Surgery Service, University Hospital Nuestra Señora de Candelaria, 38010 Santa Cruz de Tenerife,
    Spain; adelgadomolinos@gmail.com (A.D.-M.); angellopezcastillo24@gmail.com (Á.L.-C.)
[4] Transplant Coordination, University Hospital Nuestra Señora de Candelaria, 38010 Santa Cruz de Tenerife,
    Spain; sergiotomasr@hotmail.com (S.R.-R.); pcerlop@gobiernodecanarias.org (P.C.-L.)
[5] Pathology Service, University Hospital Nuestra Señora de Candelaria, 38010 Santa Cruz de Tenerife, Spain;
    vcaslop@gmail.com
[6] Human Anatomy and Histology Department, University of Salamanca, 37008 Salamanca, Spain;
    marevalo@usal.es
[7] Clinical Analysis Service, University Hospital Nuestra Señora de Candelaria, 38010 Santa Cruz de Tenerife,
    Spain; nperdel@gobiernodecanarias.org
[8] Nephrology Service, University Hospital Nuestra Señora de Candelaria, 38010 Santa Cruz de Tenerife, Spain
[9] Institute of Biomedical Technologies, University of La Laguna, 38200 San Cristóbal de La Laguna, Spain
* Correspondence: jdonate@ull.es (J.D.-C.); jnavgon@gobiernodecanarias.org (J.F.N.-G.); Tel.: +34-922-602-921
    (J.D.-C. & J.F.N.-G.); Fax: +34-922-600-562 (J.D.-C. & J.F.N.-G.)
† These authors contributed equally to this work.
‡ These authors share senior authorship.

Received: 26 February 2019; Accepted: 1 April 2019; Published: 3 April 2019

Abstract: Background: Diabetic foot syndrome (DFS) is a prevalent complication in the diabetic population and a major cause of hospitalizations. Diverse clinical studies have related alterations in the system formed by fibroblast growth factor (FGF)-23 and Klotho (KL) with vascular damage. In this proof-of-concept study, we hypothesize that the levels of FGF23 and Klotho are altered in DFS patients. Methods: Twenty patients with limb amputation due to DFS, 37 diabetic patients without DFS, and 12 non-diabetic cadaveric organ donors were included in the study. Serum FGF23/Klotho and inflammatory markers were measured by enzyme-linked immunosorbent assay (ELISA). Protein and gene expression levels in the vascular samples were determined by immunohistochemistry and quantitative real-time PCR, respectively. Results: Serum Klotho is significantly reduced and FGF23 is significantly increased in patients with DFS ($p < 0.01$). Vascular immunoreactivity and gene expression levels for Klotho were decreased in patients with DFS ($p < 0.01$). Soluble Klotho was inversely related to serum C-reactive protein ($r = -0.30$, $p < 0.05$). Vascular immunoreactivities for Klotho and IL6 showed an inverse association ($r = -0.29$, $p < 0.04$). Similarly, vascular gene expression of *KL* and *IL6* were inversely associated ($r = -0.31$, $p < 0.05$). Logistic regression analysis showed that higher Klotho serum concentrations and vascular gene expression levels were related to a lower risk of DFS, while higher serum FGF23 was associated with a higher risk for this complication.

Conclusion: FGF23/Klotho system is associated with DFS, pointing to a new pathophysiological pathway involved in the development and progression of this complication.

Keywords: diabetic foot syndrome; vascular disease; fibroblast growth factor 23; Klotho; inflammation

1. Introduction

Type 2 diabetes mellitus (T2DM) has become a critical health problem, with more than 450 million people living with this disease worldwide [1]. Diabetic foot syndrome (DFS) is a prevalent complication in this population and a major cause of hospitalizations. This syndrome carries the risk of limb amputation, which represents the most prevalent non-traumatic amputation surgery in the hospital setting [2]. Mortality associated with DFS is similar to that of breast, prostate, or colon cancer [3]. Hence, there is an overwhelming need for a better understanding of the molecular mechanisms underlying the development of DFS and foremost to detect early those at the highest risk of this complication and develop more specific and effective therapies.

DFS has a complex and multifactorial pathogenesis where an underlying vascular atherogenic process affecting both the endothelium and the smooth muscle cell layer is a critical factor. Atherosclerosis is a chronic inflammatory phenomenon involving several pathways and molecules, including inflammatory cytokines [4,5]. Closely related with the atherogenic phenomena, the alterations of the mineral metabolism have gained special prominence as pathogenic factors in the development and progression of vascular damage [6,7].

The newly described system formed by the fibroblast growth factor (FGF)-23 and Klotho (KL) proteins is recognized as one of the main regulators of mineral metabolism. FGF23 is a phosphaturic hormone that is synthesized in the bone in response to dietary phosphate intake in order to keep a normal phosphate homeostasis [8]. To this end, it binds to cognate fibroblast growth factor receptors (FGFRs; primarily FGFR1 and 3) in the renal tubular cells in the presence of the obligated co-receptor Klotho, a type 1 trans-membrane protein predominantly expressed in the kidneys [8]. In addition, a soluble form of Klotho is generated by the shedding of the ectodomain by the A Disintegrin and A Metalloproteinase 17 (ADAM17), or by an alternative RNA splicing [9–11]. Beyond this physiological role, the FGF23/Klotho system has been related to various processes associated with cardiovascular damage.

Diverse clinical studies have directly related the increase in serum concentrations of FGF23 with cardiovascular damage and with an increased mortality [12,13]. Moreover, recent works suggest that high levels of FGF23 can exert deleterious effects on the cardiovascular system by establishing low affinity bonds with FGFRs, independently of Klotho [14,15]. On the other hand, reduced concentrations of soluble Klotho have been involved in processes related to premature aging, including the appearance of vascular damage [16]. This soluble form of Klotho is detectable in urine, serum, and cerebrospinal fluid, acting as a humoral factor with multiple functions such as anti-oxidation, modulation of ion channels, anti-Wnt signaling or anti-apoptosis, and anti-senescence effects [17]. Importantly, the expression of *KL* has been recently demonstrated in human vascular tissue [16,18] and both the levels of systemic and vascular Klotho have been related with cardiovascular complications [16,19–21].

In this proof-of-concept study, we hypothesize that the levels of FGF23 and Klotho are altered in DFS patients. We determined the serum concentrations and the vascular expression levels of FGF23 and Klotho in a group of diabetic patients, with and without DFS. We also measured the levels of inflammatory mediators, including the cytokines tumor necrosis factor (TNF)-α, interleukin (IL)-6 and IL10, and analyzed the relationship among these variables and the presence of DFS.

2. Materials and Methods

2.1. Patients

Eighty-five diabetic patients undergoing an elective open vascular surgery procedure due to established clinical atherosclerotic artery disease were considered for enrollment in this study. Exclusion criteria included hemodynamic instability; history of chronic inflammatory, immunologic, or tumoral disease; positive serology to hepatitis B, hepatitis C, or HIV; acute inflammatory or intercurrent infectious episodes in the previous month; institutionalization; treatment with immunotherapy or immunosuppressive drugs; previous organ transplantation and advanced renal disease (estimated glomerular filtration rate (eGFR) lower than 30 mL/min/1.73m^2).

For comparative purposes in the immunohistochemical and gene expression analyses, femoral samples from 12 cadaveric non-diabetic organ donors matched by age, sex, and eGFR, and without any medical history of cardiovascular disease were recovered during organ retrieval surgery.

The study protocol was approved by the Institutional Ethics Committee of the University Hospital Nuestra Señora de Candelaria (Santa Cruz de Tenerife, Spain) and complied with ethical standards of the Declaration of Helsinki. Written informed consent was obtained from all participants before they participated in the study.

2.2. Samples and Biochemical Markers

During the surgical procedure, a sample of the carotid or femoral arteries, according to the affected vessel, was obtained from the participants. At the same time, serum samples were drawn, aliquoted, and immediately stored at -80 °C for further analysis. Routine biochemical parameters were determined using standard methods. Serum levels of intact FGF23 were measured by specific enzyme-linked immunosorbent assay (ELISA) (EMD Millipore Corporation, Milford, MA, USA), which detects only the intact form of FGF23 with a sensitivity of 3.5 pg/mL and intra- and inter-assay coefficients of 9.5% and 6.85%, respectively. Serum Klotho was measured using the human Klotho ELISA kit (Immuno-Biological Laboratories, Takasaki, Japan), with a sensitivity of 6.15 pg/mL and intra- and inter-assay coefficients of variation of 3.1 % and 6.9 %, respectively. High-sensitivity serum C-reactive protein (hsCRP) was measured by a high-sensitivity particle enhanced immunoturbidimetric fully automated assay (Roche Diagnostics GmbH, Mannheim, Germany) in a Cobas 6000 analyzer from the same manufacturer with a sensitivity of 0.3 mg/L and intra- and inter-assay coefficients of variation of 1.6 and 8.4, respectively. Levels of the inflammatory cytokines TNFα, IL6 and IL10 were also measured by high-sensitivity ELISA methods (Quantikine®, R&D Systems, Abingdon, UK). Minimum detectable concentrations were 0.10 pg/mL, 0.70 pg/mL, and 0.50 pg/mL, respectively. Intra- and inter-assay coefficients of variability were <10.8%.

2.3. Immunohistochemistry

Sections of blood vessels were fixed in 4% buffered formalin for 24 h and subsequently dehydrated in ascending concentrations of ethanol, cleared in xylene, and embedded in paraffin. Blocks were trimmed and 3 µm sections were processed for immunohistochemistry. Primary antibodies used were: mouse monoclonal anti-FGFR1, 1:400 dilution (Abcam, Cambridge, UK); rabbit polyclonal anti-FGFR3, 1:300 dilution (Abcam); rabbit polyclonal anti-Klotho, 1:100 dilution (Abcam); mouse monoclonal anti-IL6, 1:300 dilution (Santa Cruz Biotechnology Inc., Dallas, TX, USA); and rabbit monoclonal anti-TNFα, 1:100 dilution (Santa Cruz Biotechnology Inc.). For the quantification analysis, a total of five images of each slide that include intima and media layers were captured and processed with a high-resolution video camera (Sony, DF-W-X710, Kōnan, Japan) connected to a light microscope (Nikon Eclipse 50i). The areas of tissue stained by the antibodies were quantified by using ImageJ software (Rasband, W.S., ImageJ, National Institutes of Health, Bethesda, MD, USA). Results are expressed in square microns.

2.4. Quantitative Real-Time PCR

After surgery, vascular tissue fragments were immediately placed in RNAlater® solution (Ambion (Europe) Limited, Cambridge, UK) and stored at 4 °C for subsequent RNA extraction. Total RNA was isolated by homogenization in TRI Reagent® (Sigma-Aldrich, Saint Louis, MO, USA) employing TissueRuptor (QIAGEN, Hilden, Germany). Further purification was performed using RNeasyMini kit (QIAGEN), according to manufacturer's specifications, and stored at −80 °C. RNA was quantified using a Thermo Scientific NanoDrop Lite Spectrophotometer (Thermo Scientific, Waltham, MA, USA) and its quality was assessed by the A260/A280 ratio measured in this equipment. cDNA was obtained using a High Capacity RNA-to-cDNA kit (Applied Biosystems, Foster City, CA, USA) for further use in quantitative RT-PCR (qRTPCR). Transcripts encoding for *KL, ADAM17, TNF, IL6, IL10,* and *glyceraldehyde 3-phosphate dehydrogenase* (*GAPDH*) were measured by TaqMan quantitative PCR with TaqMan Fast Universal PCR Master Mix (Applied Biosystems, Foster City, CA, USA). TaqMan gene expression assays for each transcript (Hs00183100_m1 (KLOTHO), Hs01041915_m1 (ADAM17), Hs00174128_m1 (TNFα), Hs00985639_ml (IL6), Hs0961622_m1 (IL10), and Hs99999905_m1 (GAPDH)) were analyzed in a 7500 Fast Real-Time PCR System (Applied Biosystems). The level of target mRNA was estimated by relative quantification using the comparative method ($2^{-\Delta\Delta Ct}$) by normalizing to *GAPDH* expression. mRNA levels were expressed as arbitrary units (a.u.). Quantification of each cDNA sample was tested in triplicate.

2.5. Statistical Analysis

All analyses were performed using SPSS software version 25 (IBM Corp. Armonk, NY, USA). Continuous variables are reported as mean ± SD or median with interquartile range (IQR), and categorical data as number and percent frequency. Continuous variables were checked for the normal distribution assumption using the Kolmogorov–Smirnov statistics, and those that did not satisfy the criteria were log-transformed. Comparisons between groups were performed by Chi-square test, Mann–Whitney U test, or Kruskal-Wallis test as appropriate. The Spearman correlation coefficient was calculated to assess the relationship between variables. Partial correlation analysis was performed to measure the association of serum and vascular Klotho expression with inflammatory parameters whilst controlling for the effect of covariates that were selected based on clinical relevance and the results of comparison and correlation analysis. A multiple logistic regression was performed to assess independent predictors of the presence of DFS. For this purpose, we adopted three models: In model 1, we introduced age, sex, uric acid, eGFR, and hsCRP. In model 2, we additionally included serum Klotho, FGF23, IL6, and IL10. Finally, in model 3 we adjusted the analysis for the immunoreactivity and gene expression levels of *KL*. Regarding gene expression analysis, quantification of each sample was tested in triplicate and data are expressed as arbitrary units. A two tailed *p*-value < 0.05 was considered statistically significant.

3. Results

3.1. Characteristics of the Patients and Biochemical Parameters

Eighty-five diabetic patients with established vascular disease that underwent elective revascularization surgical or lower limb amputation were initially evaluated, with 28 subjects being excluded due to exclusion criteria. Finally, 57 patients (46 males and 11 females) with a mean age of 69.8 ± 9.7 were included in the study: 20 patients were subjected to elective limb amputation by DFS, and the remaining 37 underwent elective shunt vascular surgery in the lower extremities. The specific indications for amputation included unsuccessful previous revascularization, extensive non-healing ulcers or non-healing wounds, non-reconstructable disease with persistent tissue loss, unrelenting rest pain due to muscle ischemia and gangrene. At the time of amputation, no patient had foot infection. In all cases a standard below knee transtibial amputation was performed. The clinical, biochemical, and demographic characteristics of the 57 patients included in the study are presented in Table 1.

There were no differences in any demographic characteristic between the group of patients with DFS and the remaining patients included in the study. Percentages of subjects with smoking habits, alcoholism, and hypertension were also similar in both groups. Similarly, no differences were observed in general laboratory measurements including total cholesterol, high-density lipoprotein (HDL), low-density lipoprotein (LDL), triglycerides, HbA1c, glucose, eGFR, albumin, calcium, phosphorous, and alkaline phosphatase. However, the levels of pro-inflammatory markers were higher in patients with DFS, although only reached statistical significance for hsCRP (6 (2.3–10.5) vs. 5.3 (2.4–13.6) pg/mL, $p < 0.05$) and IL6 (22 (51–28.6) vs. 6.42 (0.7–26.7) pg/mL, $p < 0.05$). Conversely, the level of anti-inflammatory cytokine IL10 was diminished in the DFS group (1.1 (0.6–6.7) vs. 3.9 (0.5–6.8) pg/mL, $p < 0.01$). Interestingly, the serum levels of Klotho were significantly diminished in patients with DFS (397.4 (318–515) vs. 617 (484–779) pg/mL, $p < 0.01$), whereas the FGF23 concentration was significantly higher (23.8 (17–32.2) vs. 15.5 (10.1–24.5) pg/mL, $p < 0.01$). On comparison with the respective levels in patients without DFS, these differences represent a median percent decrease of 37.4% for Klotho and 30.3% increase for iFGF23 concentrations in patients with DFS (Table 1 and Figure 1).

Table 1. Clinical characteristics and biochemical assessments of the diabetic patients included in the study.

Variable	Overall	DFS	No. DFS	*p*–Value
n	57	20	37	
Age (years)	69.8 ± 9.7	71.8 ± 10.2	68.7 ± 9.4	NS
Male gender, *n* (%)	46 (80)	15 (75)	31 (84)	NS
BMI (kg/m^2)	30.1 ± 3.8	30.8 ± 4.7	29.7 ± 3.3	NS
Hypertension, *n* (%)	51 (89.4)	17 (85)	34 (91.9)	NS
Alcoholism, *n* (%)	22 (38.6)	7 (35)	15 (40.5)	NS
Smoking habits, *n* (%)	36 (63.2)	12 (60)	24 (64.8)	NS
Total Cholesterol (mg/dL)	159.2 ± 38.9	159 ± 45.8	159.3 ± 35.3	NS
HDL (mg/dL)	42.4 ± 10.2	40 ± 9.8	43.7 ± 10.4	NS
LDL (mg/dL)	84.8 ± 29.4	87.5 ± 33.4	83.3 ± 27.3	NS
Triglycerides, mg/dL	155 ± 80.2	156.6 ± 100	154.2 ± 68.4	NS
HbA1c, %	7.3 ± 1.4	7.3 ± 1.9	7.2 ± 1.3	NS
Glucose	132.9 ± 42.8	126.3 ± 40.1	136.6 ± 44.4	NS
eGFR, mL/min/1.73 m^2	77.5 ± 24.7	76.6 ± 24	77.8 ± 25.4	NS
Albumin (g/L)	3.8 ± 0.7	3.7 ± 0.8	3.8 ± 0.6	NS
Calcium (mg/dl)	9.1 ± 0.6	9 ± 0.6	9.1 ± 0.6	NS
Phosphorous (mg/dl)	3.6 ± 0.5	3.6 ± 0.5	3.6 ± 0.5	NS
AP (mU/mL)	78.8 ± 58.1	77.9 ± 43.3	79.2 ± 65.3	NS
hsCRP (mg/dL)	5.8 (2.4–11.9)	6 (2.3–10.5)	5.3 (2.4–13.6)	<0.05
IL6, pg/mL	11.2 (1.1–28.1)	22 (5.1–28.6)	6.42 (0.7–26.7)	<0.05
IL10, pg/mL	1.3 (0.5–6.7)	1.1 (0.6–6.7)	3.9 (0.5–6.8)	<0.01
TNFα, pg/mL	1.3 (0.8–1.3)	1.4 (0.9–2.1)	1.1 (0.7–1.7)	NS
KLOTHO, pg/mL	532.8 (375–677)	397.4 (318–515)	617.3 (484–779)	<0.01
iFGF23, pg/mL	17.5 (12–27)	23.8 (17–32)	15.5 (10.1–24.5)	<0.01

DFS, diabetic foot syndrome; BP, blood pressure; BMI, body mass index; HDL, high-density lipoprotein; LDL, low-density lipoprotein; HbA1c, hemoglobin A1c; eGFR, estimated glomerular filtration rate; AP, alkaline phosphatase; hsCRP, high-sensitivity C-reactive protein; IL, interleukin; TNF, tumor necrosis factor; FGF23, fibroblast growth factor 23; NS, not significant. *p*-value reflects differences between with DFS and without DFS.

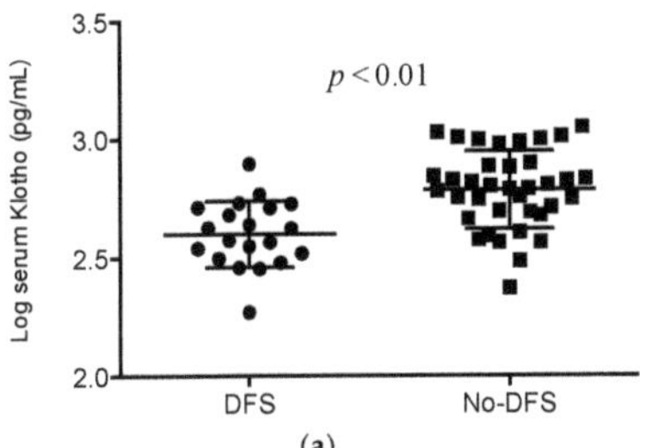

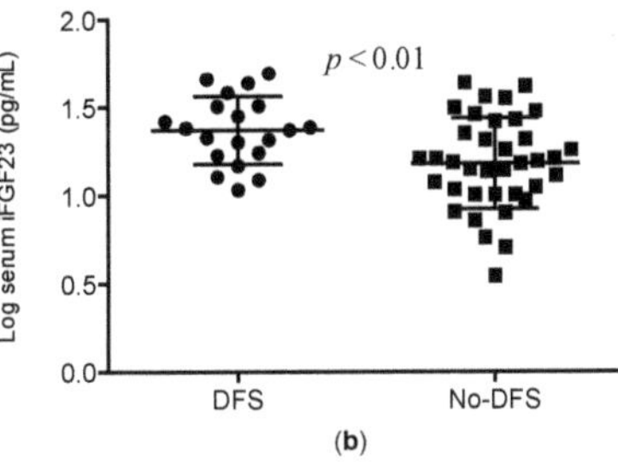

Figure 1. Differences in the log-transformed blood levels of (**a**) Klotho and (**b**) FGF23 between patients with and without diabetic foot syndrome (DFS).

3.2. Immunohistochemical Analysis

For immunohistochemical determinations, serial sections of femoral or carotid were obtained from the participants. All the vascular samples obtained from the 20 patients with DFS were excised from the femoral artery. Fragments obtained from the group of 37 patients without DFS included carotid and femoral samples (21 and 16, respectively). All the fragments from the donor group were obtained from the femoral artery.

The results obtained show the presence of Klotho protein, as well as the receptors for FGF23 (FGFR1 and FGFR3), in all the samples studied (Figure 2A). Mean immunoreactivity levels for Klotho were significantly decreased in vascular sections obtained from patients with DFS as compared to sections from diabetic patients without DFS ($p < 0.05$) and from the control group ($p < 0.01$) (Figure 2B). No differences were observed in Klotho levels between patients without DFS and the control group. The difference in Klotho values between the patients with and without DFS remained when only the fragments of femoral arteries were considered in the analysis ($p < 0.05$). No differences were observed in the immunoreactivity levels for FGFR1 and FGFR3 between the two groups of diabetic patients. However, the levels of both proteins were higher in diabetic patients than in controls ($p < 0.01$). Finally, mean immunoreactivity levels of the pro-inflammatory cytokines IL6 and TNFα were higher in the group of diabetic patients than in controls ($p < 0.05$), with no differences regarding the presence or absence of DFS (Figure 2B).

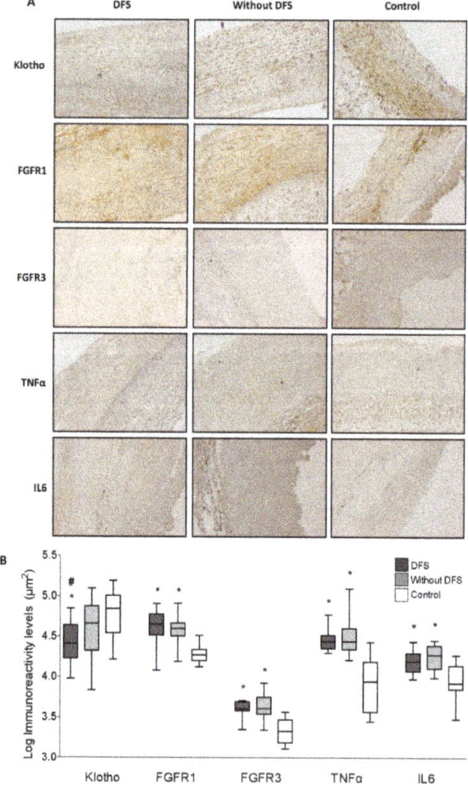

Figure 2. (a) Immunohistochemical staining for Klotho, FGFR1 (fibroblast growth factor receptor (1), FGFR3, TNFα (tumor necrosis factor α), and IL6 (interleukin 6) in arterial sections of patients with and without diabetic foot syndrome (DFS), and control donors (magnification 4×). (b) Mean immunoreactivity levels. * $p < 0.01$ vs. control group; # $p < 0.05$ vs. patients without DFS.

3.3. Vascular Gene Expression

Gene expression levels of *KL, ADAM17,* and the cytokines *IL6, IL10,* and *TNF* were analyzed in all the vascular samples. Results showed that expression levels of *Klotho* were significantly reduced in DFS patients (2.04 (0.79–3.36)) when comparing to either diabetic patients without DFS (4.21 (1.98–6.61)) or to the control group (6.81 (8.51–5.12)); $p < 0.01$, for both comparisons (Figure 3). These differences represent a mean 51.4% and 70% lower expression of *KL* in the vascular wall of DFS patients compared to patients without DFS and to control individuals, respectively. *ADAM17* expression levels were similar among all the groups. Concerning inflammatory cytokines, the vascular gene expression levels of *TNF* and *IL6* were higher in the group of patients with DFS when compared to control subjects ($p < 0.05$ and $p < 0.01$, respectively), although only the *IL6* levels differed between patients with and without DFS (2.04 (0.79–3.36) vs. 1.41 (0.34–7.83), $p < 0.01$). This difference represented a mean 30.9% higher expression of *IL6* in the vascular wall of patients with DFS. Finally, the expression levels of the anti-inflammatory cytokine *IL10* in both groups of diabetic patients were reduced with respect to the control subjects ($p < 0.05$).

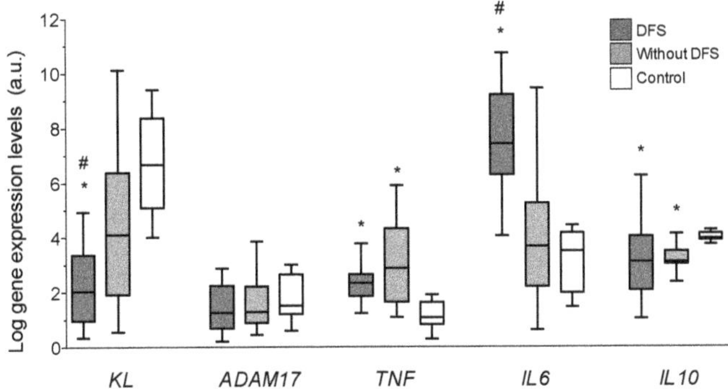

Figure 3. Gene expression levels of *KL, ADAM17* (A Disintegrin and A Metalloproteinase 17), *TNF* (tumor necrosis factor α), *IL6* (interleukin 6), and *IL10* determined by quantitative RT-PCR in vascular walls of patients with diabetic foot syndrome (DFS), without DFS, and controls. * $p < 0.01$ vs. control group; # $p < 0.05$ vs. patients without DFS.

3.4. Correlations and Multivariate Analysis

A correlation analysis was performed between Klotho levels and the other parameters determined in the study (Table 2). A statistically significant positive correlation was found between eGFR and serum and vascular expression levels of Klotho ($r = 0.329$ and $r = 0.354$, respectively, $p < 0.01$). Vascular Klotho levels were also directly related to serum Klotho concentrations ($r = 0.288$, $p < 0.05$) and inversely related to serum uric acid ($r = 0.342$, $p < 0.01$), serum creatinine ($r = -0.389$, $p < 0.01$), and vascular protein levels of IL6 ($r = -0.337$, $p < 0.01$). Finally, vascular expression levels of *KL and ADAM17* showed a positive correlation ($r = 0.369$, $p < 0.01$).

Partial correlation analysis was performed to measure the associations of serum and vascular Klotho expression with inflammatory parameters after controlling for covariates (age, eGFR, HbA1c, hemoglobin, hematocrit, HDL, LDL, triglycerides, and uric acid) (Table 3). In the first analysis, soluble Klotho was significantly related to serum CRP ($r = -0.30$, $p < 0.05$), with a marginal association with serum IL6. In the second analysis, vascular mRNA expression of *KL* was directly associated with the expression of *ADAM17* ($r = 0.41$, $p = 0.01$), whilst a significant inverse association was observed with *IL6* expression ($r = -0.31$, $p < 0.05$). Regarding vascular immunoreactivity for Klotho, this parameter showed an inverse association with vascular immunoreactivity for IL6 ($r = -0.29$, $p < 0.04$).

Finally, the multivariate logistic regression analysis using the presence/absence of DFS as the dependent variable showed that higher serum Klotho concentrations and gene expression levels in the vascular wall were associated with a lower risk for DFS, with immunoreactivity values for vascular Klotho showing a marginal association. On the contrary, higher serum FGF23 was related to a higher risk for this complication (Table 4).

Table 2. Correlations of serum and vascular gene expression levels of Klotho in diabetic patients.

Variable	Log Serum KL		Log Vascular *KL*	
	r	*p*-Value	*r*	*p*-Value
BMI (kg/m^2)	0.129	NS	−0.062	NS
Total Cholesterol (mg/dL)	0.055	NS	−0.139	NS
HDL (mg/dL)	0.255	<0.05	−0.131	NS
LDL (mg/dL)	0.021	NS	−0.152	NS
Triglycerides, mg/dL	−0.058	NS	0.134	NS
HbA1c, %	−0.034	NS	0.131	NS
Hemoglobin	0.344	<0.05	0.104	NS
Albumin (g/L)	−0.038	NS	0.065	NS
Hematocrit	0.341	<0.05	0.109	NS
Calcium (mg/dl)	0.206	NS	−0.51	NS
Phosphorous (mg/dl)	0.071	NS	0.17	NS
AP (mU/mL)	−0.62	NS	0.03	NS
Uric acid (mg/dL)	0.096	NS	−0.342	<0.01
eGFR, mL/min/1.73 m^2	0.329	<0.01	0.354	<0.01
hsCRP (mg/dL)	−0.196	<0.05	0.076	NS
Serum IL6 (pg/mL)	−0.204	NS	−0.094	NS
Serum IL10 (pg/mL)	0.172	NS	−0.099	NS
Serum TNFα (pg/mL)	−0.159	NS	−0.114	NS
Serum Klotho (pg/mL)			0.125	NS
Serum iFGF23 (pg/mL)	0.042	NS	−0.064	NS
VIR Klotho (μm^2)	0.288	<0.05	0.097	NS
VIR FGFR1 (μm^2)	−0.15	NS	0.158	NS
VIR FGFR3 (μm^2)	0.153	NS	−0.106	NS
VIR IL6 (μm^2)	−0.124	NS	−0.337	<0.01
VIR TNFα (μm^2)	0.076	NS	−0.177	NS
Log *KL* mRNA (a.u.)	0.125	NS		
Log *ADAM17* mRNA (a.u.)	0.183	NS	0.369	<0.01
Log *IL6* mRNA (a.u.)	−0.81	NS	0.133	NS
Log *IL10* mRNA (a.u.)	−0.069	NS	0.223	NS
Log *TNF* mRNA (a.u.)	0.069	NS	0.183	NS

BMI, body mass index; HDL, high-density lipoprotein; LDL, low-density lipoprotein; HbA1c, hemoglobin A1c; AP, alkaline phosphatase; eGFR, estimated glomerular filtration rate; hsCRP, high-sensitivity C-reactive protein; IL, interleukin; TNF, tumor necrosis factor; KL, Klotho; a.u., arbitrary units; FGF23, fibroblast growth factor 23; VIR, vascular immunoreactivity; NS, not significant.

Table 3. Association between serum and vascular expression levels of Klotho with inflammatory parameters assessed by partial correlation analysis.

Serum Klotho Concentrations		
Serum hsCRP	$r = -0.30$	$p < 0.05$
Serum IL6	$r = -0.23$	$p = 0.10$
Serum IL10	$r = 0.20$	$p = 0.17$
Serum TNFα	$r = -0.12$	$p = 0.39$
Vascular mRNA *KL* Expression Levels		
Expression *ADAM17*	$r = 0.41$	$p = 0.001$
Expression *IL6*	$r = -0.31$	$p < 0.05$
Expression *IL10*	$r = 0.19$	$p = 0.20$
Expression *TNF*	$r = 0.03$	$p = 0.81$
Vascular Klotho Immunoreactivity Levels		
Immunoreactivity IL6	$r = -0.29$	$p < 0.05$
Immunoreactivity TNFα	$r = -0.06$	$p = 0.64$

hsCRP, high-sensitivity C-reactive protein; IL6, interleukin 6; IL10, interleukin 10; TNFα, tumor necrosis factor α; ADAM17, A Disintegrin and A Metalloproteinase 17. Covariates: age, estimated glomerular filtration rate, glycated hemoglobin, hemoglobin, hematocrit, HDL, LDL, triglycerides, and uric acid.

Table 4. Multivariate logistic regression analysis for the presence of DFS.

Model 1		
Independent Variable	**Odds Ratio (95% CI)**	***p*-Value**
Age	1.03 (0.97 to 1.10)	0.28
Gender	0.59 (0.13 to 2.78)	0.51
Uric acid	0.96 (0.64 to 1.45)	0.87
eGFR	1.05 (0.97 to 1.03)	0.72
hsCRP	0.99 (0.93 to 1.06)	0.95
Model 2 (Model 1 + Serum Klotho, FGF23, IL6 and IL10)		
Independent Variable	**Odds Ratio (95% CI)**	***p*-Value**
Age	1.00 (0.91 to 1.10)	0.82
Gender	0.75 (0.08 to 7.04)	0.80
Uric acid	1.09 (0.57 to 2.08)	0.78
eGFR	1.01 (0.97 to 1.05)	0.49
hsCRP	0.97 (0.89 to 1.07)	0.64
Serum Klotho	0.99 (0.98 to 0.99)	<0.01
Serum FGF23	1.10 (1.05 to 1.21)	<0.05
Serum IL6	1.00 (0.98 to 1.01)	0.88
Serum IL10	1.06 (0.85 to 1.32)	0.60
Model 3 (Model 2 + Soluble Klotho)		
Independent Variable	**Odds Ratio (95% CI)**	***p*-Value**
Age	0.91 (0.78 to 1.06)	0.23
Gender	4.98 (0.17 to 14.63)	0.35
Uric acid	0.64 (0.27 to 1.50)	0.31
eGFR	1.04 (0.98 to 1.12)	0.15
hsCRP	1.16 (0.93 to 1.33)	0.23
Serum Klotho	0.99 (0.98 to 0.99)	<0.05
Serum FGF23	1.22 (1.02 to 1.47)	<0.05
Serum IL6	0.99 (0.97 to 1.01)	0.53
Serum IL10	0.97 (0.77 to 1.23)	0.83
VIR Klotho	1.00 (0.99 to 1.00)	0.06
KL gene expression	0.66 (0.43 to 0.99)	<0.05

eGFR, estimated glomerular filtration rate; hsCRP, high-sensitivity C-reactive protein; FGF23, fibroblast growth factor 23; IL, interleukin; VIR, vascular immunoreactivity.

4. Discussion

The main results of the present study show that patients with DFS have lower serum concentrations of soluble Klotho and elevated concentrations of FGF23, while at the vascular level, both immunoreactivity and gene expression levels of *KL* were also diminished. Moreover, these alterations in the components of the FGF23/Klotho system are independently associated with the presence of DFS. These findings have been not reported previously and suggest an intriguing possibility about the potential role of FGF23/Klotho in the development and/or progression of DFS.

Vascular disease is a critical factor underlying the pathogenesis of DFS [22]. The FGF23/Klotho system has been decisively related to various processes associated with vascular damage. Recent works suggest that FGF23 can exert direct effects on the cardiovascular system. It has been proven that at high concentrations, FGF23 is able to establish low affinity bonds with its receptors, independently of the presence of Klotho, which could cause deleterious effects on multiple organs and tissues [14]. From a clinical point of view, elevated levels of FGF23 have been associated with the presence of vascular dysfunction [15], as well as with a greater severity of atherosclerotic vascular damage [23]. A direct effect of FGF23 on the integrity of vascular tissue has been proposed, a hypothesis that has been reinforced by the recent demonstration of the expression of its receptors, as well as Klotho, in the human vascular wall [16,18]. Our results show that patients with DFS present not only higher serum levels of FGF23, but also increased vascular expression of FGFRs. These findings point out the possibility that high FGF23 could exert direct pathogenic actions on the vasculature even though vascular levels of Klotho are diminished.

On the other hand, Klotho deficiency has been also associated with vascular damage. In murine models, absence of Klotho causes a syndrome of accelerated aging, which included arteriosclerosis and vascular calcifications [24], alterations that could be reversed by administration of the *Kl* gene or by parabiosis with the wild mice [25]. More recent studies have confirmed the protective effects of Klotho on the vascular system, including its participation in the maintenance of endothelial homeostasis and vascular functionality [26]. In the clinical setting, low serum Klotho concentrations in adult subjects without known risk factors for cardiovascular disease have been related to a reduced capacity of flow-mediated dilation of the brachial artery and higher values of epicardial fat thickness and carotid artery intima-media thickness, suggesting that Klotho deficiency may be an early predictor of subclinical atherosclerosis [27]. Moreover, recent works have related the presence of established clinical atherosclerotic disease with low serum Klotho levels [19,21,28]. To the best of our knowledge, only one previous study has evaluated the serum Klotho concentrations in T2DM in relation to diabetic complications. In that study, Zhang et al. [29] studied 102 T2DM patients with DFS and observed that these individuals presented a 44.79% reduction in serum Klotho as compared with non-diabetic controls, with no differences with respect to patients with other complications but no diabetic foot. In the present work, patients with DFS presented a median percent decrease of 37.4% in serum Klotho concentrations as compared with diabetic subjects without DFS. These differences could be explained by the distinct subjects' characteristics between the studies, since in the work by Zhang et al. the patients were younger (mean age, 56.1 vs. 71.8 years), with a greater female proportion (43.1% vs. 25%), lower mean BMI (27.4 vs. 30.8 kg/m^2), and more importantly, all of our patients had severe DFS with lower limb amputation, whereas in the study by Zhang et al. the criteria for the DFS group included foot ulcers, deformity, infection, or gangrene. Therefore, it is possible to speculate that reduction in serum Klotho may reflect a progressive process that is associated with the severity of DFS.

Importantly, the endogenous expression of *KL* in human arteries has been recently demonstrated, an issue of special relevance since the vascular wall may constitute a source of soluble Klotho, with putative autocrine and/or paracrine protective beneficial effects on vascular homeostasis [16,18]. Clinical data regarding the relationship between vascular *KL* expression and atherosclerotic disease are scarce. In previous works, our group reported that low vascular *KL* gene expression was associated with the presence and severity of coronary artery disease [21] and clinical atherosclerotic disease [19] independently of established cardiovascular risk factors. The present study reports for the first-time data about vascular expression of Klotho, both at gene and protein levels, in patients with DFS. In particular, we have detected a 51.4% lower gene expression and a 52.1% lower protein expression of Klotho in vascular samples of patients with DFS as compared with diabetic subjects without this complication. Overall, these findings may be relevant from a mechanistic perspective since Klotho has been involved in maintaining the health of the vasculature, and thus, disruption of Klotho homeostasis may be an important factor in the development and progression of atherosclerotic disease in general, and DFS in particular.

Finally, it is important to note the interplay between inflammation and Klotho, since inflammation induces the reduction of *KL* expression, both in the kidney as well as in the vessels [16,20,30]. Our results show that subjects with DFS presented higher serum concentrations of hsCRP and IL6, and reduced levels of the anti-inflammatory cytokine IL10. In addition, the immunoreactivity levels of TNFα and IL6 were increased in the two groups of diabetic patients as compared with non-diabetic subjects. Regarding gene expression, diabetic subjects presented a higher expression of *TNF* and lower levels of *IL10* compared to control subjects, whereas IL6 expression was significantly higher in patients with DFS as compared with non-diabetic individuals and diabetic subjects without DFS. Importantly, after adjusting for the effect of other variables, arterial *KL* mRNA levels showed an independent negative relationship with *IL6* expression, while vascular Klotho immunoreactivity was also negatively associated with immunoreactivity levels for IL6, supporting the role of inflammation as an inducer of vascular Klotho downregulation in DFS.

Although this study provides novel information about the relationship between DFS and FGF23/Klotho, several limitations need to be acknowledged. First, the small sample size since this work was designed as a proof-of-concept study, which might imply that some of our results did not reach statistical significance and limit the ability to adequately describe the closeness of the relationship between the variables; however, the results indicate that the initial assumptions can be valid for generating new hypotheses. Second, given the cross-sectional nature of the study, we can only demonstrate associations without definitive inferences on their direction or causality. Third, serum levels of other factors related to Klotho and FGF23, such as vitamin D, which may potentially impact DFS, were not measured, and therefore, their potential influence on the results may not be completely ruled out.

5. Conclusions

In conclusion, this paper describes that patients with DFS present lower serum concentrations of Klotho and elevated FGF23, and reports the reduced expression of Klotho, both at the protein and gene level, in their vascular beds. Importantly, these parameters were independently associated with DFS. Overall, these findings point to a new pathophysiological pathway potentially involved in DFS. Further studies are imperative to elucidate the role of the FGF23/Klotho system in the development and progression of this complication.

Author Contributions: J.F.N.-G., C.M.-F., J.D.-C. and E.M.-N. equally contributed to the concept and design of the study. C.H.-C. obtained the clinical information from the participants. A.D.-M. and A.L.-C. obtained the blood and vascular samples from the diabetic patients. S.R.-R. and P.C.-L. obtained the blood and vascular samples from the organ donors. C.F., V.G.T., V.C.L.-T. and N.P.-D. processed the blood and vascular samples and performed serum measurements and gene expression analysis. M.A.A.-G performed immunohistochemistry analysis. J.F.N.-G. and C.M.-F. performed the statistical analyses and wrote the manuscript. All authors approved the final version for publication. J.F.N.-G. is the guarantor of this work and, as such, had full access to all the data in the study and takes responsibility for the integrity of the data and the accuracy of the data analysis.

Funding: This work was supported by Fundación DISA (Ref. DISA012) and by Fondo de Investigación Sanitaria, Instituto de Salud Carlos III (ISCIII) (PI13/01726, PI16/00024) y REDINREN (Red de Investigación Renal) ISCIII (RD16/0009/0022), Sociedad Española de Nefrología (Ayuda SEN2015), and ACINEF. The authors acknowledge cofunding by Fondo Europeo de Desarrollo Regional, Unión Europea ("Una forma de hacer Europa"). J.D.-C. is recipient of a Sara Borrell Contract (CD16/00165). E.M.-N. is recipient of a research contract from the ISCIII (FI14/00033). C.F. is recipient of a fellowship from Agencia Canaria de Investigación, Innovación y Sociedad de la Información, Consejería de Economía, Industria, Comercio y Conocimiento, Gobierno de Canarias (ACIISI, TESIS2018010110), cofunded by Fondo Social Europeo (FSE) Programa Operativo Integrado de Canarias 2014-2020, Eje 3 Tema Prioritario 74 (85%). V.G.T. is recipient of a research grant "Stop Fuga de Cerebros" from Roche España and FUNCANIS (C16/004).

Conflicts of Interest: The authors declare no conflict of interest. The funders played no role in the study design, data collection and analysis, decision to publish, or preparation of the manuscript.

References

1. International Diabetes Federation. *IDF Diabetes Atlas*, 8th ed.; International Diabetes Federation: Brussels, Belgium, 2017.

2. Bild, D.E.; Selby, J.V.; Sinnock, P.; Browner, W.S.; Braveman, P.; Showstack, J.A. Lower-extremity amputation in people with diabetes. Epidemiology and prevention. *Diabetes Care* **1989**, *12*, 24–31. [CrossRef]

3. Armstrong, D.G.; Wrobel, J.; Robbins, J.M. Guest Editorial: Are diabetes-related wounds and amputations worse than cancer? *Int. Wound J.* **2007**, *4*, 286–287. [PubMed]

4. Ross, R. Atherosclerosis—An inflammatory disease. *N. Engl. J. Med.* **1999**, *340*, 115–126. [CrossRef] [PubMed]

5. van der Wal, A.C.; Becker, A.E.; van der Loos, C.M.; Das, P.K. Site of intimal rupture or erosion of thrombosed coronary atherosclerotic plaques is characterized by an inflammatory process irrespective of the dominant plaque morphology. *Circulation* **1994**, *89*, 36–44. [CrossRef]

6. Navarro-González, J.F.; Mora-Fernández, C.; Muros de Fuentes, M.; Donate-Correa, J.; Cazaña-Pérez, V.; García-Pérez, J. Effect of phosphate binders on serum inflammatory profile, soluble CD14, and endotoxin levels in hemodialysis patients. *Clin. J. Am. Soc. Nephrol.* **2011**, *6*, 2272–2279. [CrossRef] [PubMed]

7. Navarro-González, J.F.; Mora-Fernández, C.; Muros, M.; Herrera, H.; García, J. Mineral metabolism and inflammation in chronic kidney disease patients: A cross−sectional study. *Clin. J. Am. Soc. Nephrol.* **2009**, *4*, 1646–1654. [CrossRef]

8. Yu, X.; Ibrahimi, O.A.; Goetz, R.; Zhang, F.; Davis, S.I.; Garringer, H.J.; Linhardt, R.J.; Ornitz, D.M.; Mohammadi, M.; White, K.E. Analysis of the biochemical mechanisms for the endocrine actions of fibroblast growth factor-23. *Endocrinology* **2005**, *146*, 4647–4656. [CrossRef]

9. Imura, A.; Iwano, A.; Tohyama, O.; Tsuji, Y.; Nozaki, K.; Hashimoto, N.; Fujimori, T.; Nabeshima, Y.-I. Secreted Klotho protein in sera and CSF: Implication for post−translational cleavage in release of Klotho protein from cell membrane. *FEBS Lett.* **2004**, *565*, 143–147. [CrossRef]

10. Akimoto, T.; Yoshizawa, H.; Watanabe, Y.; Numata, A.; Yamazaki, T.; Takeshima, E.; Iwazu, K.; Komada, T.; Otani, N.; Morishita, Y.; et al. Characteristics of urinary and serum soluble Klotho protein in patients with different degrees of chronic kidney disease. *BMC Nephrol.* **2012**, *13*, 155.

11. Matsumura, Y.; Aizama, H.; Shiraki-lida, T.; Naqai, R.; Kuro-o, M.; Nabeshima, Y. Identification of the human klotho gene and its two transcripts encoding membrane and secreted klotho protein. *Biochem Biophys. Res. Commun.* **1998**, *242*, 626–630. [CrossRef]

12. Kanbay, M.; Nicoleta, M.; Selcoki, Y.; Ikizek, M.; Aydin, M.; Eryonucu, B.; Duranay, M.; Akcay, A.; Armutcu, F.; Covic, A. Fibroblast growth factor 23 and fetuin A are independent predictors for the coronary artery disease extent in mild chronic kidney disease. *Clin. J. Am. Soc. Nephrol.* **2010**, *5*, 1780–1786. [CrossRef] [PubMed]

13. Gutiérrez, O.M.; Mannstadt, M.; Isakova, T.; Rauh-Hain, J.A.; Tamez, H.; Shah, A.; Smith, K.; Lee, H.; Thadhani, R.; Jüppner, H.; et al. Fibroblast growth factor 23 and mortality among patients undergoing hemodialysis. *N. Engl. J. Med.* **2008**, *359*, 584–592. [CrossRef] [PubMed]

14. Faul, C.; Amaral, A.P.; Oskouei, B.; Hu, M.-C.; Sloan, A.; Isakova, T.; Gutierrez, O.M.; Aguillon-Prada, R.; Lincoln, J.; Hare, J.M.; et al. FGF23 induces left ventricular hypertrophy. *J. Clin. Investig.* **2011**, *121*, 4393–4408. [CrossRef]

15. Mirza, M.A.; Larsson, A.; Lind, L.; Larsson, T.E. Circulating fibroblast growth factor-23 is associated with vascular dysfunction in the community. *Atherosclerosis* **2009**, *205*, 385–390. [CrossRef]

16. Lim, K.; Lu, T.S.; Molostvov, G.; Lee, C.; Lam, F.T.; Zehnder, D.; Hsiao, L.L. Vascular Klotho deficiency potentiates the development of human artery calcification and mediates resistance to FGF-23. *Circulation* **2012**, *125*, 2243–2255. [CrossRef] [PubMed]

17. Hu, M.C.; Kuro-o, M.; Moe, O.W. Renal and extrarenal actions of Klotho. *Semin. Nephrol.* **2013**, *33*, 118–129. [CrossRef]

18. Donate-Correa, J.; Mora-Fernández, C.; Martínez-Sanz, R.; Muros-De-Fuentes, M.; Perez, H.; Meneses-Pérez, B.; Cazaña-Pérez, V.; Navarro-Gonzalez, J.F. Expression of FGF23/KLOTHO system in human vascular tissue. *Int. J. Cardiol.* **2013**, *165*, 179–183. [CrossRef] [PubMed]

19. Martín-Núñez, E.; Donate-Correa, J.; López-Castillo, Á.; Delgado-Molinos, A.; Ferri, C.; Rodríguez-Ramos, S.; Cerro, P.; Pérez-Delgado, N.; Castro, V.; Hernández-Carballo, C.; et al. Soluble levels and endogenous vascular gene expression of *KLOTHO* are related to inflammation in human atherosclerotic disease. *Clin. Sci. (Lond.)* **2017**, *131*, 2601–2609. [CrossRef]

20. Donate-Correa, J.; Martín-Núñez, E.; Martínez-Sanz, R.; Muros-de-Fuentes, M.; Mora-Fernández, C.; Pérez-Delgado, N.; Navarro-González, J.F. Influence of Klotho gene polymorphisms on vascular gene expression and its relationship to cardiovascular disease JF. *J. Cell Mol. Med.* **2016**, *20*, 128–133. [CrossRef]

21. Navarro-González, J.F.; Donate-Correa, J.; Muros de Fuentes, M.; Pérez-Hernández, H.; Martínez-Sanz, R.; Mora-Fernández, C. Reduced Klotho is associated with the presence and severity of coronary artery disease. *Heart* **2014**, *100*, 34–40. [CrossRef] [PubMed]

22. Schaper, N.C.; Nabuurs-Franssen, M.H. The diabetic foot: Pathogenesis and clinical evaluation. *Semin. Vasc. Med.* **2002**, *2*, 221–228. [CrossRef]

23. Mirza, M.A.; Hansen, T.; Johansson, L.; Ahlström, H.; Larsson, A.; Lind, L. Relationship between circulating FGF23 and total body atherosclerosis in the community. *Nephrol. Dial. Transplant.* **2009**, *24*, 3125–3131. [CrossRef]

24. Kuro-O, M.; Matsumura, Y.; Aizawa, H.; Suga, T.; Utsugi, T.; Ohyama, Y.; Kurabayashi, M.; Kaname, T.; Kume, E.; Iwasaki, H.; et al. Mutation of the mouse klotho gene leads to a syndrome resembling ageing. *Nature* **1997**, *390*, 45–51. [CrossRef] [PubMed]

25. Saito, Y.; Nakamura, T.; Ohtama, Y.; Suzuki, T.; Iida, A.; Shiraki-Iida, T.; Kuro-O, M.; Nabeshima, Y.-I.; Kurabayashi, M.; Nagai, R. In vivo klotho gene delivery protects against endothelial dysfunction in multiple risk factor syndrome. *Biochem. Biophys. Res. Commun.* **2000**, *276*, 767–772. [CrossRef] [PubMed]

26. Nagai, R.; Saito, Y.; Ohyama, Y.; Aizawa, H.; Suga, T.; Nakamura, T.; Kurabayashi, M.; Kuroo, M. Endothelial dysfunction in the klotho mouse and downregulation of klotho gene expression in various animal models of vascular and metabolic diseases. *Cell. Mol. Life Sci.* **2000**, *57*, 738–746. [CrossRef] [PubMed]

27. Keles, N.; Caliskan, M.; Dogan, B.; Keles, N.N.; Kalcik, M.; Aksu, F.; Kostek, O.; Aung, S.M.; Isbilen, B.; Oguz, A. Low serum level of Klotho is an early predictor of atherosclerosis. *Tohoku J. Exp. Med.* **2015**, *237*, 17–23. [CrossRef] [PubMed]

28. Semba, R.D.; Cappola, A.R.; Sun, K.; Bandinelli, S.; Dalal, M.; Crasto, C.; Guralnik, J.M.; Ferrucci, L. Plasma klotho and cardiovascular disease. *J. Am. Geriatr. Soc.* **2011**, *59*, 1596–1601. [CrossRef]

29. Zhang, L.; Liu, T. Clinical implication of alterations in serum Klotho levels in patients with type 2 diabetes mellitus and its associated complications. *J. Diabetes Complicat.* **2018**, *32*, 922–930. [CrossRef]

30. Ruiz-Andrés, O.; Sánchez-Niño, M.D.; Moreno, J.A.; Ruiz-Ortega, M.; Ramos, A.M.; Sanz, A.B.; Ortiz, A. Downregulation of kidney protective factors by inflammation: Role of transcription factors and epigenetic mechanisms. *Am. J. Physiol. Renal Physiol.* **2016**, *311*, 1329–1340. [CrossRef]

Journal of
Clinical Medicine

Article

Ultrasound Tissue Characterization of Carotid Plaques Differs Between Patients with Type 1 Diabetes and Subjects without Diabetes

Esmeralda Castelblanco [1,2], Àngels Betriu [3], Marta Hernández [3,4], Minerva Granado-Casas [1,3], Emilio Ortega [5], Berta Soldevila [1,2,6], Anna Ramírez-Morros [1], Josep Franch-Nadal [2,7], Manel Puig-Domingo [1], Elvira Fernández [3], Angelo Avogaro [8], Núria Alonso [1,2,3,6,*] and Dídac Mauricio [2,3,9,*]

[1] Department of Endocrinology and Nutrition, University Hospital and Health Science Research Institute Germans Trias i Pujol, 08916 Badalona, Spain; esmeraldacas@gmail.com (E.C.); mgranado@igtp.cat (M.G.-C.); bsolde@hotmail.com (B.S.); aramirez@igtp.cat (A.R.-M.); mpuigd@gmail.com (M.P.-D.)

[2] Center for Biomedical Research on Diabetes and Associated Metabolic Diseases (CIBERDEM), Instituto de Salud Carlos III, 08907 Barcelona, Spain; josep.franch@gmail.com

[3] Biomedical Research Institute of Lleida, 25198 Lleida, Spain; angels.betriu.bars@gmail.com (A.B.); martahernandez@gmail.com (M.H.); efernandez@irblleida.cat (E.F.)

[4] Department of Endocrinology and Nutrition, University Hospital Arnau de Vilanova, 25198 Lleida, Spain

[5] Department of Endocrinology and Nutrition, Institut d'Investigacions Biomèdiques August Pi Suñer, CIBEROBN, Hospital Clinic, 08036 Barcelona, Spain; eortega1@clinic.cat

[6] Department of Medicine, Barcelona Autonomous University (UAB), 08916 Barcelona, Spain

[7] Primary Health Care Center Raval Sud, Gerència d'Atenció Primaria, Institut Català de la Salut, 08001 Barcelona, Spain

[8] Department of Medicine, University of Padova, 35128 Padova, Italy; angelo.avogaro@unipd.it

[9] Department of Endocrinology and Nutrition, University Hospital de la Santa Creu i Sant Pau & Institut d'Investigació Biomèdica Sant Pau (IIB Sant Pau), 08041 Barcelona, Spain

[*] Correspondence: didacmauricio@gmail.com (D.M.); nalonso32416@yahoo.es (N.A.); Tel.: +34-935565661 (D.M.); +34-934-978-860 (N.A.)

Received: 15 March 2019; Accepted: 22 March 2019; Published: 28 March 2019

Abstract: The aim of the study was to investigate ultrasound tissue characterization of carotid plaques in subjects with and without diabetes type 1 (T1D). B-mode carotid ultrasound was performed to assess the presence and type of plaque in a group of 340 subjects with and 304 without T1D, all of them without cardiovascular disease. One hundred and seven patients with T1D (49.5% women; age 54 ± 9.8 years) and 67 control subjects without diabetes who had at least one carotid plaque were included in the study. The proportion of subjects who had only echolucent plaques was reduced in the group of patients with T1D (48.6% vs. 73.1%). In contrast, the proportion with only echogenic (25.2% vs. 7.5%) and calcified plaques (9.4% vs. 1.5%) was increased compared with subjects without diabetes. Moreover, having at least one echogenic plaque was more frequent in T1D patients compared with subjects without diabetes (49.5% vs. 26.9% $p = 0.005$). In addition to diabetes (OR 2.28; $p = 0.026$), age (OR 1.06, $p = 0.002$) was the other variable associated with echogenic plaque existence in multiple regression analysis. Patients with T1D exhibit a differential pattern of carotid plaque type compared with subjects without diabetes, with an increased frequency of echogenic and extensively calcified plaques.

Keywords: type 1 diabetes; plaque characteristics; carotid plaque; echogenic plaque

---

1. Introduction

Atherosclerotic cardiovascular disease (CV) is a major cause of mortality in patients with type 1 diabetes [1,2]. Its main clinical manifestations include myocardial infarction, stroke and peripheral vascular disease. The common nexus of this group of cardiovascular events is atherosclerosis, which is characterized as a chronic lipid deposition process in the sub-endothelial space that develops insidiously throughout life and is typically in an advanced stage when its symptoms manifest clinically in the form of a cardiovascular event.

Ultrasonography is a non-invasive, safe, reproducible and well-validated method for visualizing and quantifying atherosclerotic lesions [3]. The presence and burden of subclinical carotid atherosclerotic plaque is strongly associated with future arterial CV events [4,5]. A high prevalence of carotid plaques has been described in patients with type 2 diabetes without evidence of clinical CV disease [6]. Data regarding atherosclerotic plaques in patients with type 1 diabetes are very scarce. In this sense, our group has recently described an increased proportion and burden of carotid plaques in a group of 340 patients with T1D without CV events compared with a group of subjects without diabetes [7]. On the other hand, plaque echogenicity as assessed by carotid ultrasound reliably predicts the content of soft tissue and the amount of calcification in carotid plaques [8]. In this regard, echolucent plaques exhibit more lipid, inflammatory cells and haemorrhage compared with echogenic plaques, which contain more calcification and fibrous tissue; characteristics that have been validated by histopathology [8–10]. Otherwise, calcified plaques in the coronary arteries determined by computed tomography scanning coronary artery calcium (CAC) are accepted as a measure of CV disease burden both in patients with [11] and without diabetes [4]. Vascular calcification beyond coronary arteries (i.e., carotid and aortic plaques) has also been described to be associated with all-cause and CV mortality both in patients with [12] and without type 2 diabetes [13].

Studies that have evaluated plaques characteristics by carotid ultrasonography demonstrated that lipid rich plaques (echolucent) are associated with an increased risk of CV events in patients with symptomatic stenotic ($\geq$50%) plaques [14,15]. Most studies that have analyzed the presence and ultrasound characteristics of carotid plaques in patients with diabetes have been performed in asymptomatic subjects with T2D. In these patients both echolucent [16,17] and echogenic [18] plaques were shown to be associated with an increased risk of future CV events. Data regarding subclinical atherosclerotic plaques characteristics assessed by carotid ultrasound in patients with type 1 diabetes are lacking. To our knowledge, no studies to date have analyzed plaque characteristics by carotid ultrasound in patients with type 1 diabetes.

The present study aimed to analyze ultrasound non-stenotic carotid plaque echogenicity in a group of patients with type 1 diabetes and to compare the results obtained with a group of subjects without diabetes, all of whom do not have previous history of clinical CV disease and exhibit normal renal function.

2. Experimental Section

2.1. Subjects

This study is a sub-analysis of a previous study from our research group [7]. From that study, 174 subjects with at least one atherosclerotic plaque, including 107 patients with type 1 diabetes and 67 without diabetes, were selected. Our previous study was a cross-sectional study conducted in 340 patients with type 1 diabetes and 304 subjects without diabetes matched by sex and age. All participants were free from previous CV disease. Subjects with type 1 diabetes were recruited from the diabetic outpatient clinics at University Hospital Germans Trias i Pujol and University Hospital Arnau de Vilanova in the north-western region of Spain (Catalonia). All potential participants were identified from the electronic clinical records from the two participating institutions that belong to the same health care organization.

For type 1 diabetic subjects, the inclusion criteria were as follows: age >18 years; diabetes for at least 1 year; normal renal function (estimated glomerular filtration rate (eGFR >60 mL/min)); no previous CV disease defined as any form of clinical coronary heart disease, stroke or peripheral vascular disease (including any form of diabetic foot disease). Moreover, we excluded patients with a urine albumin excretion ratio >300 mg/g. The selection of subjects without diabetes was based on the same criteria, except for the criteria that concerned the diabetic specific microvascular complications. Additionally, subjects in the control group had fasting glucose and HbA1c values less than 100 mg/dL and 5.7% (39 mmol/mol), respectively.

For each subject, age, sex, body mass index and waist circumference were obtained by standardized methods. Subjects were considered to have hypertension or dyslipidaemia if they were under anti-hypertensive or lipid-lowering agent treatment, respectively. Serum and spot urine samples were collected in the fasting state, and all serum and urine tests were performed using standard laboratory methods as previously described [19].

2.2. Carotid Ultrasound Imaging

The detailed protocol used to evaluate the presence of carotid plaques by ultrasound has been previously described [19]. Briefly, carotid ultrasonography imaging was performed using a LOGIQ® E9 (General Electric, Wauwatosa, WI, USA) equipped with a 15-MHz linear array probe or a Sequoia 512 (Siemens, North Rhine, Westphalia, Germany) equipped with a 15-MHz linear array probe. Plaques were identified using B-mode and colour Doppler examinations in both the longitudinal and transverse planes to consider circumferential asymmetry and were defined as a "focal structure that encroaches into the arterial lumen of at least 0.5 mm or 50% of the surrounding carotid intima media thickness value or demonstrates a thickness of 1.5 mm, as measured from the media-adventitia interference to the intima-lumen surface" according to the Mannheim consensus [20]. Further, the LOGIQ E9 ultrasound system had a 3D image acquisition system with a low depth sweep box with a 45° angle and a slow acquisition time to scan the volume. Volume acquisition used 3 orthogonal sectional plans (A, B, C) and started with the sectional image (A), which gave a 2D image showing the longitudinal view of the carotid arteries and bulb. Images B and C showed the transverse and horizontal axes, respectively. The volume sweep was recorded as raw data and used for volume calculations. Volume calculation was made by delineating the plaque contours and with the transverse plane (B) as described previously [21]. Plaque volume was evaluated in 96 subjects (67 with type 1 diabetes and 29 without diabetes. Volume plaque was measured as total and mean plaque volume. Total plaque volume was defined as the sum of all plaque volume, and mean plaque volume is the sum of the mean plaque volume of each subject.

Atherosclerotic plaques were classified using the well-known five-type classification system based on visual assessment of echogenicity with vessel lumen and adventitia as reference structures: Uniformly echolucent, predominantly echolucent, uniformly echogenic, predominantly echogenic and extensively calcified plaques [19]. We additionally reclassified individuals into 4 clinical categories that have been shown to be useful in terms of future outcomes [16]: Echolucent (lipid- and haemorrhage-rich plaques), mixed (when patients had both echolucent and echo-rich plaques), echogenic (fibrotic or fibro-fatty plaque); and extensively calcified plaques (echo-shadowing from calcifications) as previously described [9]. This classification has been validated against histopathology [9,22] and grey scale medium measured on ultrasound images [23]. The arterial territories explored included the common and internal carotid territories and the bifurcation from the left and right carotid arteries. All participants in the study underwent the same carotid ultrasound examination, and all measures and ultrasound studies were assessed at each participating hospital by the same researcher.

The Local Ethics Committee of both participating centers approved the protocol (PI11/11 and PI-13-095) in accordance with the Declaration of Helsinki, and all participants signed informed consent forms.

2.3. Statistical Analysis

The descriptive statistics of the mean (standard deviation) or median [interquartile range] were estimated for quantitative variables with a normal or non-normal distribution, respectively. For the qualitative variables, absolute and relative frequencies were used. Normally distributed data were analyzed using the Shapiro–Wilk test. The differences between groups were assessed by Student's test, analysis of variance or Mann–Whitney test, and Kruskal–Wallis test depending on the distributions of the quantitative variables. The significance of the differences in qualitative variables was assessed by Chi-squared test or Fisher's exact test. A conditional logistic regression model was performed to study the echogenic plaque and its association with other variables. All variables of the bivariate analysis with a p-value < 0.2 were used. Only the main effects with a significant contribution to the final model according to the likelihood ratio test were included in the final model. Goodness of fit logistic model assumptions were evaluated using the Hosmer–Lemeshow test. Statistical significance was established at a p-value < 0.05. Data management and analyses were performed with the free software environment R version 3.3.1 and SPSS software (version 22, SPSS, Chicago, IL, USA).

3. Results

Clinical characteristics of the 174 subjects with carotid plaque included in this study are presented in Table 1. Patients with type 1 diabetes exhibited an increased prevalence of hypertension and dyslipidaemia. Among them, up to 60 (56.1%) had diabetic retinopathy. The duration of diabetes was 24.7 (±12) years. In addition, 76 (71%) patients were under statin treatment, and 53 (49.5%) were under antiplatelet treatment. Patients with type 1 diabetes also exhibited reduced waist circumference, diastolic blood pressure, and plasma triglyceride, total cholesterol and low-density lipoprotein (LDL) cholesterol concentrations compared with the non-diabetic group. Moreover, patients with type 1 diabetes exhibited increased systolic blood pressure, heart rate, and high-density lipoprotein (HDL) cholesterol concentrations compared with the group of subjects without diabetes (Table 1).

Table 1. Baseline characteristics of the participants.

	Non-Diabetic Group (n = 67)	Type 1 Diabetes (n = 107)	p-Value
Age, years	52.6 (9.13)	54 (9.8)	0.351
Sex, women, n (%)	28 (41.8%)	53 (49.5%)	0.401
Race, non-Caucasian, n (%)	2 (3%)	0 (0%)	0.147
Tobacco exposure, n (%)	35 (53%)	70 (65.4%)	0.144
Antiplatelet treatment, n (%)	3 (10.3%)	53 (49.5%)	<0.001
Dyslipidaemia, n (%)	20 (29.9%)	78 (72.9%)	<0.001
Hypertension, n (%)	13 (19.4%)	60 (56.1%)	<0.001
Systolic blood pressure, mmHg	131 (14.2)	137 (18)	0.010
Diastolic blood pressure, mmHg	79.6 (8)	74.7 (11.6)	0.001
Statin treatment, n (%)	8 (11.9%)	76 (71%)	<0.001
Heart rate, beats/min	71 (11.1)	76.8 (11.9)	0.002
Body mass index, kg/m^2	27.8 [25; 30.3]	26.4 [24.3; 29.6]	0.124
Waist, cm	99.6 (12.8)	92.9 (12.7)	0.001
HbA1c, %	5.5 [5.3; 5.7]	7.6 [7.15; 8.15]	<0.001
HbA1c, mmol/mol	36.6 [34; 39]	60 [54; 65.5]	<0.001
Total cholesterol, mg/dL	203 [185; 228]	173 [155; 198]	<0.001
HDL cholesterol, mg/dL	54 [46; 63]	59 [52; 70]	0.005
LDL cholesterol, mg/dL	124 [112; 148]	97.4 [80.6; 113]	<0.001
Triglycerides, mg/dL	99 [65; 142]	71 [57; 90]	0.001
Creatinine, mg/dL	0.81 [0.7; 0.9]	0.79 [0.7; 0.9]	0.166
GFR, mL/min/1.73 m^2	97.9 [85.6; 105]	96.1 [88.3; 105]	0.899
Diabetes duration, years	-	24.7 (12)	-
Plaque			0.581
One plaque	35 (52.2%)	50 (46.7%)	
Multiple plaques ($\geq$2)	32 (47.8%)	57 (53.3%)	

Data are mean (SD) or median [interquartile range] or n (%). GFR, Glomerular filtration rate estimate based on the CKD-EPI (Chronic Kidney Disease Epidemiology Collaboration) equation; HDL, high-density lipoprotein; HDL, low-density lipoprotein.

3.1. Ultrasound Examination

No differences were observed in the proportion of subjects with either one or more carotid plaque between patients with type 1 diabetes and subjects without diabetes (Table 1). Regarding plaque echogenicity, differences were observed between study groups ($p = 0.001$) (Figure 1A). The proportion of patients with type 1 diabetes who had only echolucent, i.e., lipid-rich plaques, was reduced in patients with type 1 diabetes compared with subjects without diabetes (48.6% vs. 73.1%, respectively). In contrast, the proportion of patients with only echogenic plaques, i.e., those who contain more calcification and fibrous tissue, was increased in the group of patients with type 1 diabetes (25.2% vs. 7.5%). In addition, the proportion of patients with at least one echogenic plaque was increased in patients with type 1 diabetes compared with subjects without diabetes (49.5% vs. 26.9%, $p = 0.005$) (Figure 1B). When stratifying the groups by sex, differences in the proportion of patients with echogenic plaques were observed. In subjects without diabetes, the proportion of those with echogenic plaques was increased in men (12.8% vs. 0%). In the group of patients with type 1 diabetes no differences were found between sexes (24.1% in men and 26.4% in women).

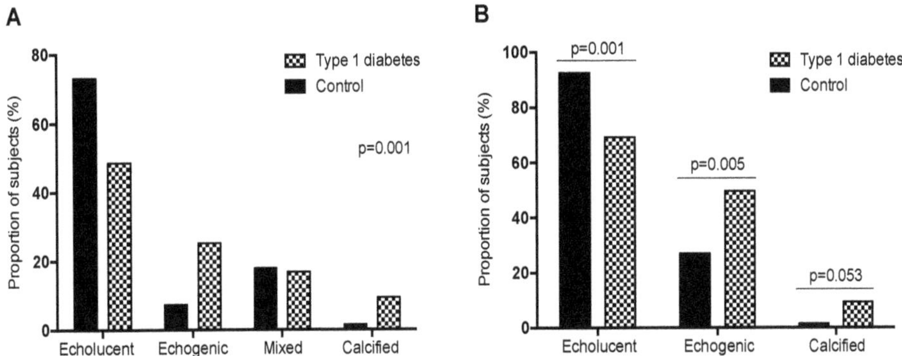

Figure 1. Atherosclerotic plaque type. (**A**) Proportion of patients with different types of atherosclerotic plaques (only echolucent, only echogenic, mixed of echolucent and echogenic plaques, and calcified with or without another plaque type). (**B**) Proportion of patients with different types of atherosclerotic plaques (at least one echolucent plaque, at least one echogenic plaque, and at least one calcified plaque).

Finally, no differences were noted in the proportion of patients with mixed plaque types (echolucent and echogenic) between the study groups (16.8% vs. 17.9%). Additionally, the proportion of patients with at least one extensively calcified plaque was increased in patients with type 1 diabetes compared with subjects without diabetes (9.4% vs. 1.5%) (Figure 1B). None of the study subjects had high-grade (>50%) carotid artery stenosis.

3.2. Measurements of the Plaque Volume

Plaque volume was assessed in 96 subjects; 67 type 1 diabetes patients and 29 control subjects. There were no differences of total plaque volume between subjects with and without diabetes, median [IQR], (60 [30; 130] mm^3 vs. 50 [20; 80] mm^3; $p = 0.260$). However, the mean plaque volume was higher in type 1 diabetes patients compared to control subjects (40 [30; 70] mm^3 vs. 30 [20; 40] mm^3; $p = 0.028$). On the other hand, plaque volume (total and mean) were higher in individuals with echogenic plaque compared with individuals with non-echogenic plaque 100 [60; 160] mm^3 vs. 40 [20; 69] mm^3; $p < 0.001$ and 55 [34.2; 80] mm^3 vs. 30 [20; 50] mm^3; $p = 0.003$, respectively).

3.3. Characteristics of Patients with Echogenic Plaques

Seventy-one study subjects exhibited echogenic plaques (including those with or without type 1 diabetes). Among them, up to 53 (74.6%) had type 1 diabetes and up to 41 (57.7%) were under statin treatment. Subjects with echogenic plaques were older, exhibited an increased frequency of hypertension and dyslipidaemia, had higher systolic blood pressure and were more frequently under antiplatelet treatment. On the other hand, total and LDL cholesterol concentrations were reduced in those patients with echogenic plaques compared with the group without echogenic plaques. Finally, the proportion of patients with more than one plaque was increased in the group of subjects with echogenic plaques (Table A1 in Appendix A).

3.4. Logistic Regression Model of Echogenic Plaques

The multiple regression model was constructed using the candidate parameters identified in the bivariate analysis and the clinically relevant variables to explain the risk of having echogenic plaques. The model for having echogenic plaques vs. not having echogenic plaques revealed that the variables associated with the presence of echogenic plaques were older age (OR = 1.06; p = 0.002) and type 1 diabetes (OR = 2.28; p = 0.026); this model had a good discriminatory ability (AUCROC = 0.73; p < 0.001) (Table 2).

Table 2. Logistic regression model for the presence of any echogenic plaques.

Presence of any Echogenic Plaque vs. No Echogenic Plaque [a]			
	Odds Ratio	**95% CI**	p
Age	1.062	1.023–1.103	0.002
Sex, women	0.539	0.272–1.068	0.076
BMI	0.949	0.872–1.032	0.221
Diabetes	2.276	1.104–4.690	0.026
sBP	1.019	0.998–1.040	0.075

BMI, body mass index; sBP, systolic blood pressure. [a] The logistic model showed a good discrimination with an AUC of 0.73 (95% CI: [0.65–0.80] and no significant lack of calibration (Hosmer–Lemeshow test p-value = 0.21).

4. Discussion

In the present study that included a large sample of patients with type 1 diabetes and subjects without diabetes, all of whom had at least one carotid atherosclerotic plaque, we report for the first time that the ultrasound tissue characterization of carotid plaques differs between patients with type 1 diabetes and subjects without diabetes. The results obtained show that the proportion of subjects with echogenic or extensively calcified plaques is increased in patients with type 1 diabetes compared with subjects without diabetes, whereas the proportion of echolucent plaques is lower in the former.

In the general population, non-invasive measures of subclinical atherosclerosis, such as CAC burden and atherosclerotic plaque presence and burden in the carotid arteries, are features associated with an increased risk of CV events [4,8,12,13]. In addition, plaque characteristics are associated with the future risk of CV events. Regarding carotid plaques, ultrasonographic plaque characterization studies performed in the general population have demonstrated that echolucent plaques are independently associated with an increased risk of cerebrovascular and other CV events [24] in symptomatic patients with stenotic plaques ($\geq$50%) [15] and those with non-stenotic plaques [25]. On the other hand, plaque volume, as well as the number of atherosclerotic plaques (i.e., plaque burden), have both been associated with an increased risk of future CV events [26]. First, in our study, the mean plaque volume in type 1 diabetic subjects was higher, although this preliminary finding should be interpreted with caution as the number of subjects is quite low. This prevented us from further statistical work-up using these data. Interestingly, the proportion of patients with more than one plaque was increased in the group of subjects with echogenic plaques. In concordance with this finding, these patients had a higher total plaque volume than the non-echogenic group.

Thus, the presence of echogenic plaques in a given patient is associated with an increased burden of atherosclerosis (total number and volume of plaques). Therefore, future studies should address the issue of whether subjects with echogenic plaques, especially if they have type 1 diabetes, are at a higher risk of cardiovascular events than those with echolucent plaques. Additionally, in the general population, carotid plaque calcification, which is defined as that with an acoustic shadowing, is a strong and an independent predictor of vascular events in asymptomatic subjects [11,27]. In type 2 diabetes, the presence of both echolucent [16,17,28] and echogenic plaques [12,18] is also independently associated with an increased risk of CV events. In contrast to echolucent plaques, which are lipid rich, echogenic plaques contain more calcification and fibrous tissue. In the general population and patients with type 2 diabetes carotid plaque calcification, as assessed by computed tomography, is independently associated with a future risk of cardiovascular events [12]. Not only the presence of calcium in the atherosclerotic plaque but also the pattern of the calcium depots has been suggested to be involved in the vulnerability of the atherosclerotic lesion. In this sense, microcalcification patterns, i.e., spotty calcium depots, are associated with vulnerability of coronary atherosclerotic plaques in contrast to extensive calcifications, which are associated with stable plaques [29–31]. These findings suggest that microcalcification may increase wall stress in the fibrous cap, resulting in plaque instability as a characteristic of vulnerable plaques [31,32]. In the present study, presence and characteristics of carotid plaques were evaluated with carotid B-mode ultrasound. Actually, as contrast-enhanced modalities that are the best method to analyze the characteristics of vulnerable atherosclerotic plaques such as inflammation, intraplaque haemorrhage and ulceration, the use of this ultrasound method might have clearly improved the adequate characterization of carotid atherosclerotic plaques of type 1 diabetic patients; unfortunately, this is a rather time-consuming and expensive procedure that we could not use in the current study.

Few studies have analyzed the presence and tissue characteristics of atherosclerotic plaques in type 1 diabetes. Atherosclerotic CV disease is the leading cause of death in people with type 1 diabetes whose risk of CV disease is two- to seven-fold increased and not completely accounted for by traditional CV risk factors [33]. For example, women exhibit a two-fold increased risk of fatal and nonfatal vascular events compared with men with type 1 diabetes [1]. Our group has recently described an increased proportion and burden of carotid plaques in type 1 diabetic patients without CV events compared with subjects without diabetes [7]. In this group of patients, the presence of advanced stages of diabetic retinopathy is independently associated with the presence and the burden of subclinical carotid atherosclerosis. Plaque tissue characteristics in patients with type 1 diabetes have been mainly studied in coronary arteries by computed tomography. Studies demonstrate that diabetes is associated with an increased prevalence of any degree of calcification compared with control subjects [32]. Moreover, coronary artery calcification is greatly increased in women, and the gender difference in calcification observed in the general population is lost in subjects with type 1 diabetes [34,35]. In addition, Djaberi et al., used multislice computed tomography to assess coronary atherosclerosis and found no differences in average CAC scores and the prevalence of coronary atherosclerosis between asymptomatic patients with type 1 and type 2 diabetes, with the exception of an increased percentage of noncalcified plaques in patients with type 2 diabetes [36].

To the best of our knowledge, no studies to date have analyzed the ultrasound characteristics of carotid plaques in patients with type 1 diabetes. The results obtained in the present study demonstrated that the proportion of subjects with echogenic plaques is increased in patients with type 1 diabetes compared with subjects without diabetes without differences between gender in the proportion of subjects who have echogenic plaques. These results point to a greater effect of calcification in women compared with men given that echogenic plaques contain more calcified tissue compared with echolucent plaques. These results are similar to those that have described a reduction in the gender difference in coronary artery calcification given that the prevalence of CAC is similar in men and women [35]. The analysis performed in the present study to analyze those variables associated with the risk of having echogenic plaques noted that older age and type 1 diabetes were the only ones that were independently associated with the risk of having this type of plaque. However, we must acknowledge that although we found an association between each of these two variables with the presence of echogenic plaques, the low number of subjects studied may reduce the chance of finding an association with other cardiovascular disease-related or unexplored factors. Finally, regarding extensive plaque calcification, our results demonstrate that the proportion of patients with at least one extensively calcified plaque was increased in patients with type 1 diabetes compared with subjects without diabetes, which is similar to the increased coronary artery calcification described in patients with type 1 diabetes compared with nondiabetic subjects.

In the general population, treatment with statins promotes coronary artery calcification [37]. Few studies have analyzed the effect of statin treatment on calcium content in coronary arteries in patients with diabetes mellitus. These studies have been performed in patients with type 2 diabetes. The results indicate that patients who receive statins exhibit an increased degree of calcification as assessed by CAC compared with those who are not receiving statins; however, the progression of CAC scores is faster in those who are not receiving statins [38,39]. In this sense, some authors have suggested that statins render microcalcifications more confluent, which could be associated with a reduction in vessel wall stress [40]. In fact, patients with diabetes and patients with a previous CV event are frequently treated with statins given that this treatment decreases the incidence of CV events in these populations. Our study reported no differences in the plaque type distribution between the complete series and patients without statin treatment (Figure A1 in Appendix A), or between subjects under statin treatment with and without echogenic plaques.

Our study has several limitations. First, although we adjusted for risk factors known to be associated with atherosclerosis, it is possible that some confounding factors exist and were incompletely accounted for. Second, the cross-sectional design precludes conclusions about causality; therefore, a larger prospective study is needed to establish the usefulness of plaque type characterization as a predictive factor for CV events in patients with type 1 diabetes. Finally, an additional limitation is that only a limited number of patients with echogenic plaques were studied; therefore, larger studies are warranted as no final conclusions can be drawn from our findings.

5. Conclusions

In summary, our study demonstrates that in the absence of previous CV disease, non-stenotic carotid plaque ultrasound characteristics differ between patients with type 1 diabetes and subjects without diabetes. The proportion of echogenic plaques, which comprise calcified tissue, is increased in patients with type 1 diabetes, and no differences exist in the proportion of subjects with echogenic plaques between genders. This finding is consistent with the notion that diabetes is associated with vascular calcification and the loss of gender differences in calcification. Whether the increase in the proportion of plaques with this ultrasonographic pattern observed in patients with type 1 diabetes will translate into an increased risk of future cardiovascular events will be determined in the prospective follow-up of this cohort of patients.

Author Contributions: Conceptualization, E.C., N.A. and D.M.; methodology, A.B., M.H., M.G.-C., and B.S.; data curation, E.C. and A.R.-M.; writing—original draft preparation, E.C., N.A.; writing—review and editing, E.O., E.F., J.F.-N., M.P.-D., D.M. and A.A.; supervision, N.A. and D.M.; project administration, D.M.; funding acquisition, D.M.

Funding: This research was funded by grants from the Carlos III National Institute of Health, ISCIII (PI12/0183 and PI15/0625) with co-funding from the European Regional Development Fund (ERDF). CIBER for Diabetes and Associated Metabolic Diseases (CIBERDEM) is an initiative of ISCIII, Spain. The Health Sciences Research Institute Germans Trias i Pujol is part of the CERCA Programme/Generalitat de Catalunya.

Acknowledgments: The authors thank Montserrat Martinez (Statistics Unit, IRB Lleida) for her assistance in conducting the statistical analysis and Nuria Villalmanzo for their assistance in conducting the laboratory technics (she holds a fellowship from Pla Estratègic i Innovació en Salut -PERIS- del Departament de Salut de la Generalitat de Catalunya, SLT002/16/00171). We want to particularly acknowledge the patients, IGTP-HUGTP and IRBLleida (B.0000682) Biobanks integrated in the Spanish National Biobanks Network of Instituto de Salud

Carlos III (PT17/0015/0045 and PT17/0015/0027, respectively) and Tumour Bank Network of Catalonia for its collaboration.

Conflicts of Interest: The authors declare no conflict of interest.

Appendix A

Table A1. Baseline characteristics of the study group according to the presence of echogenic plaques.

Variable	Non-Echogenic Plaque	Echogenic	*p*-Value Plaque
	n = 103	*n* = 71	
Age, years	51.2 (9.22)	56.7 (9.11)	<0.001
Sex, women, *n* (%)	52 (50.5%)	29 (40.8%)	0.272
Race, non Caucasian, *n* (%)	2 (1.94%)	0 (0.00%)	0.514
Diabetes mellitus type 1, *n* (%)	54 (52.4%)	53 (74.6%)	0.005
Diabetes duration, years	22.2 (11.8)	27.2 (11.8)	0.030
Tobacco exposure, *n* (%)	58 (56.9%)	47 (66.2%)	0.281
Hypertension, *n* (%)	33 (32.0%)	40 (56.3%)	0.002
Dyslipidaemia, *n* (%)	47 (45.6%)	43 (60.6%)	0.075
BMI, kg/m^2	27 [25;29.9]	26.7 [24.2;29.8]	0.424
Waist, cm	96.0 (12.0)	94.5 (14.6)	0.469
Systolic blood pressure, mmHg	132 (17.4)	139 (15.5)	0.004
Diastolic blood pressure, mmHg	76.3 (10.2)	76.9 (11.3)	0.730
Heart rate, beat/min	74.1 (12.8)	75.6 (10.6)	0.391
Antiplatelet treatment, *n* (%)	23 (29.1%)	33 (57.9%)	0.001
Statin treatment, *n* (%)	43 (41.7%)	41 (57.7%)	0.055
Glucose, mg/dL	99 [87.5; 140]	127 [88; 211]	0.034
HbA1c, %	6.7 [5.5; 7.6]	7.40 [6.3; 7.9]	0.006
HbA1c, mmol/mol	49.7 [36.6; 60]	57 [45; 63]	0.006
Triglycerides, mg/dL	74 [59; 106]	81 [60.5; 126]	0.313
Total cholesterol, mg/dL	192 [170; 218]	178 [160; 202]	0.016
HDL, mg/dL	58 [49.5; 70]	56 [47.5; 68.5]	0.296
LDL, mg/dL	112 [92.5; 141]	107 [82.5; 120]	0.036
Creatinine, mg/dL	0.8 [0.67; 0.9]	0.79 [0.66; 0.92]	0.756
GFR, mL/min/1.73 m^2	97.6 [87.4; 106]	96 [86.8; 104]	0.485
Plaque, *n* (%)			<0.001
One plaque	67 (65%)	18 (25.4%)	
Multiple plaques	36 (35%)	53 (74.6%)	

Data are mean (SD) or median [interquartile range] or *n* (%). GFR, Glomerular filtration rate estimate by the CKD-EPI (Chronic Kidney Disease Epidemiology Collaboration) equation.

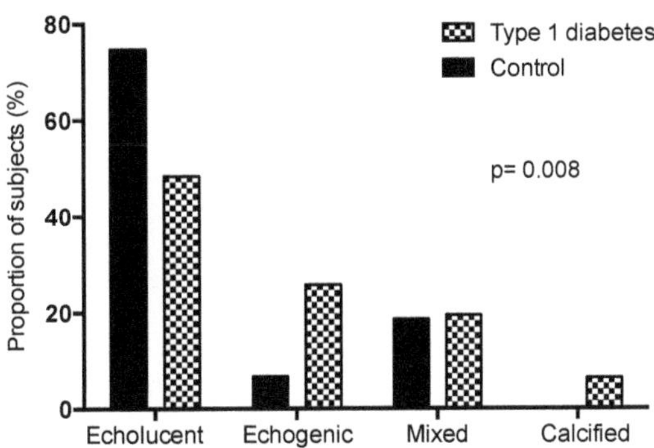

Figure A1. Proportion of patients without statin treatment with different types of atherosclerotic plaque (only echolucent, only echogenic, mixed of echolucent and echogenic plaques, and calcified with or without another plaque type), *n* = 90.

References

1. Huxley, R.R.; Peters, S.A.; Mishra, G.D.; Woodward, M. Risk of all-cause mortality and vascular events in women versus men with type 1 diabetes: A systematic review and meta-analysis. *Lancet Diabetes Endocrinol.* **2015**, *3*, 198–206. [CrossRef]
2. Peters, S.A.; Huxley, R.R.; Woodward, M. Articles Diabetes as a risk factor for stroke in women compared with men: A systematic review and meta-analysis of 64 cohorts, including 775,385 individuals and 12,539 strokes. *Lancet* **2014**, *383*, 1973–1980. [CrossRef]
3. Stein, J.H.; Korcarz, C.E.; Hurst, R.T.; Lonn, E.; Kendall, C.B.; Mohler, E.R.; Najjar, S.S.; Rembold, C.M.; Post, W.S. American Society of Echocardiography Carotid Intima-Media Thickness Task Force. Use of carotid ultrasound to identify subclinical vascular disease and evaluate cardiovascular disease risk: A consensus statement from the American Society of Echocardiography carotid intima-media thickness task force endorsed by the Society for Vascular Medicine. *J. Am. Soc. Echocardiogr.* **2008**, *21*, 93–111. [PubMed]
4. Baber, U.; Mehran, R.; Sartori, A.; Schoos, M.M.; Sillesen, H.; Muntendam, P.; Garcia, M.J.; Gregson, J.; Pocock, S.; Falk, E.; et al. Prevalence, impact, and predictive value of detecting subclinical coronary and carotid atherosclerosis in asymptomatic adults. *J. Am. Coll. Cardiol.* **2015**, *65*, 1065–1074. [CrossRef]
5. Inaba, Y.; Chen, J.A.; Bergmann, S.R. Carotid plaque, compared with carotid intima-media thickness, more accurately predicts coronary artery disease events: A meta-analysis. *Atherosclerosis* **2012**, *220*, 128–133. [CrossRef]
6. Catalan, M.; Herreras, Z.; Pinyol, M.; Sala-Vila, A.; Amor, A.J.; de Groot, E.; Gilabert, R.; Ros, E.; Ortega, E. Prevalence by sex of preclinical carotid atherosclerosis in newly diagnosed type 2 diabetes. *Nutr. Metab. Cardiovasc. Dis.* **2015**, *25*, 742–748. [CrossRef] [PubMed]
7. Carbonell, M.; Castelblanco, E.; Valldeperas, X.; Betriu, À.; Traveset, A.; Granado-Casas, M.; Hernández, M.; Vázquez, F.; Martín, M.; Rubinat, E.; et al. Diabetic retinopathy is associated with the presence and burden of subclinical carotid atherosclerosis in type 1 diabetes. *Cardiovas. Diabetol.* **2018**, *17*, 66. [CrossRef] [PubMed]
8. Grønholdt, M.-L.M. Ultrasound and lipoproteins as predictors of lipid-rich, rupture-prone plaques in the carotid artery. *Arter. Thromb. Vasc. Biol.* **1999**, *19*, 2–13. [CrossRef]
9. Gray-Weale, A.C.; Graham, J.C.; Burnett, J.R.; Byrne, K.; Lusby, R.J. Carotid artery atheroma: Comparison of preoperative B-mode ultrasound appearance with carotid endarterectomy specimen pathology. *J. Cardiovasc. Surg.* **1988**, *29*, 676–681.
10. Mitchell, C.C.; Stein, J.H.; Cook, T.D.; Salamat, S.; Wang, X.; Varghese, T.; Jackson, D.C.; Sandoval Garcia, C.; Wilbrand, S.M.; Dempsey, R.J. Histopathological Validation of Grayscale Carotid Plaque Characteristics Related to Plaque Vulnerability. *Ultrasound Med. Biol.* **2017**, *43*, 129–137. [CrossRef]
11. Mathiesen, E.B.; Bønaa, K.H.; Joakimsen, O. Echolucent plaques are associated with high risk of ischemic cerebrovascular events in carotid stenosis: The tromsø study. *Circulation* **2001**, *103*, 2171–2175. [CrossRef] [PubMed]
12. Grønholdt, M.-L.M.; Nordestgaard, B.G.; Schroeder, T.; Vorstrup, S.; Sillesen, H. Ultrasonic echolucent carotid plaques predict future strokes. *Circulation* **2001**, *104*, 68–73. [CrossRef] [PubMed]
13. Agarwal, S.; Morgan, T.; Herrington, D.M.; Xu, J.; Cox, A.J.; Freedman, B.I.; Carr, J.J.; Bowden, D.W. Coronary calcium score and prediction of all-cause mortality in diabetes: The diabetes heart study. *Diabetes Care* **2011**, *34*, 1219–1224. [CrossRef] [PubMed]
14. Cox, A.J.; Hsu, F.-C.; Agarwal, S.; Freedman, B.I.; Herrington, D.M.; Carr, J.J.; Bowden, D.W. Prediction of mortality using a multi-bed vascular calcification score in the Diabetes Heart Study. *Cardiovasc. Diabetol.* **2014**, *13*, 160. [CrossRef]
15. Allison, M.A.; Hsi, S.; Wassel, C.L.; Morgan, C.; Ix, J.H.; Wright, C.M.; Criqui, M.H. Calcified atherosclerosis in different vascular beds and the risk of mortality. *Arter. Thromb. Vasc. Biol.* **2011**, *32*, 140–146. [CrossRef]
16. Katakami, N.; Takahara, M.; Kaneto, H.; Sakamoto, K.; Yoshiuchi, K.; Irie, Y.; Kubo, F.; Katsura, T.; Yamasaki, Y.; Kosugi, K.; et al. Ultrasonic tissue characterization of carotid plaque improves the prediction of cardiovascular events in diabetic patients: A pilot study. *Diabetes Care* **2012**, *35*, 2640–2646. [CrossRef] [PubMed]
17. Irie, Y.; Katakami, N.; Kaneto, H.; Takahara, M.; Nishio, M.; Kasami, R.; Sakamoto, K.; Umayahara, Y.; Sumitsuji, S.; Ueda, Y.; et al. The utility of ultrasonic tissue characterization of carotid plaque in the prediction of cardiovascular events in diabetic patients. *Atherosclerosis* **2013**, *230*, 399–405. [CrossRef]

18. Vigili de Kreutzenberg, S.; Fadini, G.P.; Guzzinati, S.; Mazzucato, M.; Volpi, A.; Coracina, A.; Avogaro, A. Carotid plaque calcification predicts future cardiovascular events in type 2 diabetes. *Diabetes Care* **2015**, *38*, 1937–1944. [CrossRef]

19. Alonso, N.; Traveset, A.; Rubinat, E.; Ortega, E.; Alcubierre, N.; Sanahuja, J.; Hernández, M.; Betriu, A.; Jurjo, C.; Fernández, E.; et al. Type 2 diabetes-associated carotid plaque burden is increased in patients with retinopathy compared to those without retinopathy. *Cardiovasc. Diabetol.* **2015**, *14*, 33–39. [CrossRef]

20. Touboul, P.-J.; Hennerici, M.G.; Meairs, S.; Adams, H.; Amarenco, P.; Bornstein, N.; Csiba, L.; Desvarieux, M.; Ebrahim, S.; Hernandez, R.; et al. Mannheim Carotid Intima-Media Thickness and Plaque Consensus (2004–2006–2011). *Cerebrovasc. Dis.* **2012**, *34*, 290–296. [CrossRef]

21. Ludwig, M.; Zielinski, T.; Schremmer, D.; Stumpe, K.O. Reproducibility of 3-dimensional ultrasound readings of volume of carotid atherosclerotic plaque. *Cardiovas. Ultrasound* **2008**, *26*, 42. [CrossRef]

22. European Carotid Plaque Study Group. Carotid Artery Plaque Composition—Relationship to Clinical Presentation and Ultrasound B-mode Imaging. *Eur. J. Vasc. Endovasc. Surg.* **1995**, *10*, 23–30. [CrossRef]

23. Mayor, I.; Momjian, S.; Lalive, P.; Sztajzel, R. Carotid plaque: Comparison between visual and grey-scale median analysis. *Ultrasound Med. Biol.* **2003**, *29*, 961–966. [CrossRef]

24. Larsen, J.R.; Tsunoda, T.; Tuzcu, E.M.; Schoenhagen, P.; Brekke, M.; Arnesen, H.; Hanssen, K.F.; Nissen, S.E.; Dahl-Jorgensen, K. Intracoronary ultrasound examinations reveal significantly more advanced coronary atherosclerosis in people with type 1 diabetes than in age- and sex-matched non-diabetic controls. *Diab. Vasc. Dis. Res.* **2007**, *4*, 62–65. [CrossRef] [PubMed]

25. Costacou, T.; Edmundowicz, D.; Prince, C.; Conway, B.; Orchard, T.J. Progression of coronary artery calcium in type 1 diabetes mellitus. *Am. J. Cardiol.* **2007**, *100*, 1543–1547. [CrossRef] [PubMed]

26. Polak, J.F.; Szklo, M.; Kronmal, R.A.; Burke, G.L.; Shea, S.; Zavodni, A.E.; O'Leary, D.H. The value of carotid artery plaque and intima-media thickness for incident cardiovascular disease: The multi-ethnic study of atherosclerosis. *J. Am. Heart Assoc.* **2013**, *2*, e000087. [CrossRef] [PubMed]

27. Djaberi, R.; Schuijf, J.D.; Boersma, E.; Kroft, L.J.; Pereira, A.M.; Romijn, J.A.; Scholte, A.J.; Jukema, J.W.; Bax, J.J. Differences in atherosclerotic plaque burden and morphology between type 1 and 2 diabetes as assessed by multislice computed tomography. *Diabetes Care* **2009**, *32*, 1507–1512. [CrossRef] [PubMed]

28. Zavodni, A.E.H.; Wasserman, B.A.; McClelland, R.L.; Gomes, A.S.; Folsom, A.R.; Polak, J.F.; Lima, J.A.; Bluemke, D.A. Carotid artery plaque morphology and composition in relation to incident cardiovascular events: The Multi-Ethnic Study of Atherosclerosis (MESA). *Radiology* **2014**, *271*, 381–389. [CrossRef]

29. Prabhakaran, S.; Singh, R.; Zhou, X.; Ramas, R.; Sacco, R.L.; Rundek, T. Presence of calcified carotid plaque predicts vascular events: The Northern Manhattan Study. *Atherosclerosis* **2007**, *195*, e197–e201. [CrossRef]

30. Hunt, K.J.; Evans, G.W.; Folsom, A.R.; Sharrett, A.R.; Chambless, L.E.; Tegeler, C.H.; Heiss, G. Acoustic shadowing on B-mode ultrasound of the carotid artery predicts ischemic stroke: The Atherosclerosis Risk in Communities (ARIC) study. *Stroke* **2001**, *32*, 1120–1126. [CrossRef]

31. Wilson, P.W.; Kauppila, L.I.; O'Donnell, C.J.; Kiel, D.P.; Hannan, M.; Polak, J.M.; Cupples, L.A. Abdominal aortic calcific deposits are an important predictor of vascular morbidity and mortality. *Circulation* **2001**, *103*, 1529–1534. [CrossRef] [PubMed]

32. Nonin, S.; Iwata, S.; Sugioka, K.; Fujita, S.; Norioka, N.; Ito, A.; Nakagawa, M.; Yoshiyama, M. Plaque surface irregularity and calcification length within carotid plaque predict secondary events in patients with coronary artery disease. *Atherosclerosis* **2017**, *256*, 29–34. [CrossRef] [PubMed]

33. McClelland, R.L.; Jorgensen, N.W.; Budoff, M.; Budoff, M.; Blaha, M.J.; Post, W.S.; Kronmal, R.A.; Bild, D.E.; Shea, S.; Liu, K.; et al. 10-year coronary heart disease risk prediction using coronary artery calcium and traditional risk factors: Derivation in the MESA (Multi-Ethnic Study of Atherosclerosis) with validation in the HNR (Heinz Nixdorf Recall) Study and the DHS (Dallas Heart Study). *J. Am. Coll. Cardiol.* **2015**, *66*, 1643–1653. [PubMed]

34. Lin, R.; Chen, S.; Liu, G.; Xue, Y.; Zhao, X. Association between carotid atherosclerotic plaque calcification and intraplaque hemorrhage A Magnetic Resonance Imaging Study. *Arterioscler. Thromb. Vasc. Biol.* **2017**, *37*, 1228–1233. [CrossRef]

35. Beckman, J.A.; Ganz, J.; Creager, M.A.; Ganz, P.; Kinlay, S. Relationship of clinical presentation and calcification of culprit coronary artery stenoses. *Arterioscler. Thromb. Vasc. Biol.* **2001**, *21*, 1618–1622. [CrossRef] [PubMed]

36. Shemesh, J.; Stroh, C.I.; Tenenbaum, A.; Hod, H.; Boyko, V.; Fisman, E.Z.; Motro, M. Comparison of coronary calcium in stable angina pectoris and in first acute myocardial infarction utilizing double helical computerized tomography. *Am. J. Cardiol.* **1998**, *81*, 271–275. [CrossRef]
37. Ehara, S. Spotty Calcification typifies the culprit plaque in patients with acute myocardial infarction: An Intravascular Ultrasound Study. *Circulation* **2004**, *110*, 3424–3429. [CrossRef] [PubMed]
38. Maldonado, N.; Kelly-Arnold, A.; Vengrenyuk, Y.; Laudier, D.; Fallon, J.T.; Virmani, R.; Cardoso, L.; Weinbaum, S. A mechanistic analysis of the role of microcalcifications in atherosclerotic plaque stability: Potential implications for plaque rupture. *Am. J. Physiol. Heart Circ. Physiol.* **2012**, *303*, H619–H628. [CrossRef]
39. Lee, S.E.; Chang, H.J.; Sung, J.M.; Park, H.B.; Heo, R.; Rizvi, A.; Lin, F.Y.; Kumar, A.; Hadamitzky, M.; Kim, Y.J.; et al. Effects of Statins on Coronary Atherosclerotic Plaques: The PARADIGM (Progression of AtheRosclerotic PlAque DetermIned by Computed TomoGraphic Angiography Imaging) Study. *JACC Cardiovasc. Imaging* **2018**, *11*, 1475–1484. [CrossRef]
40. Puri, R.; Libby, P.; Nissen, S.E.; Wolski, K.; Ballantyne, C.M.; Barter, P.J.; Chapman, M.J.; Erbel, R.; Raichlen, J.S.; Uno, K.; et al. Long-term effects of maximally intensive statin therapy on changes in coronary atheroma composition: Insights from SATURN. *Eur. Heart J. Cardiovasc. Imaging* **2014**, *15*, 380–388. [CrossRef]

Journal of
Clinical Medicine

Article

Poorer Quality of Life and Treatment Satisfaction is Associated with Diabetic Retinopathy in Patients with Type 1 Diabetes without Other Advanced Late Complications

Minerva Granado-Casas [1,2], Esmeralda Castelblanco [1,3], Anna Ramírez-Morros [1], Mariona Martín [4], Nuria Alcubierre [2], Montserrat Martínez-Alonso [5], Xavier Valldeperas [6], Alicia Traveset [7], Esther Rubinat [8], Ana Lucas-Martin [4], Marta Hernández [2,9], Núria Alonso [1,3] and Didac Mauricio [2,3,10,*]

[1] Department of Endocrinology and Nutrition, Health Sciences Research Institute & University Hospital Germans Trias i Pujol, 08916 Badalona, Spain; mgranado@igtp.cat (M.G.-C.); esmeraldacas@gmail.com (E.C.); aramirez@igtp.cat (A.R.-M.); nalonso32416@yahoo.es (N.A.)

[2] Lleida Institute for Biomedical Research Dr. Pifarré Foundation, IRBLleida, University of Lleida, 25198 Lleida, Spain; cubito@lleida.org (N.A.); martahernandezg@gmail.com (M.H.)

[3] Centre for Biomedical Research on Diabetes and Associated Metabolic Diseases (CIBERDEM), Instituto de Salud Carlos III, 08907 Barcelona, Spain

[4] Department of Endocrinology & Nutrition, University Hospital Germans Trias i Pujol, 08916 Badalona, Spain; marionamarting@gmail.com (M.M.); alucas.germanstrias@gencat.cat (A.L.-M.)

[5] Systems Biology and Statistical Methods for Biomedical Research, IRBLleida, University of Lleida, 25198 Lleida, Spain; mmartinez@irblleida.cat

[6] Department of Ophthalmology, University Hospital Germans Trias i Pujol, 08916 Badalona, Spain; xvalldeperas@gmail.com

[7] Department of Ophthalmology, University Hospital Arnau de Vilanova, 25198 Lleida, Spain; atravesetm@gmail.com

[8] Department of Nursing and Physiotherapy, University of Lleida, 25198 Lleida, Spain; rubinatesther@gmail.com

[9] Department of Endocrinology & Nutrition, University Hospital Arnau de Vilanova, 25198 Lleida, Spain

[10] Department of Endocrinology & Nutrition, Hospital de la Santa Creu i Sant Pau, 08041 Barcelona, Spain

* Correspondence: didacmauricio@gmail.com; Tel.: +34-935-565-661

Received: 16 January 2019; Accepted: 14 March 2019; Published: 18 March 2019

Abstract: Diabetic retinopathy (DR) may potentially cause vision loss and affect the patient's quality of life (QoL) and treatment satisfaction (TS). Using specific tools, we aimed to assess the impact of DR and clinical factors on the QoL and TS in patients with type 1 diabetes. This was a cross-sectional, two-centre study. A sample of 102 patients with DR and 140 non-DR patients were compared. The Audit of Diabetes-Dependent Quality of Life (ADDQoL-19) and Diabetes Treatment Satisfaction Questionnaire (DTSQ-s) were administered. Data analysis included bivariate and multivariable analysis. Patients with DR showed a poorer perception of present QoL ($p = 0.039$), work life ($p = 0.037$), dependence ($p = 0.010$), and had a lower average weighted impact (AWI) score ($p = 0.045$). The multivariable analysis showed that DR was associated with a lower present QoL ($p = 0.040$), work life ($p = 0.036$) and dependence ($p = 0.016$). With regards to TS, DR was associated with a higher perceived frequency of hypoglycaemia ($p = 0.019$). In patients with type 1 diabetes, the presence of DR is associated with a poorer perception of their QoL. With regard to TS, these subjects also show a higher perceived frequency of hypoglycaemia.

Keywords: diabetic retinopathy; type 1 diabetes; quality of life; treatment satisfaction; patient-reported outcomes

---

1. Introduction

Diabetic retinopathy (DR) is a significant diabetic complication in patients with type 1 diabetes [1]. This complication is a potential cause of vision loss in the diabetic population that can negatively affect their quality of life (QoL) [2]. Moreover, DR is also associated with an increased risk for all-cause mortality and cardiovascular disease in patients with type 1 diabetes [3].

QoL is a multidimensional, subjective and dynamic construct comprising the individual's subjective perception of the physical, psychological and social well-being aspects of his or her life [4]. Treatment satisfaction (TS) is an individual's subjective appraisal of his or her experience of the treatment, including both the process and results [4]. From the patient's point of view, QoL and TS as patient-reported outcomes (PROs) are two important measures because these are often stronger determinants of medical outcomes such as hospitalization, mortality and the presence of complications [5,6]. Furthermore, an assessment of PROs has been accepted to complement visual acuity information in clinical trials, interventions or research [7]. Regarding QoL and TS, recent studies have found that having DR negatively impacts the QoL and TS in patients with type 2 diabetes [8–14].

To our knowledge, only one study with a cross-sectional design has assessed the relationship between QoL and the presence of DR in patients with type 1 diabetes using a diabetes-specific QoL questionnaire [15]. So far, all of the published scientific evidence regarding this issue used generic instruments or visual function scales to appraise the impact of DR in the QoL of patients with type 1 diabetes [16–30]. The FinnDiane study, which included a large sample of patients with type 1 diabetes, did not find any association between the presence of DR and health-related quality of life (HRQoL) using a generic instrument [16]; it should be pointed out that, in this study, DR was determined to be present if the subject had ever received treatment for DR and that 53% of the patients had also diabetic nephropathy, which could influence the results. Other cross-sectional studies with large samples of patients with type 1 diabetes did not report any association between DR and HRQoL [17,23,24]. However, other prospective cohort and cross-sectional studies observed a negative impact of DR on HRQoL in patients with type 1 diabetes using generic instruments [25,28]. Two recently published studies found a lower HRQoL in patients with type 1 diabetes and more severe DR in the presence of other diabetic complications such as neuropathy, nephropathy and cardiovascular disease [29,30]. The Wisconsin Epidemiologic Study of Diabetic Retinopathy did not find changes in the QoL with the presence of DR over time using a visual function scale [18,19]; nevertheless, researchers included a study sample of patients with other important diabetic complications that could influence the results. Furthermore, the Diabetes Control and Complications Trial/Epidemiology of Diabetes Interventions and Complications (DCCT/EDIC) Study designed to assess the effects of intensive insulin treatment and risk factors on the patient-reported visual function in patients with type 1 diabetes did not find changes with DR [20]. Although researchers used a diabetes-specific QoL questionnaire and a visual function scale, these patients also showed other diabetic complications. Other studies performed in patients with type 1 and type 2 diabetes using a visual function scale related lower scores with DR [21,22]; however, the study subjects had also other diabetic complications and comorbidities such as cardiovascular diseases, cancer, arthritis or rheumatism and hearing problems. Regarding DR and PROs, it has been pointed out that generic instruments have a lack of sensitivity to assess specific domains of QoL that are related to diabetic complications [7]. This is an important point because the scientific evidence has shown that patients can report a good patient-reported health status, often measured with generic questionnaires, even though their diabetes-specific QoL is negative [8,9,31]. Besides, visual function scales cannot be regarded as sufficient to assess the diabetes-related QoL due to the lack of specific quality of life domains related to diabetes. For this reason, the disease-specific questionnaires, but not the generic tools, are the more adequate instruments to measure specific disease-related QoL [31].

To the best of our knowledge, there have not been any studies primarily designed to assess the impact of DR on QoL and TS in patients with type 1 diabetes. In addition, there are not any studies that have assessed the impact of DR on QoL and TS that included a well-defined sample of patients with type 1 diabetes and that used diabetes-specific questionnaires. Therefore, we hypothesized that patients with type 1 diabetes who have been diagnosed with DR have a lower perception of their QoL and TS in comparison to their non-DR counterparts. The aim of the study was to assess the association of DR with QoL and TS in adult type 1 diabetic patients with DR as compared to a group of patients without this condition. Additionally, we also determined which clinical factors that could be related to QoL and TS.

2. Materials and Methods

This was a cross-sectional, two-centre study. Participants were patients with type 1 diabetes regularly cared for at their reference hospital from two health care regions (Lleida and Badalona). The recruitment took place between January 2013 and May 2015. From a total number of 330 patients with type 1 diabetes that were invited to participate, 276 accepted to be included in the study, while 17 could not be included because of exclusion criteria. Finally, a total sample of 259 patients with type 1 diabetes was included in the previously published study [32]. From this sample, 16 patients were excluded because they had not had their eyes assessed; furthermore, one patient was excluded because he did not answer the questionnaires. A final sample of 102 patients with DR and 140 without DR were included. A detailed description of the characteristics of the study are provided elsewhere [32]. The inclusion criteria were: having the diagnosis of type 1 diabetes, a current age greater than 18 years, and a disease duration of more than 1 year. We intended to avoid the inclusion of other advanced late diabetic complications to exclude their negative impact on top of DR. Thus, the exclusion criteria were as follows: psychological or cognitive deterioration (e.g., mental diseases or dementia); being a healthcare professional; the presence of previous clinical cardiovascular diseases (ischemic heart disease, cerebrovascular disease, peripheral arterial disease and heart failure) or diabetic foot disease; the presence of any eye disease that could influence the results (macular oedema, media opacity that hindered assessment of the retina, other concomitant retinal pathology, ophthalmological surgery in the previous year and laser treatments during the six months before the study visit); pregnancy; and the presence of other advanced diabetic late complications such as macroalbuminuria (urine albumin/creatinine ratio > 299 mg/g) and renal insufficiency (estimated glomerular filtration rate < 60 mL/min). In the DR group, three patients had glaucoma and one patient had myopia superior to 5 dioptres. In the non-DR group, three patients had myopia superior to 5 dioptres. The assessment of DR was performed and classified by an expert ophthalmologist according to the international clinical classification system [33]. Optic coherence tomography measurements were performed using spectral-domain OCT (SD-OCT) (Cirrus HD-OCT, model 4000, Carl Zeiss Meditec, Dublin, CA, USA) and deep Enhanced Imaging (EDI-OCT), with HD 5 Line Raster scan pattern. An ophthalmologist assessed and classified DR as follows: no apparent retinopathy, mild non-proliferative retinopathy, moderate non-proliferative retinopathy, severe non-proliferative retinopathy and proliferative retinopathy. The ethics committee of each of the two participating centres (Ethics Committee of the University Hospital Arnau de Vilanova, and Ethics Committee of the University Hospital Germans Trias i Pujol) approved the study. Written informed consent form was obtained from all participants.

2.1. Clinical Variables

Anthropometric measures, including waist circumference, weight, height and body mass index (BMI), blood pressure and laboratory variables were determined according to standard procedures. Cardiovascular disease was excluded based on detailed anamnesis and careful review of all clinical records. Besides, predefined questionnaires specially designed for the study were used for conduction of personal interviews and data extraction for collection of all other variables: medication use, physical activity, educational level, ethanol consumption and smoking habit. Hypertension and dyslipidaemia were defined by the use of specific medication for treatment of these two conditions. Microalbuminuria was defined as an albumin creatinine ratio above 30 mg/g. Physical activity was assessed according to two previously validated methods [34,35]. This was classified as regular physical activity (if the patients spent more than 25 min/day in any activity that requires 4 METS (the metabolic equivalent), or as sedentary if the participants spent up to 25 min/day. Ethanol consumption (g per day) was estimated from frequency, beverage type, and average amount according to previously validated methods [32].

2.2. Quality of Life

The Audit of Diabetes-Dependent Quality of Life (ADDQoL-19) questionnaire in Spanish was used to assess the QoL of the patients. This is an instrument specifically designed for its use in subjects with diabetes, and validated in the diabetic Spanish population [36–38]. The questionnaire consists of 21 items, 19 of which are related to specific life domains and are scored multiplying the impact rating (from −3 to +1) by the importance rating (from +3 to 0); these produce scores ranging from +3 to −9 points. The overall score is the mean of 19 specific domains and is assigned as the average weighted impact score (AWI); this ranges from +3 (highest QoL) to −9 (poorest QoL). Additionally, there are two overview items that are scored separately; they measure present QoL ranging from +3 (excellent) to −3 (very bad) and diabetes-specific QoL, which is scored from −3 (maximum negative impact) to +1 (maximum positive impact). Two trained researchers (MG-C and MM) conducted individual interviews with all of the patients. All questionnaires have been recommended by the World Health Organization and the International Diabetes Federation to assess the PROs [6].

2.3. Treatment Satisfaction

The Diabetes Treatment Satisfaction Questionnaire-status version (DTSQ-s) in Spanish, a diabetes-specific questionnaire validated for Spanish diabetic subjects was administered [39,40]. This questionnaire consists of 8 items scored on a 6-point scale from 0 (very unsatisfied) to 6 (very satisfied). The final score is calculated by summing up the individual scores from six items (current treatment, convenience, flexibility, understanding, recommend to others and continue with); this final score ranges from 36 (very satisfied) to 0 (very unsatisfied). Additionally, two items are calculated separately and measure the perceived frequency of hyperglycaemia and hypoglycaemia, respectively; they are scored from 0 (never) to 6 (always).

2.4. Sample Size

The sample size was determined by using the standard deviation for the ADDQoL items from a previously published study on DR and QoL that was performed in patients with type 2 diabetes by our group in the same region [8]. Then, we assumed that the differences would be similar in patients with type 1 diabetes for present QoL, diabetes-specific QoL and AWI items. A sample of 54 patients with DR and 54 without DR was deemed necessary to detect significant differences between both groups. We used the Mann-Whitney test and a statistical power of 80% with a significance level of 5%. From our previous study, a sufficiently large number of patients with type 1 diabetes was available to fulfil the sample size [32].

2.5. Statistical Methods

The descriptive analysis of quantitative variables included mean and standard deviation for the normally distributed variables, as well as median and interquartile intervals. Qualitative variables were summarized by absolute and relative frequencies. Bivariate analysis comparing the groups of patients with and without diabetic retinopathy included the *t*-test (or the Mann–Whitney test for non-normally distributed variables) and the chi-squared test for quantitative and qualitative patient characteristics as well as QoL and TS outcomes (including overall and subscale scores), respectively. Simple linear regression analysis was performed to assess the univariate association of patients' characteristics with overall QoL and TS, as well as each of their domains. Those domains showing a significant univariate association with DR were evaluated in a linear multivariable regression analysis to assess the statistical significance of their association after having adjusted by other patients' characteristics. For this purpose, in a first step and separately for each domain associated with DR, we estimated a multivariable linear (or ordinary least squares) regression model including all variables with a *p*-value <0.25 according to the likelihood ratio test (LRT). In the next step, we used the same method (LRT) to simplify the model by dropping non-significant variables, starting with the one with the highest and non-significant *p*-value, until obtaining a model with variables showing a significant contribution according to the LRT. Afterwards, we checked the possible additional significant contribution of any other patient characteristics. The final proposed regression models included any possible interactions involving the patient characteristics that were included in the corresponding regression model. We used a significance level of 0.05 and the open-source program R for all statistical analysis [41].

3. Results

The clinical and demographic characteristics of the participants are shown in Table 1. Patients with DR were older (p = 0.004), had higher systolic blood pressures (p = 0.003), higher glycated haemoglobin (HbA1c) (p < 0.001), longer diabetes duration (p < 0.001) and lower high density lipoprotein (HDL) concentrations (p = 0.015) in comparison with non-DR patients. Furthermore, the former had a higher frequency of hypertension (p < 0.001) and microalbuminuria (p = 0.014).

Table 1. Clinical and demographic characteristics of the study groups.

Characteristics	No DR (n = 140)	DR (n = 102)	p
Age (years)	42.1 ± 10.3	46.2 ± 10.8	0.004
Sex (men)	62 (44.3)	47 (46.1)	0.884
Race (Caucasian)	137 (97.9)	101 (99.0)	0.640
Study site			0.168
Badalona	85 (60.7)	52 (51.0)	
Lleida	55 (39.3)	50 (49.0)	
Educational level			0.069
Less than primary school	8 (5.9)	4 (4.1)	
Completed primary school	34 (25.2)	35 (36.5)	
Secondary/high school	49 (36.3)	39 (40.6)	
Graduate school or higher	44 (32.6)	18 (18.8)	
Smoking			0.282
Non or former smoker	106 (75.7)	70 (68.6)	
Yes	34 (24.3)	32 (31.4)	
Regular physical activity	98 (70.0)	77 (75.5)	0.425
Insulin dose (UI/kg/day)	0.57 [0.41; 0.75]	0.65 [0.45; 0.86]	0.027
Waist circumference (cm)	86.0 [79.0; 95.0]	90.0 [81.0; 100.0]	0.050
Systolic blood pressure (mmHg)	120.0 [110.0; 134.0]	130.0 [118.0; 140.0]	0.003
Diastolic blood pressure (mmHg)	74.2 ± 9.6	73.6 ± 9.3	0.666
Body mass index (kg/m^2)	24.7 [22.4; 27.3]	25.9 [23.1; 28.3]	0.091
Hypertension	20 (14.3)	40 (39.2)	<0.001
Dyslipidaemia	49 (35.0)	48 (48.0)	0.056
Microalbuminuria	5 (3.6)	13 (12.9)	0.014
Diabetes duration (years)	17.0 [11.0; 22.0]	26.5 [19.2; 33.0]	<0.001
Glaucoma	–	3 (2.9)	–
Myopia over 5 dioptres	3 (2.1)	1 (0.9)	–
HbA1c (%)	7.3 [6.8; 7.8]	7.7 [7.2; 8.5]	<0.001
HbA1c (mmol/mol)	56.3 [50.8; 61.7]	60.7 [55.2; 69.4]	<0.001
Total cholesterol (mg/dL)	177.0 [164.0; 201.0]	176 [158.0; 200.0]	0.560
HDL cholesterol (mg/dL)	65.5 [55.0; 75.0]	59.5 [51.0; 71.0]	0.015
LDL cholesterol (mg/dL)	102.0 [87.0; 115.0]	100.0 [83.2; 119.0]	0.777
Triglycerides (mg/dL)	64.0 [49.8; 80.0]	68.0 [53.0; 89.8]	0.084
Ethanol consumption (g/day)	5.6 ± 9.2	6.5 ± 11.6	0.686

Data are shown as median [interquartile], means ± SD or n (%). DR, diabetic retinopathy; HbA1c, glycated haemoglobin; HDL, high density lipoprotein; LDL, low density lipoprotein.

3.1. Diabetes-Related Quality of Life

Patients with DR showed lower present QoL (p = 0.039) in comparison with their counterparts without DR (Table 2). Work life and dependence scores were also lower in patients with DR (p = 0.037 and p = 0.010, respectively). Although no differences between the groups in the diabetes-specific QoL item (p = 0.069) were observed, the AWI score showed a poorer QoL in patients with DR (p = 0.045). In the multivariable analysis, DR was associated with a low score in present QoL (β = −0.25; p = 0.040) and with a poorer perception of QoL items such as work life (β = −0.78; p = 0.036) and dependence (β = −0.83; p = 0.016) (Table 3). Together with smoking habit, these variables only explained a 2.0%, 2.3% and 3.8% of the variability of those scores, respectively. On the other hand, no association was found between DR and the AWI score (p = 0.111) after adjusting for smoking status, age, insulin dose and physical activity (Supplemental Table S1). However, there was a negative correlation between older age and the AWI score (p = 0.015). Conversely, physical activity and the daily insulin dose showed a positive association with a higher AWI score (p = 0.005 and p = 0.028, respectively).

Table 2. Bivariate analysis for the Audit of Diabetes Dependent Quality of Life (ADDQoL-19) and the Diabetes Treatment Satisfaction Questionnaire-status (DTSQ-s) by diabetic retinopathy status.

Items	No DR (n = 140)	DR (n = 102)	p
ADDQoL-19			
Present QoL	1.00 [1.00; 2.00]	1.00 [0.00; 1.00]	0.039
Diabetes-specific QoL	−1.00 [−2.00; −1.00]	−2.00 [−2.00; −1.00]	0.069
Leisure	−1.00 [−3.00; 0.00]	−2.00 [−3.75; 0.00]	0.198
Work life	0.00 [−3.00; 0.00]	−2.00 [−4.00; 0.00]	0.037
Travel	−2.00 [−4.00; 0.00]	−1.00 [−3.75; 0.00]	0.242
Holidays	−2.00 [−3.00; 0.00]	−2.00 [−3.00; 0.00]	0.987
Physical ability	−2.00 [−4.00; 0.00]	−2.00 [−4.00; 0.00]	0.306
Family life	0.00 [−3.00; 0.00]	0.00 [−3.00; 0.00]	0.333
Friends/social life	0.00 [0.00; 0.00]	0.00 [0.00; 0.00]	0.556
Personal relationships	0.00 [−2.00; 0.00]	0.00 [−2.00; 0.00]	0.122
Sex life	0.00 [−2.00; 0.00]	0.00 [−3.00; 0.00]	0.054
Physical appearance	0.00 [−1.00; 0.00]	0.00 [−2.00; 0.00]	0.144
Self-confidence	0.00 [−2.25; 0.00]	0.00 [−3.75; 0.00]	0.315
Motivation	0.00 [−2.00; 0.00]	0.00 [−2.00; 0.00]	0.950
Society/people's reactions	0.00 [0.00; 0.00]	0.00 [0.00; 0.00]	0.301
Future	−2.00 [−4.00; 0.00]	−2.00 [−6.00; 0.00]	0.144
Finances	0.00 [0.00; 0.00]	0.00 [0.00; 0.00]	0.070
Living conditions	0.00 [0.00; 0.00]	0.00 [0.00; 0.00]	0.853
Dependence	0.00 [−3.00; 0.00]	−2.00 [−4.00; 0.00]	0.010
Freedom to eat	−4.00 [−6.00; −2.00]	−4.00 [−9.00; −2.00]	0.447
Freedom to drink	−2.00 [−4.50; 0.00]	−2.50 [−6.00; 0.00]	0.270
AWI	−1.32 [−2.05; −0.68]	−1.61 [−2.66; −0.94]	0.045
DTSQ-s			
Hyperglycaemia frequency perception	3.00 [2.00; 4.00]	3.00 [2.00; 4.00]	0.385
Hypoglycaemia frequency perception	2.00 [2.00; 3.00]	3.00 [2.00; 4.00]	0.022
Final score	28.00 [23.00; 31.00]	28.00 [24.00; 31.80]	0.987

Data are shown as median [interquartile]. AWI, average weighted impact score; DR, diabetic retinopathy; QoL, quality of life.

To further explore the effect of advanced DR stages on QoL and TS, we analyzed differences between study groups: without DR, with mild DR (grade 1), and more than mild DR (grades 2–4) (Supplemental Table S2). The comparison among the 3 groups revealed no statistical differences. However, head to head comparison between groups yielded some differences. For instance, patients with mild DR showed a poorer perception of the present QoL (p = 0.040), dependence (p = 0.005), and an AWI score (p = 0.038) in comparison with non-DR patients (Supplemental Table S2). Furthermore, patients with advanced (more than mild) DR had a lower AWI score (p = 0.032) in comparison with the non-DR group. No statistical differences were observed between groups with the other items (Supplemental Table S2).

Table 3. Multivariable linear regression models for present quality of life, work life and dependence as reported on the Audit of Diabetes Dependent Quality of Life (ADDQoL-19) questionnaire.

	Present QoL [1]		Work Life [2]		Dependence [3]	
	Estimate β (95% CI)	*p*	Estimate β (95% CI)	*p*	Estimate β (95% CI)	*p*
(Intercept)	1.00 (0.83; 1.17)	<0.001	−1.81 (−2.31; −1.32)	<0.001	−2.45 (−1.51; −1.73)	<0.001
Retinopathy	−0.25 (−0.49; −0.01)	0.040	−0.78 (−1.50; −0.05)	0.036	−0.83 (−1.51; −0.16)	0.016
Smoker, current	0.12 (−0.14; 0.39)	0.361	0.07 (−0.74; 0.87)	0.867	0.49 (−0.26; 1.24)	0.197

No significant contribution of the variables according to the likelihood ratio test: age, sex, race, educational level, insulin dose, physical activity, hypertension, dyslipidaemia, diabetes duration, body mass index, waist circumference, microalbuminuria, glycated haemoglobin, systolic and diastolic blood pressure, HDL and LDL cholesterol and triglycerides. Model estimates (β) refers to the overall mean (intercept) and the estimated change in the score mean between patient with and without the corresponding characteristic. [1] Coefficient of determination, r-squared: 2.0%. [2] Coefficient of determination, r-squared: 2.3%. [3] Coefficient of determination, r-squared: 3.8%. QoL, quality of life.

3.2. Treatment Satisfaction

The perception of the frequency of hypoglycaemia was higher in the DR group (p = 0.022) (Table 2). However, no significant differences were observed between the groups in terms of hyperglycaemia frequency perception and the DTSQ-s final score. We did not find any difference among groups defined by the status of DR in any of the DTSQ-s items (Supplemental Table S2). In the multivariable analysis for the hypoglycaemia frequency perception item of the DTSQ-s, we observed a higher hypoglycaemia frequency perception in subjects with DR (β = 0.019; p = 0.019) (Table 4). In addition, male sex was associated with a lower hypoglycaemia frequency perception (β = −0.60; p = 0.001). DR together with male sex only explained 6.3% of the variability. In Supplemental Table S3, the multivariable analysis of the DTSQ-s final score did not show any relationship with DR (p = 0.244). Only sex and diabetes duration were associated with the DTSQ-s final score.

Table 4. Multivariable linear regression model for hypoglycaemia frequency perception of the Diabetes Treatment Satisfaction Questionnaire-status (DTSQ-s).

	Estimate β (95% CI)	*p*
(Intercept)	2.91 (2.62; 3.20)	<0.001
Retinopathy	0.44 (0.07; 0.80)	0.019
Sex, male	−0.60 (−0.96; −0.24)	0.001

No significant contribution of the variables according to the likelihood ratio test: age, race, educational level, insulin dose, physical activity, hypertension, dyslipidaemia, diabetes duration, body mass index, waist circumference, smoking, systolic and diastolic blood pressure, triglycerides, glycated haemoglobin, microalbuminuria, HDL and LDL cholesterol. Model estimates (β) refers to the overall mean (intercept) and the estimated change in the score mean between patient with and without the corresponding characteristic. Coefficient of determination, r-squared: 6.3%.

4. Discussion

Our results indicate that patients with type 1 diabetes and DR had a poorer QoL than those without DR. In addition, DR is associated with poorer individual items related to QoL, i.e., present QoL, work life and dependence. Additionally, older age was also associated with a poorer QoL; however, physical activity and insulin dose were associated with a higher QoL. In terms of overall TS, although no difference between the two groups was observed, DR was associated with a higher hypoglycaemia frequency perception. In addition, a longer diabetes duration in men was associated with higher TS.

The current study showed a poorer perception of the QoL in the presence of DR; this could be due to the fact that DR is a condition that more frequently develops in those patients with poorer management of the disease and less healthy lifestyle behaviour. This could explain, at least in part, the poorer perceived QoL of patients with DR in comparison with those without DR. However, there is only one previous cross-sectional study that has assessed the QoL in patients with type 1 diabetes using a diabetes-specific QoL instrument [15]. That study did not find any association between DR and QoL, which is in contrast with our results. Nevertheless, this could be because patients in that study had a high frequency of advanced late complications, such as diabetic foot disease, which could be potential confounders. In addition, the previous study was not specifically designed to assess the relationship between DR and QoL; the authors studied a sample of patients with type 1 and type 2 diabetes together, with a small sample of patients with DR (n = 74) [15]. Other cross-sectional studies that focused on patients with type 1 diabetes and used generic instruments to measure QoL also did not find an association between DR and QoL [16,17,24]. Fenwick et al. [23] performed a cross-sectional study to assess the impact of DR on the QoL with a sample of patients that included both type 1 and type 2 diabetes. They did not find any association between DR and QoL because they used a generic instrument; besides, these cross-sectional studies were not specifically designed to assess the relationship between the QoL and DR in this population. Nevertheless, a prospective study performed on patients with type 1 diabetes using generic instruments found a negative relationship between the presence of late diabetic complications and HRQoL [25]. However, these patients had a higher frequency of comorbidities and other complications apart from DR. Peasgood et al. [28] conducted a post hoc analysis from an interventional study and observed a decreased HRQoL in patients with type 1 diabetes who also had DR. These results are consistent with the results of the current study. Although they also used generic instruments to measure a specific condition associated with diabetes, the sample showed other diabetic complications, which were self-reported; all of these characteristics could influence the results. Furthermore, another cross-sectional study involving patients with type 1 diabetes observed a lower HRQoL in patients with severe DR; this is similar to our findings, although their results were not adjusted for other complications [29]. There are two cross-sectional studies performed with a sample of patients with type 1 and type 2 diabetes that found a lower HRQoL in the presence of DR [26,27]. These are similar findings to the current study, even though the authors used generic instruments and data from both types of diabetes were pooled. A previous study performed on patients with type 2 diabetes by our group showed the negative impact of the presence and severity of DR on the QoL [8]. Additionally, the PANORAMA study, which was conducted with a large sample of patients with type 2 diabetes, also showed that a negative perception of QoL was associated with the presence of DR [9]; this is similar to the current results in patients with type 1 diabetes, even though the two types of diabetes have remarkably different characteristics. There are other studies that yielded results that are discordant with our findings. First, the Wisconsin Epidemiologic Study for Diabetic Retinopathy did not find changes in the QoL with the presence and severity of DR [18,19]. We should point out that in that study the authors used a visual function scale to assess the diabetes-specific QoL. Furthermore, a relevant proportion of the study subjects had other diabetic complications that could significantly impact on QoL, such as nephropathy, neuropathy, cardiovascular diseases and limb amputations. Finally, the DCCT/EDIC study did not observe a relationship of QoL with DR [20]; although they used a diabetes-specific instrument and a visual function scale, again the study subjects showed other complications that could influence QoL.

The factor associated with a poorer QoL was advanced age; this is in line with our previous results in subjects with type 2 diabetes [42]. Nyanzi et al. [15] found a negative relationship between age and some aspects of QoL, which is similar to our results; moreover, in line with our study, they did not find any association between smoking status and perceived QoL. Other studies performed in patients with type 1 and type 2 diabetes also found that a reduced QoL was associated with older age [9,24,43,44]. However, a cross-sectional study of patients with type 2 diabetes did not find any association between QoL and these factors, despite using a diabetes-specific QoL instrument [14]. On the other hand, physical activity was associated with a higher QoL in patients with type 1 diabetes, which is in line with our previously published results about patients with type 2 diabetes [8,42].

However, we could not identify any study that has specifically assessed this issue in the diabetic population. The association of the daily insulin dose with a higher QoL has not been described in type 1 diabetes. We have no clear explanation for this finding. Lastly, emotional factors, such as the lack of optimism and a depressive style, contribute to impairing the QoL in women with type 1 diabetes [44]. However, in the current study we did not find any statistical difference between men and women in the emotional items of the QoL questionnaire.

In the current study, the DTSQ-s final score was neither different between the two study groups nor correlated with the presence of DR; this is similar to the results of the recent PANORAMA study and other studies performed on patients with type 2 diabetes [8,9,45]. However, no studies have assessed the relationship between DR and TS in patients with type 1 diabetes specifically. Furthermore, we found that DR was associated with a higher hypoglycaemia frequency perception. This increased perception of hypoglycaemia is probably associated with poorer glycaemic control, which is associated with a high risk for the development of diabetic complications such as DR [3]. Another factor that may influence this perception in patients with type 1 diabetes is ethanol consumption. However, there were no differences between both study groups in ethanol intake. A longer diabetes duration in men was associated with higher TS and a lower hypoglycaemia frequency perception; this is also in line with our previous results [8,42] and with other cross-sectional studies performed in patients with type 2 diabetes [9,45].

This study has several limitations. A causality association between QoL, TS and DR in patients with type 1 diabetes cannot be established due to the cross-sectional study design; additionally, changes over time in terms of QoL and TS cannot be determined by this study. In addition, hypoglycaemia unawareness and the frequency of self-blood glucose monitoring are factors that may influence the hypoglycaemia frequency perception in patients with type 1 diabetes. The absence of data on these aspects in our study is another limitation. Furthermore, it is well-known that insulin pump therapy is a treatment option for those patients with recurrent episodes of severe hypoglycaemia. Therefore, this treatment option may also have an impact on the perception of hypoglycaemia frequency; unfortunately, although the proportion of patients treated with pump therapy is low in our region, we did not collect data on this mode of therapy. Unfortunately, in the current study we did not assess the visual function of the study participants. In this regard, we should point out that the impact of DR on QoL is mainly produced by impairment of the visual function. Furthermore, although we could analyse the effect of advanced DR stages, the number of patients with advanced DR (moderate, severe and proliferative DR) was very low; for this reason, we cannot draw any conclusion as to whether patients with more severe DR are those than have poorer QoL and treatment satisfaction outcomes. This also applies to conditions known to produce visual impairment (e.g., myopia) that were present in a low number of participants. Although DR was significantly associated with specific QoL domains and overall diabetic treatment satisfaction, according to our results, there is a high percentage of variability in these patient-reported outcomes that remains unexplained. Actually, some other factors or comorbidities associated with DR, that were not assessed in the current study, could have a relevant contribution to explain our findings on QoL and TS. For instance, although we used diabetic foot disease as an exclusion criterion, we did not assess diabetic neuropathy in this sample of patients, a complication that could affect the QoL and TS of the patients. All the limitations mentioned above call for caution when interpreting the current results. Further studies are clearly needed to address all the aforementioned issues, especially those in relation to eye-related burden, i.e., the contribution of the impairment of visual function and the impact of advanced stages of DR.

On the other hand, this study has several strengths. This is the first study that has assessed the potential relationship between QoL and TS in adult patients with type 1 diabetes and DR by using diabetes-specific QoL and TS questionnaires. Many studies have used generic instruments to assess QoL. In point of fact, generic instruments cannot assess the impact of a specific condition on the aspects of patient's life due to a demonstrated lack of sensitivity of these instruments; this has been described in scientific literature [5–7]. Additionally, we recruited a relevant number of patients with type 1 diabetes without other diabetic complications and comorbid conditions in this two-centre study and included a representative sample of subjects.

5. Conclusions

In conclusion, DR was associated with poorer QoL and TS in patients with type 1 diabetes. Moreover, DR was associated with a higher hypoglycaemia frequency perception. Finally, the current results have implications for potential future research because there has been no other study that specifically assesses the impact of DR on the QoL and TS of adult patients with type 1 diabetes. Future research should lead to the identification of the individual contribution of each of the relevant factors on these patient-reported outcomes.

Supplementary Materials: The following are available online at http://www.mdpi.com/2077-0383/8/3/377/s1, Table S1: Multivariate linear regression for the Audit of Diabetes Dependent Quality of Life (ADDQoL-19) average weighted impact score, Table S2: Multivariate linear regression for the Diabetes Treatment Satisfaction Questionnaire-status (DTSQ-s) final score.

Author Contributions: Conceptualization, M.G.-C. and D.M.; methodology, M.G.-C., M.M.-A. and D.M.; formal analysis, M.M.-A.; investigation, M.G.-C., E.C., A.R.-M., M.M., N.A., X.V., A.T., E.R., A.L.-M., M.H. and N.A.; Writing—Original Draft preparation, M.G.-C. and D.M.; Writing—Review and Editing, M.G.-C. and D.M.; supervision, D.M.

Funding: This study was supported by the Catalan Diabetes Association (Beca d'Educació Terapèutica 2015), Spain. Additional support from grants PI12/00183 and PI15/00625 from the Instituto de Salud Carlos III (Ministry of Economy and Competitiveness, Spain) to DM is acknowledged. CIBERDEM is an initiative from the Instituto de Salud Carlos III (Plan Nacional de I + D + I and Fondo Europeo de Desarrollo Regional). M.G.-C. holds a predoctoral fellowship from the Ministerio de Educación, Cultura y Deporte, FPU15/03005.

Acknowledgments: We particularly acknowledge the patients, IGTP-HUGTP and IRB Lleida (B.0000682) Biobanks integrated in the Spanish National Biobanks Network of Instituto de Salud Carlos III (PT17/0015/0045 and PT17/0015/0027, respectively) for their collaboration.

Conflicts of Interest: The authors declare no conflict of interest.

References

1. Cho, N.H.; Kirigia, J.; Mbanya, J.C.; Ogurstova, K.; Guariguata, L.; Rathmann, W.; Roglic, G.; Forouhi, N.G.; Dajani, R.; Esteghamati, A.; et al. IDF Diabetes Atlas. Available online: https://www.idf.org/e-library/epidemiology-research/diabetes-atlas/134-idf-diabetes-atlas-8th-edition.html (accessed on 3 September 2018).

2. Hendrick, A.M.; Gibson, M.V.; Kulshreshtha, A. Diabetic Retinopathy. *Prim. Care* **2015**, *42*, 451–464. [CrossRef] [PubMed]

3. Van Hecke, M.V.; Dekker, J.M.; Stehouwer, C.D.A.; Polak, B.C.P.; Fuller, J.H.; Sjolie, A.K.; Kofinis, A.; Rottiers, R.; Porta, M.; Chaturvedi, N. Diabetic retinopathy is associated with mortality and cardiovascular disease incidence: The EURODIAB prospective complications study. *Diabetes Care* **2005**, *28*, 1383–1389. [CrossRef] [PubMed]

4. Speight, J.; Reaney, M.D.; Barnard, K.D. Not all roads lead to Rome—A review of quality of life measurement in adults with diabetes. *Diabet. Med.* **2009**, *26*, 315–327. [CrossRef]

5. Rubin, R.R.; Peyrot, M. Quality of life and diabetes. *Diabetes Metab. Res. Rev.* **1999**, *15*, 205–218. [CrossRef]

6. Sánchez-Lora, F.J.; Santana, T.T.; Trigueros, A.G. Instrumentos específicos de medida de la calidad de vida relacionada con la salud en la diabetes mellitus tipo 2 disponibles en España. *Med. Clin.* **2010**, *135*, 658–664. [CrossRef] [PubMed]

7. Fenwick, E.K.; Pesudovs, K.; Rees, G.; Dirani, M.; Kawasaki, R.; Wong, T.Y.; Lamoureux, E.L. The impact of diabetic retinopathy: Understanding the patient's perspective. *Br. J. Ophthalmol.* **2010**, *95*, 774–782. [CrossRef]

8. Alcubierre, N.; Rubinat, E.; Traveset, A.; Martinez-Alonso, M.; Hernandez, M.; Jurjo, C.; Mauricio, D. A prospective cross-sectional study on quality of life and treatment satisfaction in type 2 diabetic patients with retinopathy without other major late diabetic complications. *Health Qual. Life Outcomes* **2014**, *12*, 131. [CrossRef] [PubMed]

9. Bradley, C.; Eschwège, E.; de Pablos-Velasco, P.; Parhofer, K.G.; Simon, D.; Vandenberghe, H.; Gönder-Frederick, L. Predictors of quality of life and other patient-reported outcomes in the PANORAMA multinational study of people with type 2 diabetes. *Diabetes Care* **2018**, *41*, 267–276. [CrossRef]

10. Man, R.E.K.; Fenwick, E.K.; Sabanayagam, C.; Li, L.J.; Tey, C.S.; Soon, H.J.T.; Cheung, G.C.M.; Tan, G.S.W.; Wong, T.Y.; Lamoureux, E.L. Differential impact of unilateral and bilateral classifications of diabetic retinopathy and diabetic macular edema on vision-related quality of life. *Investig. Ophthalmol. Vis. Sci.* **2016**, *57*, 4655–4660. [CrossRef]

11. Pereira, D.M.; Shah, A.; D'Souza, M.; Simon, P.; George, T.; D'Souza, N.; Suresh, S.; Baliga, M.S. Quality of life in people with diabetic retinopathy: Indian study. *J. Clin. Diagnost. Res.* **2017**, *11*, NC01–NC06. [CrossRef]

12. Sepúlveda, E.; Poínhos, R.; Constante, M.; Pais-Ribeiro, J.; Freitas, P.; Carvalho, D. Relationship between chronic complications, hypertension, and health—Related quality of life in Portuguese patients with type 2 diabetes. *Diabetes Metab. Syndr. Obes.* **2015**, *8*, 535–542. [CrossRef] [PubMed]

13. Das, T.; Wallang, B.; Semwal, P.; Basu, S.; Padhi, T.R.; Ali, M.H. Changing clinical presentation, current knowledge-attitude-practice, and current vision related quality of life in self-reported type 2 diabetes patients with retinopathy in eastern India: The LVPEI eye and diabetes study. *J. Ophthalmol.* **2016**, *2016*, 1–9. [CrossRef]

14. Timar, R.; Velea, I.; Timar, B.; Lungeanu, D.; Oancea, C.; Roman, D.; Mazilu, O. Factors influencing the quality of life perception in patients with type 2 diabetes mellitus. *Patient Prefer. Adher.* **2016**, *10*, 2471–2477. [CrossRef]

15. Nyanzi, R.; Wamala, R.; Atuhaire, L.K. Diabetes and quality of life: A Ugandan perspective. *J. Diabetes Res.* **2014**, *2014*, 1–9. [CrossRef]

16. Ahola, A.J.; Saraheimo, M.; Forsblom, C.; Hietala, K.; Sintonen, H.; Groop, P.-H.; FinDiane Study, G. Health-related quality of life in patients with type 1 diabetes-association with diabetic complications (the FinDiane Study). *Nephrol. Dial. Transplant.* **2010**, *25*, 1903–1908. [CrossRef] [PubMed]

17. Alva, M.; Gray, A.; Mihaylova, B.; Clarke, P. The effect of diabetes complications of health-related quality of life: The importance of longitudinal data to address patient heterogeneity. *Health Econ.* **2014**, *23*, 487–500. [CrossRef] [PubMed]

18. Hirai, F.E.; Tielsch, J.M.; Klein, B.E.K.; Klein, R. Ten-year change in vision-related quality of life in type 1 diabetes: Wisconsin epidemiologic study of diabetic retinopathy. *Ophthalmology* **2011**, *118*, 353–358. [CrossRef] [PubMed]

19. Hirai, F.E.; Tielsch, J.M.; Klein, B.E.K.; Klein, R. Ten-year change in self-rated quality of life in a type 1 diabetes population: Wisconsin epidemiologic study of diabetic retinopathy. *Qual. Life Res.* **2013**, *22*, 1245–1253. [CrossRef]

20. Gubitosi-Klug, R.A.; Sun, W.; Cleary, P.A.; Braffett, B.H.; Aiello, L.P.; Das, A.; Tamborlane, W.; Klein, R. Effects of prior intensive insulin therapy and risk factors on patient-reported visual function outcomes in the Diabetes Control and Complications Trial/Epidemiology of Diabetes Interventions and Complications (DCCT/EDIC) cohort. *JAMA Ophthalmol.* **2016**, *134*, 137–145.

21. Matza, L.S.; Rousculp, M.D.; Malley, K.; Boye, K.S.; Oglesby, A. The longitudinal link between visual acuity and health-related quality of life in patients with diabetic retinopathy. *Health Qual. Life Outcomes* **2008**, *6*, 1–10. [CrossRef]

22. Cusick, M.; SanGiovanni, J.P.; Chew, E.Y.; Csaky, K.G.; Hall-Shimel, K.; Reed, G.F.; Caruso, R.C.; Ferris, F.L. Central visual function and the NEI-VFQ-25 near and distance activities subscale scores in people with type 1 and 2 diabetes. *Am. J. Ophthalmol.* **2005**, *139*, 1042–1050. [CrossRef] [PubMed]

23. Fenwick, E.K.; Xie, J.; Ratcliffe, J.; Konrad, P.; Finger, R.P.; Wong, T.Y.; Lamoureux, E.L. The impact of diabetic retinopathy and diabetic macular edema on health-related quality of life in type 1 and type 2 diabetes. *Investig. Ophthalmol. Vis. Sci.* **2012**, *53*, 677–684. [CrossRef] [PubMed]

24. Hart, H.E.; Bilo, H.J.G.; Redekop, W.K.; Stolk, R.P.; Assink, J.H.; Jong, B.M. Quality of life of patients with type I diabetes mellitus. *Qual. Life Res.* **2003**, *12*, 1089–1097. [CrossRef] [PubMed]

25. Hart, H.E.; Redekop, W.K.; Berg, M.; Bilo, H.J.G.; Meyboom-De Jong, B. Factors that predicted change in health-related quality of life were identified in a cohort of diabetes mellitus type 1 patients. *J. Clin. Epidemiol.* **2005**, *58*, 1158–1164. [CrossRef]

26. Mata, A.R.; Álvares, J.; Diniz, L.M.; Ruberson Ribeiro da Silva, M.; Alvernaz dos Santos, B.R.; Guerra Júnior, A.A.; Cherchiglia, M.L.; Andrade, E.I.; Godman, B.; Acurcio, F.D. Quality of patients with diabetes mellitus types 1 and 2 from a referal health care center in Minas Gerais, Brazil. *Expert Rev. Clin. Pharmacol.* **2016**, *9*, 739–746. [CrossRef]

27. Solli, O.; Stavem, K.; Kristiansen, I.S. Health-related quality of life in diabetes: The associations of complications with EQ-5D scores. *Health Qual. Life Outcomes* **2010**, *8*, 1–8. [CrossRef]

28. Peasgood, T.; Brennan, A.; Mansell, P.; Elliott, J.; Basarir, H.; Kruger, J. The impact of diabetes-related complications on preference-based measures of health-related quality of life in adults with type I diabetes. *Med. Decis. Mak.* **2016**, *36*, 1020–1033. [CrossRef]

29. Jansson, R.W.; Hufthammer, K.O.; Krohn, J. Diabetic retinopathy in type 1 diabetes patients in Western Norway. *Acta Ophthalmol.* **2018**, *96*, 465–474. [CrossRef]

30. Weisman, A.; Lovblom, L.E.; Keenan, H.A.; Tinsley, L.J.; D'Eon, S.; Boulet, G.; Farooqi, M.A.; Lovshin, J.A.; Orszag, A.; Lytvyn, Y.; et al. Diabetes care disparities in long-standing type 1 diabetes in Canada and the U.S.: A Cross-sectional comparison. *Diabetes Care* **2018**, *41*, 88–95. [CrossRef]

31. Mauricio, D. Quality of life and treatment satisfaction are highly relevant patient-reported outcomes in type 2 diabetes mellitus. *Ann. Transl. Med.* **2018**, *6*, 220. [CrossRef]

32. Granado-Casas, M.; Alcubierre, N.; Martín, M.; Real, J.; Ramírez-Morros, A.M.; Cuadrado, M.; Alonso, N.; Falguera, M.; Hernández, M.; Aguilera, E.; et al. Improved adherence to Mediterranean Diet in adults with type 1 diabetes mellitus. *Eur. J. Nutr.* **2018**, 1–9. [CrossRef] [PubMed]

33. Wilkinson, C.P.; Ferris, F.L.; Klein, R.E.; Lee, P.P.; Agardh, C.D.; Davis, M.; Dills, D.; Kampik, A.; Pararajasegaram, R.; Verdaguer, J.T.; et al. Proposed international clinical diabetic retinopathy and diabetic macular edema disease severity scales. *Ophthalmology* **2003**, *110*, 1677–1682. [CrossRef]

34. Bernstein, M.S.; Morabia, A.; Sloutskis, D. Definition and prevalence of sedentarism in an urban population. *Am. J. Public Health* **1999**, *89*, 862–867. [CrossRef]

35. Cabrera de León, A.; Rodríguez-Pérez, M.D.C.; Rodríguez-Benjumeda, L.M.; Anía-Lafuente, B.; Brito-Díaz, B.; Muros de Fuentes, M.; Almeida-González, D.; Batista-Medina, M.; Aguirre-Jaime, A. Sedentary lifestyle: Physical activity duration versus percentage of energy expenditure. *Rev. Esp. Cardiol.* **2007**, *60*, 244–250.

36. Bradley, C.; De Pablos-Velasco, P.; Parhofer, K.G.; Eschwge, E.; Gönder-Frederick, L.; Simon, D. PANORAMA: A European study to evaluate quality of life and treatment satisfaction in patients with type-2 diabetes mellitus—Study design. *Prim. Care Diabetes* **2011**, *5*, 231–239. [CrossRef]

37. Bradley, C.; Todd, C.; Gorton, T.; Symonds, E.; Martin, A.; Plowright, R. The development of an individualised questionnaire measure of perceived impact of diabetes on quality of life: The ADDQoL. *Qual. Life Res.* **1999**, *8*, 79–81. [CrossRef]

38. Depablos-Velasco, P.; Salguero-Chaves, E.; Mata-Poyo, J.; Derivas-Otero, B.; Garcia-Sanchez, R.; Viguera-Ester, P. Quality of life and satisfaction with treatment in subjects with type 2 diabetes: Results in Spain of the PANORAMA study. *Endocrinol. Nutr.* **2014**, *61*, 18–26. [CrossRef] [PubMed]

39. Bradley, C. Diabetes Treatment Satisfaction Questionnaire (DTSQ). In *Handbook of Psychology and Diabetes: A Guide to Psychological Measurement in Diabetes Research and Practice*; Harwood Academic Publishers: Amsterdam, The Netherlands, 1994; pp. 111–112.

40. Gomis, R.; Herrera-Pombo, J.; Calderón, A.; Rubio-Terrés, C.; Sarasa, P. Validación del cuestionario "Diabetes treatment satisfaction questionnaire" (DTSQ) en la población española. *Pharmacoeconomics* **2006**, *3*, 7–20. [CrossRef]

41. R Core Team. R: A language and environment for statistical computing. Available online: https://www.r-project.org (accessed on 20 June 2018).

42. Granado-Casas, M.; Martínez-Alonso, M.; Alcubierre, N.; Ramírez-Morros, A.; Hernández, M.; Castelblanco, E.; Torres-Puiggros, J.; Mauricio, D. Decreased quality of life and treatment satisfaction in patients with latent autoimmune diabetes of the adult. *PeerJ* **2017**, *5*, e3928. [CrossRef] [PubMed]

43. Collins, M.; O'Sullivan, T.; Harkins, V.; Perry, I. Quality of life and quality of care in patients with diabetes experiencing different models of care. *Diabetes Care* **2009**, *32*, 603–605. [CrossRef]

44. Gawlik, N.R.; Bond, M.J. The role of negative affect in the assessment of quality of life among women with type 1 diabetes mellitus. *Diabetes Metab. J.* **2018**, *42*, 130–136. [CrossRef]
45. Biderman, A.; Noff, E.; Harris, S.B.; Friedman, N.; Levy, A. Treatment satisfaction of diabetic patients: What are the contributing factors? *Fam. Pract.* **2009**, *26*, 102–108. [CrossRef]

 Journal of
Clinical Medicine

Article

Association of Glucose Fluctuations with Sarcopenia in Older Adults with Type 2 Diabetes Mellitus

Noriko Ogama [1,2,3], Takashi Sakurai [1,4,*], Shuji Kawashima [2,5], Takahisa Tanikawa [5,6], Haruhiko Tokuda [5,6,7], Shosuke Satake [2], Hisayuki Miura [8], Atsuya Shimizu [9], Manabu Kokubo [9], Shumpei Niida [7], Kenji Toba [1], Hiroyuki Umegaki [3] and Masafumi Kuzuya [3,10]

1   Center for Comprehensive Care and Research on Memory Disorders, National Center for Geriatrics and Gerontology, Obu 474-8511, Japan; n-ogama@ncgg.go.jp (N.O.); toba@ncgg.go.jp (K.T.)
2   Department of Geriatric Medicine, National Center for Geriatrics and Gerontology, Obu 474-8511, Japan; kawashu@ncgg.go.jp (S.K.); satakes@ncgg.go.jp (S.S.)
3   Department of Community Healthcare and Geriatrics, Nagoya University Graduate School of Medicine, Nagoya 466-8550, Japan; umegaki@med.nagoya-u.ac.jp (H.U.); kuzuya@med.nagoya-u.ac.jp (M.K.)
4   Department of Cognition and Behavior Science, Nagoya University Graduate School of Medicine, Nagoya 466-8550, Japan
5   Department of Endocrinology and Metabolism, National Center for Geriatrics and Gerontology, Obu 474-8511, Japan; hoisan@ncgg.go.jp (T.T.); tokuda@ncgg.go.jp (H.T.)
6   Department of Clinical Laboratory, National Center for Geriatrics and Gerontology, Obu 474-8511, Japan
7   Medical Genome Center, National Center for Geriatrics and Gerontology, Obu 474-8511, Japan; sniida@ncgg.go.jp
8   Department of Home Care Coordinators, National Center for Geriatrics and Gerontology, Obu 474-8511, Japan; hmiura@ncgg.go.jp
9   Department of Cardiology, National Center for Geriatrics and Gerontology, Obu 474-8511, Japan; ashimizu@ncgg.go.jp (A.S.); mkokubo@ncgg.go.jp (M.K.)
10   Institutes of Innovation for Future Society, Nagoya University, Nagoya 464-8601, Japan
*   Correspondence: tsakurai@ncgg.go.jp; Tel.: +81-562-46-2311

Received: 31 January 2019; Accepted: 28 February 2019; Published: 6 March 2019

Abstract: Type 2 diabetes mellitus accelerates loss of muscle mass and strength. Patients with Alzheimer's disease (AD) also show these conditions, even in the early stages of AD. The mechanism linking glucose management with these muscle changes has not been elucidated but has implications for clarifying these associations and developing preventive strategies to maintain functional capacity. This study included 69 type 2 diabetes patients with a diagnosis of cognitive impairment ($n = 32$) and patients with normal cognition ($n = 37$). We investigated the prevalence of sarcopenia in diabetes patients with and without cognitive impairment and examined the association of glucose alterations with sarcopenia. Daily glucose levels were evaluated using self-monitoring of blood glucose, and we focused on the effects of glucose fluctuations, postprandial hyperglycemia, and the frequency of hypoglycemia on sarcopenia. Diabetes patients with cognitive impairment displayed a high prevalence of sarcopenia, and glucose fluctuations were independently associated with sarcopenia, even after adjusting for glycated hemoglobin A1c (HbA1c) levels and associated factors. In particular, glucose fluctuations were significantly associated with a low muscle mass, low grip strength, and slow walking speed. Our observation suggests the importance of glucose management by considering glucose fluctuations to prevent the development of disability.

Keywords: Alzheimer's disease; type 2 diabetes mellitus; glucose fluctuations; sarcopenia

1. Introduction

The definition and diagnosis of sarcopenia were updated by the European Working Group on Sarcopenia in Older People [1]. The revised consensus focuses on the importance of promoting early detection and treatment of sarcopenia [1]. Because sarcopenia predicts adverse outcomes such as chronic disease progression, mortality, and functional disability [2], developing strategies to prevent sarcopenia in older adults is necessary.

Older adults with type 2 diabetes have a higher prevalence of sarcopenia than non-diabetic individuals [3,4]. Some studies have reported that increased glycated hemoglobin A1c (HbA1c) levels are associated with impaired muscle quality, muscle strength and physical performance [5–7]. However, HbA1c levels do not adequately reflect the mean glucose concentration and are not associated with hypoglycemic risk in diabetes patients [8,9]. Severe hypoglycemic events lead to deterioration in general health, resulting in an increased risk of mobility disabilities [10]. Therefore, continuous assessments of daily glucose excursions are needed to determine the association of glucose management with sarcopenia.

Muscle changes and physical dysfunctions are observed in the early stages of Alzheimer's disease (AD) [11,12]. Lean mass is reduced in early AD and is associated with brain atrophy [11]. Furthermore, the nutritional status is correlated with regional cerebral glucose metabolism in prodromal AD, and the prevalence of sarcopenia increases with the degree of cognitive decline [13,14]. These studies suggest that AD-related degenerative pathologies have a negative impact on muscle structure and physical function. Additionally, AD patients with diabetes have a variety of difficulties in glucose management and therefore might have a higher prevalence of sarcopenia than diabetes patients without AD. However, to date, the prevalence of sarcopenia in diabetes patients with and without cognitive impairment remains unclear. Furthermore, the association of glucose management with sarcopenia in these patients is also unknown.

The aims of this study were as follows: (1) to clarify the prevalence of sarcopenia in older type 2 diabetes patients with cognitive impairment or normal cognition, and (2) to clarify the association of glucose management with sarcopenia in these patients. We hypothesized that diabetes patients with cognitive impairment would display a high prevalence of sarcopenia and that abnormal glucose profiles would be associated with muscle changes and physical dysfunction.

2. Methods

2.1. Participants

The study was conducted in accordance with the Declaration of Helsinki, and the protocol was approved by the Ethics Committee of the National Center for Geriatrics and Gerontology (NCGG) (approval no. 682). All participants provided written informed consent before participating in the study. We included 69 outpatients (cognitive impairment: $n = 32$; normal cognition: $n = 37$) who visited the NCGG hospital from 2014 to 2016. The presence of cognitive impairment was defined as probable or possible AD and amnestic mild cognitive impairment (aMCI) according to the criteria of the National Institute on Aging–Alzheimer's Association workgroups and the definition provided by Petersen et al. [15,16]. Persons with NC attended NCGG hospital for suspected memory impairment but were assessed as having normal cognition. Patients meeting all the following criteria were included in this study: (1) outpatients with type 2 diabetes treated with antidiabetic agents; (2) aged 65 years or older; (3) living in their houses; (4) with families or caregivers who support self-monitoring of blood glucose (SMBG); and (5) a Mini-Mental State Examination (MMSE) score ≥ 10 for cognitive impairment. The exclusion criteria were as follows: (1) severe hearing loss and visual impairment; (2) severe health conditions, such as cardiac failure, renal disorder or liver dysfunction; and (3) neurological disorders other than AD or aMCI.

2.2. Assessment of Clinical Parameters and Comorbidities Associated with Diabetes

Clinical data and blood samples were distributed from the Biobank, which collects and stores biological material and associated clinical data for biomedical research. Data on the diagnosis, antidiabetic medication use, and polypharmacy (defined as taking five or more types of oral medicine) [17] were obtained from clinical charts. Global cognitive function was assessed by the MMSE [18]. Diabetes-associated complications were evaluated for the coexistence of neuropathy, retinopathy and nephropathy [19]. Diabetic neuropathy was defined as either the loss of Achilles tendon reflex or the presence of neuropathic symptoms. Diabetic retinopathy was fundoscopically assessed through dilated pupils by experienced ophthalmologists. Diabetic nephropathy was defined as an albumin-to-creatinine ratio >300 µg/mg or a urinary protein concentration >0.5 mg/dL. We obtained information on the following biochemical parameters: HbA1c, triglyceride, total cholesterol, high-density lipoprotein cholesterol, low-density lipoprotein cholesterol, estimated glomerular filtration rate, serum albumin, and urinary albumin. HbA1c levels were expressed in the National Glycohemoglobin Standardization Program units.

2.3. Measurement of Daily Glucose Levels

Daily glucose levels were recorded by SMBG (MS-FR201B; Terumo Corp., Tokyo, Japan) and were measured at five time points per day (05:00 h, before breakfast, 2 h after breakfast, before lunch, and before dinner) on eight separate days during a two-month period. Glucose levels usually nadir early in the morning and before each meal, and postprandial glucose levels usually peak at 2 h after breakfast [20]. Therefore, we evaluated the glucose levels at these time points. When diabetic patients with cognitive impairment used SMBG, their families helped with the measurement. Hypoglycemia was defined as a glucose level ≤70 mg/dL [21], and the presence or absence of hypoglycemic symptoms was recorded at every SMBG measurement point. Glucose fluctuations were determined based on the diurnal range from minimum glucose levels to maximum glucose levels.

2.4. Evaluation of Sarcopenia

Sarcopenia was defined as low muscle mass plus low muscle strength and/or low physical performance according to the Asian Working Group for Sarcopenia (AWGS) [22]. Low muscle mass was defined as a calf circumference <31 cm [23], which was measured with the patient in the supine position with the left knee raised at a right angle from the thigh. Calf circumference was correlated with appendicular skeletal muscle mass measured by dual-energy X-ray absorptiometry and has been proposed as a surrogate marker of muscle mass for diagnosing sarcopenia [24]. Low muscle strength was determined by low hand grip strength (<26 kg in men and <18 kg in women). Hand grip strength was measured with a Smedley dynamometer (Matsumiya Medical Instruments, Tokyo, Japan). Low physical performance was assessed by slow walking speed. The presence of slow walking speed was defined as individuals who answered "No" to the following question: "Can you cross the road within the green signal interval?" [25].

2.5. Statistical Analysis

All analyses were performed using SPSS for Windows version 22.0 (IBM Corp., Armonk, NY, USA). The distributions of data were assessed for normality using the Shapiro–Wilk test. Differences in demographics between diabetic patients with cognitive impairment and patients with normal cognition were examined by unpaired *t*-tests (for parametric variables) or the Mann-Whitney U test (for non-parametric variables). Categorical variables were analyzed by the chi-squared test or Fisher's exact test. Differences in glucose levels with and without sarcopenia were analyzed using unpaired *t*-tests or the Mann–Whitney U test. To identify the factors associated with sarcopenia, we performed logistic regression analysis. First, we conducted stepwise analysis to select the most influential glucose index for sarcopenia. Next, we conducted logistic regression with forward variable selection to construct a model based on the variables associated with sarcopenia, defined as those with $p < 0.05$, adjusted for factors related to the development of sarcopenia (i.e., age, HbA1c level and the presence of diabetic neuropathy) [4,5,7]. In this analysis, we calculated the adjusted odds ratio (OR) and 95% confidence interval (CI) for factors associated with sarcopenia. The dependent variable was sarcopenia, and the independent variable was the glucose index, which was selected as the most influential variable associated with sarcopenia in a first regression analysis. Statistical significance was set at $p < 0.05$.

3. Results

3.1. Clinical Characteristics of the Study Participants

The demographics of the participants are shown in Table 1. No differences in age, sex, body mass index, diabetic comorbidities or medication use were found between the diabetic patients with cognitive impairment and the normal cognition group. The cognitive impairment group had a lower MMSE score than the normal cognition group. Serum triglyceride levels were higher and albumin concentrations were lower in the cognitive impairment group, but the other metabolic markers did not differ between the groups. Regarding the average glucose levels during the 2 months of study, the glucose level before lunch was high in the cognitive impairment group. The prevalence of sarcopenia, represented by a low muscle mass, low grip strength and slow walking speed, was higher in the cognitive impairment group than that in the normal cognition group.

Table 1. Clinical characteristics of the study participants.

	Total (*n* = 69)		Cognitive Impairment (*n* = 32)		Normal Cognition (*n* = 37)		*p*-Value [†]
	Mean (SD) or *n* (%)	Min–Max	Mean (SD) or *n* (%)	Min–Max	Mean (SD) or *n* (%)	Min–Max	
Age, years	75.0 (5.3)	65–87	76.0 (5.8)	65–87	74.2 (4.7)	65–83	0.146
Male, *n* (%)	36 (52.2)		15 (46.9)		21 (56.8)		0.413
Body mass index, kg/m^2	23.8 (2.7)	17.8–31.0	23.6 (2.6)	17.8–29.4	24.0 (2.7)	19.9–31.0	0.597
Mini-Mental State Examination	24.4 (5.0)	13–30	21.0 (5.2)	13–29	27.4 (2.0)	22–30	<0.001
Diabetes and comorbidities							
Duration of diabetes, years	15.3 (10.8)	2–48	15.3 (10.6)	2–40	15.4 (11.0)	2–48	0.899
Diabetic neuropathy, *n* (%)	45 (65.2)		22 (68.8)		23 (62.2)		0.567
Diabetic retinopathy, *n* (%)	16 (23.2)		5 (15.6)		11 (29.7)		0.166
Diabetic nephropathy, *n* (%)	21 (30.4)		12 (37.5)		9 (24.3)		0.236
Coronary artery disease, *n* (%)	15 (21.7)		6 (18.8)		9 (24.3)		0.576
Hypertension, *n* (%)	53 (76.8)		24 (75.0)		29 (78.4)		0.740
Medications and antidiabetic agents, *n* (%)							
Polypharmacy	56 (81.2)		29 (90.6)		27 (73.0)		0.061
Biguanide	20 (29.0)		10 (31.3)		10 (27.0)		0.700
Thiazolidine	8 (11.6)		6 (18.8)		2 (5.4)		0.132
DPP4 inhibitor	49 (71.0)		23 (71.9)		26 (70.3)		0.884
Sulfonylurea	40 (58.0)		19 (59.4)		21 (56.8)		0.826
Insulin secretion promoter	2 (2.9)		2 (6.3)		0 (0.0)		0.211
α-Glucosidase inhibitor	16 (23.2)		7 (21.9)		9 (24.3)		0.810
Insulin	13 (18.8)		7 (21.9)		6 (16.2)		0.549
GLP-1 receptor agonists	2 (2.9)		1 (3.1)		1 (2.7)		1.000
Biochemical parameters							
HbA1c, %	7.1 (0.6)	6.2–9.3	7.3 (0.7)	6.2–9.3	7.0 (0.5)	6.3–8.6	0.107
Triglyceride, mg/dL	139.8 (69.4)	44–330	165.5 (72.1)	65–330	117.6 (57.2)	44–279	0.004
Total cholesterol, mg/dL	190.3 (41.1)	108–316	192.0 (41.1)	108–316	188.9 (41.6)	137–309	0.524
HDL cholesterol, mg/dL	53.6 (13.7)	27–92	50.8 (13.6)	27–83	56.0 (13.5)	37–92	0.112
LDL cholesterol, mg/dL	109.2 (36.0)	46–238	109.5 (37.2)	46–211	108.9 (35.4)	67–238	0.928
eGFR, mL/min/1.73 m^2	63.7 (17.6)	28.3–115.9	64.8 (16.9)	28.3–110.3	62.7 (18.3)	30.6–115.9	0.621
Albumin, g/dL	4.3 (0.4)	3.5–5.2	4.2 (0.3)	3.5–5.2	4.4 (0.3)	3.8–5.2	0.014
UACR, mg/gCr	156.7 (339.4)	1.5–1808.3	167.1 (349.2)	1.5–1705.4	147.8 (335.3)	2.7–1808.3	0.516
Daily blood glucose level							
05:00 h, mg/dL	116.6 (22.2)	57–254	113.5 (19.6)	57–205	119.4 (24.2)	58–254	0.485
Before breakfast, mg/dL	123.0 (21.7)	51–215	119.5 (20.4)	63–201	126.0 (22.6)	51–215	0.216
2 h after breakfast, mg/dL	180.7 (34.3)	68–383	184.7 (38.8)	68–349	177.3 (29.9)	72–383	0.880
Before lunch, mg/dL	126.7 (33.6)	43–313	137.3 (38.5)	43–313	117.6 (25.9)	45–243	0.011
Before dinner, mg/dL	134.9 (27.5)	55–331	139.2 (33.0)	60–331	131.2 (21.3)	55–267	0.339
Fluctuation, mg/dL	91.4 (28.5)	32–155	97.0 (29.7)	49–155	86.5 (26.8)	32–137	0.127
Frequency of hypoglycemia *	0.71 (1.3)	0–7	0.72 (1.5)	0–7	0.70 (1.2)	0–4	0.719

Table 1. *Cont.*

	Total (*n* = 69)		Cognitive Impairment (*n* = 32)		Normal Cognition (*n* = 37)		*p*-Value [†]
	Mean (SD) or *n* (%)	Min–Max	Mean (SD) or *n* (%)	Min–Max	Mean (SD) or *n* (%)	Min–Max	
Mobility function							
Sarcopenia, *n* (%)	8 (11.6)		7 (21.9)		1 (2.7)		0.021
Low muscle mass, *n* (%)	10 (14.5)		9 (28.1)		1 (2.7)		0.004
Low grip strength, *n* (%)	24 (34.8)		17 (53.1)		7 (18.9)		0.003
Slow walking speed, *n* (%)	4 (5.8)		4 (12.5)		0 (0.0)		0.042

Data are presented as the mean (standard deviation) or as numbers and percentages. * Indicates the per-patient averages at the measuring points during the study period. [†] The quantitative variables age, body mass index, eGFR, daily blood glucose levels before breakfast and fluctuation were examined by unpaired t-tests, and other quantitative variables were analyzed by the Mann–Whitney U test. The categorical variables thiazolidine, insulin secretion promoter, and GLP-1 receptor agonists were examined by Fisher's exact test, and other categorical variables were analyzed by the chi-squared test. Abbreviations: DPP4, dipeptidyl peptidase 4; eGFR, estimated glomerular filtration rate; GLP-1, glucagon-like peptide-1; HbA1c, glycated hemoglobin A1c; HDL cholesterol, high-density lipoprotein cholesterol; LDL cholesterol, low-density lipoprotein cholesterol; UACR, urine albumin-to-creatinine ratio.

3.2. Differences in Glucose Profiles According to Sarcopenia

First, we examined the differences in glucose profiles with and without sarcopenia in all subjects (Figure 1). Patients with sarcopenia displayed larger fluctuations in daily glucose levels compared to non-sarcopenia patients (117.3 mg/dL vs. 88.0 mg/dL, $p = 0.005$). In addition, patients with sarcopenia showed high glucose levels 2 h after breakfast and before lunch (209.5 mg/dL vs. 177.0 mg/dL, $p = 0.011$, and 152.2 mg/dL vs. 123.4 mg/dL, $p = 0.037$, respectively).

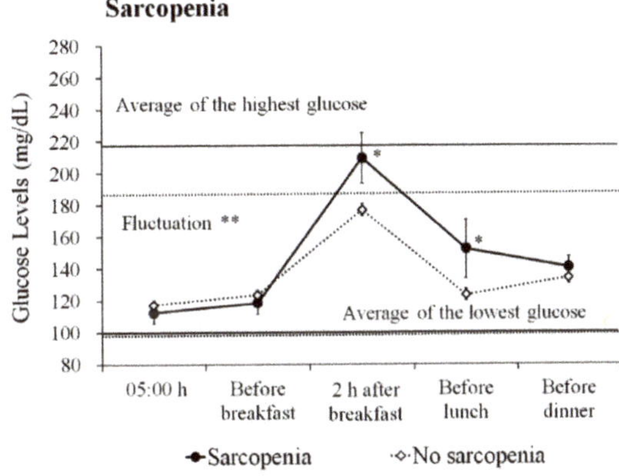

Figure 1. Daily glucose profiles based on sarcopenia. The figure shows the average ± standard error (SE) glucose level based on sarcopenia in all participants. Average of the highest glucose level: the average of the maximum glucose level of the day during the measurement period. Average of the lowest glucose level: the average of the minimum glucose level of the day during the measurement period. The solid line represents the sarcopenia group, and the dotted line indicates the no sarcopenia group. Fluctuation: the average of the diurnal range from the minimum glucose level to the maximum glucose level. ** $p < 0.01$, * $p < 0.05$.

Next, we examined the glucose profile differences with and without sarcopenia in the cognitive impairment group (Figure A1). Cognitive impairment patients with sarcopenia displayed large glucose fluctuations (119.5 mg/dL vs. 90.7 mg/dL, $p = 0.021$) and elevated glucose levels 2 h after breakfast (210.6 mg/dL vs. 177.4 mg/dL, $p = 0.044$). In the normal cognition group, because only one patient was diagnosed with sarcopenia, we could not examine the association between glucose profiles and sarcopenia.

3.3. Differences in Glucose Profiles Based on Sarcopenia Components

The differences in glucose profiles with and without sarcopenia components in all subjects are shown in Figure 2. Patients with a low muscle mass, low grip strength and slow walking speed displayed larger glucose fluctuations compared to patients without these conditions (low muscle mass: 110.8 mg/dL vs. 88.1 mg/dL, $p = 0.018$; low grip strength: 104.4 mg/dL vs. 84.4 mg/dL, $p = 0.005$; and slow walking speed: 136.6 mg/dL vs. 88.6 mg/dL, $p = 0.003$, respectively). Furthermore, patients with these conditions displayed high glucose levels 2 h after breakfast (low muscle mass: 203.8 mg/dL vs. 176.8 mg/dL, $p = 0.020$; low grip strength: 198.1 mg/dL vs. 171.5 mg/dL, $p = 0.002$; and slow walking speed: 241.8 mg/dL vs. 177.0 mg/dL, $p < 0.001$, respectively). In addition, patients with low grip strength showed high glucose levels before lunch (146.2 mg/dL vs. 116.4 mg/dL, $p = 0.001$) and before dinner (145.3 mg/dL vs. 129.4 mg/dL, $p = 0.034$), and patients with a slow walking speed showed high glucose levels before dinner (155.2 mg/dL vs. 133.7 mg/dL, $p = 0.041$).

In the cognitive impairment group, patients with a low muscle mass tended to have large glucose fluctuations and high glucose levels 2 h after breakfast, but this difference did not reach statistical significance (111.9 mg/dL vs. 91.2 mg/dL, $p = 0.076$, and 204.1 mg/dL vs. 177.1 mg/dL, $p = 0.077$, respectively) (Figure A2). Patients with low grip strength displayed large glucose fluctuations and high glucose levels 2 h after breakfast and before lunch (107.4 mg/dL vs. 85.2 mg/dL, $p = 0.033$, 197.6 mg/dL vs. 170.0 mg/dL, $p = 0.038$, and 151.4 mg/dL vs. 121.3 mg/dL, $p = 0.025$, respectively). In addition, patients with a slow walking speed displayed large glucose fluctuations and high glucose levels 2 h after breakfast (136.6 mg/dL vs. 91.4 mg/dL, $p = 0.009$, and 241.8 mg/dL vs. 176.5 mg/dL, $p = 0.001$, respectively).

3.4. Prevalence of Hypoglycemia and its Association with Sarcopenia

Hypoglycemia (glucose range: 43–70 mg/dL) was observed at 49 measurement points (1.78% of all measurement points) in all subjects. Severe hypoglycemic symptoms or events were not observed in this study. Typical hypoglycemic symptoms, such as hand tremor and palpitations, were observed in one normal cognition subject whose glucose level was 45 mg/dL. Hypoglycemia-related symptoms, such as nausea, dizziness, light-headedness, blurred vision and fatigue, were reported at four measurement points in the cognitive impairment group and four measurement points in the normal cognition group.

There was no difference in the frequency of hypoglycemia between patients with and without sarcopenia in all subjects (1.3 ± 2.5 vs. 0.6 ± 1.1, $p = 0.884$, Mann–Whitney U test). The effect of mild hypoglycemia (glucose level $\leq$80 mg/dL or $\leq$90 mg/dL) on sarcopenia was also examined. However, there was no difference in the frequency of mild hypoglycemia among the patients (glucose level $\leq$80 mg/dL; 2.4 ± 3.9 vs. 1.8 ± 2.5, $p = 0.922$, glucose level $\leq$90 mg/dL; 4.9 ± 6.8 vs. 4.6 ± 4.7, $p = 0.741$, Mann–Whitney U test).

We conducted the same analysis in the cognitive impairment group. However, no difference was found in the frequency of hypoglycemia between patients with and without sarcopenia.

3.5. Association of Glucose Indices with Sarcopenia

Because the patients with sarcopenia displayed large glucose fluctuations, high glucose levels at 2 h after breakfast and before lunch, we conducted a regression analysis using the stepwise method to extract the most influential glucose index on sarcopenia. The results revealed glucose fluctuations as the most influential variable on sarcopenia (OR = 1.041, $p = 0.012$). Therefore, we used glucose fluctuations as an independent variable for the subsequent logistic regression analysis.

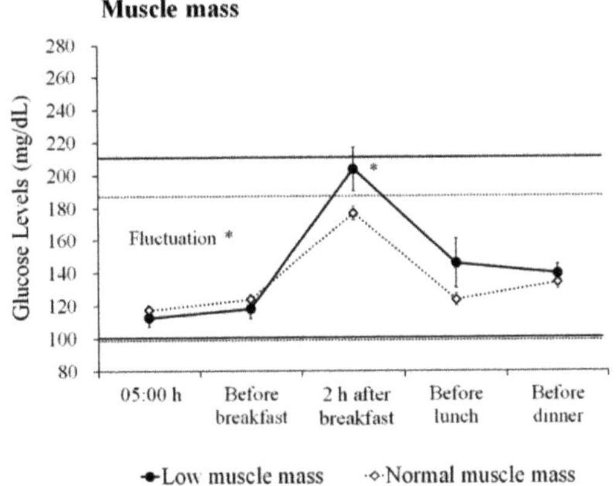

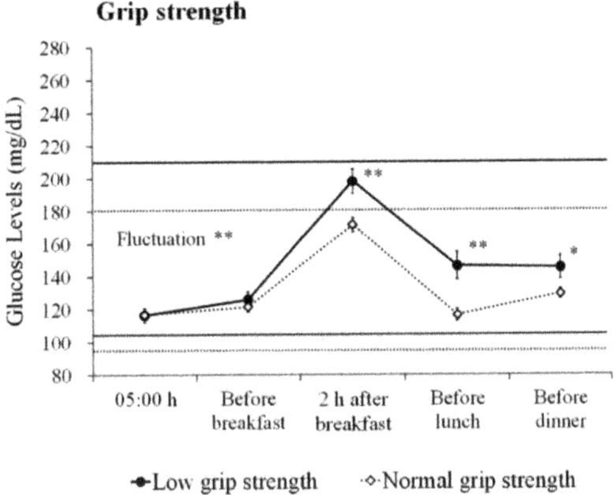

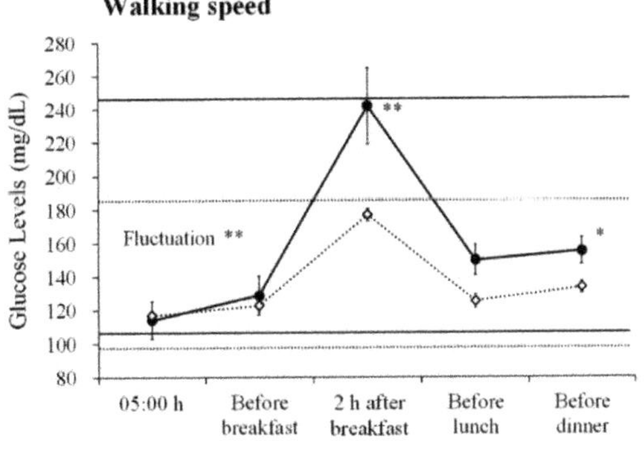

Figure 2. Daily glucose profiles based on the sarcopenia components. The figures show the average ± SE glucose level based on the components of sarcopenia in all participants. ** $p < 0.01$, * $p < 0.05$.

After adjusting for age, HbA1c level and the presence of diabetic neuropathy, glucose fluctuations were independently associated with sarcopenia (Table 2). Additionally, among the components of sarcopenia, glucose fluctuations were significantly associated with low muscle mass, low grip strength, and slow walking speed.

Table 2. Association of glucose fluctuations with sarcopenia.

	Glucose Fluctuations			
	Differences *	OR	95% CI	*p*-Value
No sarcopenia		Reference		
Sarcopenia	29.3 mg/dL	1.045	(1.007; 1.083)	0.018
Normal muscle mass		Reference		
Low muscle mass	22.7 mg/dL	1.031	(1.000; 1.064)	0.0499
Normal grip strength		Reference		
Low grip strength	20.0 mg/dL	1.029	(1.006; 1.053)	0.014
Normal walking speed		Reference		
Slow walking speed	48.0 mg/dL	1.092	(1.018; 1.172)	0.014

Logistic regression with a step-wise method. The dependent variables were sarcopenia and its components. The independent variable was glucose fluctuations. All analyses were adjusted for age, HbA1c level and the presence of diabetic neuropathy. * Indicates the differences relative to glucose levels in individuals in the no sarcopenia, normal muscle mass, normal grip strength, and normal walking speed groups. Abbreviations: CI, confidence interval; OR, odds ratio.

In the cognitive impairment group, glucose fluctuations were independently associated with sarcopenia (OR = 1.038, *p* = 0.035). Furthermore, glucose fluctuations were significantly associated with low grip strength (OR = 1.034, *p* = 0.049) and slow walking speed (OR = 1.077, *p* = 0.023). The association between glucose fluctuation and low muscle mass did not reach statistical significance, which may have been caused by the low statistical power due to the small sample size.

4. Discussion

The present study revealed that diabetes patients with cognitive impairment had a high prevalence of sarcopenia. Glucose fluctuations were independently associated with sarcopenia after adjusting for HbA1c levels and associated factors. Our observation suggests the importance of glucose management by considering glucose fluctuations in older adults with diabetes.

In the current study, the prevalence of sarcopenia was higher in the cognitive impairment group than the normal cognition group. However, diabetes comorbidities and medication were comparable between the groups. Regarding the underlying mechanisms of sarcopenia in diabetes, insulin resistance in peripheral tissues has been suggested [3,26]. Insulin resistance is associated with impaired mitochondrial function in muscles, leading to the production of oxidative damage [3]. In addition, increased levels of inflammatory cytokines, such as interleukin (IL)-1, IL-6, and tumor necrosis factor alpha (TNF-α), are observed in diabetes but are also found in AD patients [27]. These factors are closely related to decreased muscle function [28]. Furthermore, anabolic hormones play a major role in muscle integrity; testosterone decline is found in diabetes patients and also involved in the development of AD [3,29]. More recently, AD-related brain pathology has also been associated with low physical performance [30,31]. Thus, diabetes patients with cognitive impairment have multiple factors that may contribute to the development of sarcopenia.

In this investigation, we found an independent association between glucose fluctuations and sarcopenia in cognitively impaired patients. Acute hyperglycemia increases Aβ production, and altering insulin signaling lead to changes in Aβ levels in the brain [32–34]. Increased levels of Aβ have been also found in the skeletal muscle, which may cause impaired peripheral glucose metabolism [35,36]. In addition, glucose fluctuations are a greater trigger of oxidative stress than sustained hyperglycemia, and independently contribute to the development of microvascular complications [37,38]. Because brain tissue of AD show increased oxidative stress during the course of the disease [39], AD patients are more likely to be susceptible to the influence of glucose fluctuations on the brain. In fact, our previous study showed that glucose fluctuations during postprandial periods were independently associated with frontal white matter hyperintensity (WMH) in diabetic patients with AD but not in patients with normal cognition [40]. Frontal WMH is known to play a predominant role in motor dysfunctions in AD/aMCI patients [41,42]. Furthermore, an association between elevated 2 h post-load glucose levels evaluated by the oral glucose tolerance test and brain atrophy has also been reported [43]. Thus, brain structural changes associated with glucose fluctuations might further contribute to mobility disabilities in diabetes patients with cognitive impairment.

In this study, the prevalence of sarcopenia was lower than previously reported [3], particularly in the normal cognition group. This can be explained by differences in the measurement method for muscle mass in the current study. Low muscle mass is an essential condition for the AWGS definition of sarcopenia and was measured by calf circumference. A previous study showed that fat-free mass decreases and fat mass increases with aging [44]. However, in older adult with diabetes, muscle mass is even lower than in non-diabetic individuals [45]. Furthermore, higher level of insulin resistance has been associated with low muscle mass and high fat mass [46]. A study using peripheral quantitative computed tomography showed that diabetes patients have a larger intramuscular and intermuscular adipose tissue in calf muscles than individuals without diabetes [47]. Therefore, diabetic patients are considered to have deteriorated muscle quality compared to individuals without diabetes. It seems likely that these muscle changes cannot be correctly measured with calf circumference. Although, calf circumference is considered suitable for evaluating muscle mass, our observations suggest that calf circumference may be difficult to accurately estimate the muscle mass of patients with diabetes.

We found no association between hypoglycemia and sarcopenia. Diabetes patients who experienced hypoglycemic events had a high prevalence of comorbidities, disabilities, and malnutrition, and were often treated with insulin therapy and use of polypharmacy [10]. However, in this study, no difference was identified in these parameters between patients with and without hypoglycemia. The study participants had relatively well-controlled glucose levels, and severe hypoglycemic events were not observed. Therefore, we could not detect a sufficient impact of hypoglycemia. In fact, hospitalization due to severe hypoglycemia can cause further deterioration in physical function [10]. Hypoglycemic symptoms in older patients are mild or asymptomatic, and episodes are less likely to manifest [10]. We further addressed concerns about repeated mild hypoglycemia but found no association of mild hypoglycemia with sarcopenia.

Currently, sarcopenia and frailty are considered a third category of diabetes-related complications in addition to the traditional microvascular and macrovascular disease [4]. Because muscle weakness and impaired physical function resulting from sarcopenia are major physiological components of frailty, these conditions often overlap [3,48]. Some studies reported that frailty and cognitive decline are reciprocally related and suggested that they share underlying pathophysiological mechanisms, including AD pathology, chronic inflammation, mood disorder, and co-morbidities [49–51]. Thus, patients with cognitive impairment are at high risk of vulnerability due to diverse extrinsic and intrinsic conditions. Recently, some types of interventions, such as physical exercise, nutritional therapy, and multicomponent interventions, to prevent frailty have been attempted [52]. Exercise training has benefits for physical capacity and cognitive performance [53], improves skeletal muscle insulin sensitivity and increases muscle mass in frail older adults [54]. Furthermore, the benefits of resistance training in improving glycemic status and muscle strength have been reported [55]. Therefore, in addition to glucose management using medications appropriately to avoid glucose fluctuations, multimodal interventions would be beneficial to delay or prevent disability in older diabetic patients.

This study has several limitations. First, because this study had a cross-sectional design, causal relationships should be carefully considered. Second, the sample size was relatively small. Nevertheless, our results revealed the independent association between glucose fluctuations and sarcopenia, even after adjusting for several confounding factors. Furthermore, our sample size was largely comparable to that of a previous study that identified an association between altered glucose dynamics and frailty [56]. Additional studies with a large cohort estimated by calculation of the appropriate sample size are needed to confirm our findings. Third, the cognitive impairment group was assigned based on the criteria of probable or possible AD and aMCI [15,16], but biomarkers for AD pathology were not assessed in this study. A biological definition based on neuroimaging with positron emission tomography and cerebrospinal fluid biomarkers is necessary to achieve a precise diagnosis [57,58]. Fourth, SMBG was used evaluate glucose levels, but continuous glucose monitoring (CGM) might more accurately represent the glucose profiles [59]. However, CGM in patients with cognitive impairment is difficult. Because we measured glucose levels on 8 separate days during the two-month period, our data could reflect the characteristics of glucose profiles over this period. Finally, we did not use elaborate equipment to evaluate muscle mass and walking speed. It is suitable to measure muscle mass using bioimpedance analysis [22]. Regarding walking speed, our evaluation was based on self-reported answers related to road crossing. To safely cross the road while the traffic signal is green, a pedestrian walking speed of 1.2 m/s is required [60]. This value is the same as the cut-off for slow walking speed in Japanese elderly individuals [22].

5. Conclusions

The prevalence of sarcopenia was high in diabetes patients with cognitive impairment. Glucose fluctuations were independently associated with sarcopenia, particular with low muscle mass, low grip strength and slow walking speed. Our study described the characteristic glucose profiles in older diabetes patients with sarcopenia. To maintain functional ability in older diabetes patients, confirmation of our data in future longitudinal studies with a larger cohort is required.

Author Contributions: Conceptualization, N.O. and T.S.; formal analysis, N.O.; investigation and data curation, T.S., S.K., T.T., H.T., S.S., H.M., A.S., and M.K. (Manabu Kokubo); resources, S.N.; writing-original draft preparation, N.O.; writing-review and editing T.S.; supervision, K.T., H.U., and M.K. (Masafumi Kuzuya); funding acquisition, N.O. and T.S.

Funding: This work was supported by a grant from the Research Funding for Longevity Sciences (27-21, 30-1, 30-42) from the National Center for Geriatrics and Gerontology (NCGG), and the Japan Agency for Medical Research and Development (18dk0207027h0003).

Acknowledgments: The authors thank the NCGG Biobank for performing quality control of the clinical data and blood samples. We are also grateful to Kaori Inaguma (Center for Comprehensive Care and Research on Memory Disorders, NCGG) for their assistance with the collection of the clinical data.

Conflicts of Interest: The authors declare no conflict of interest.

Appendix A

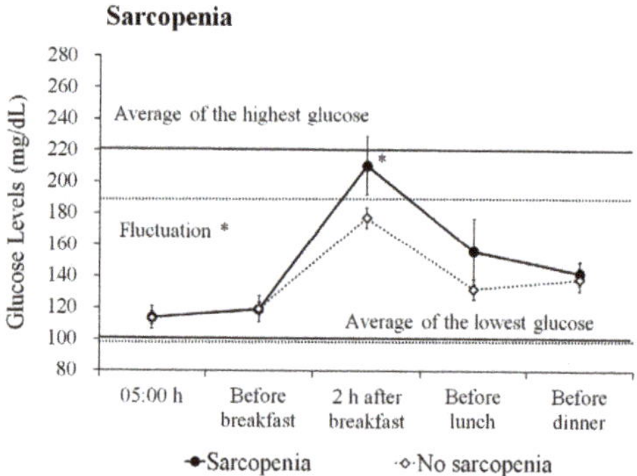

Figure A1. Daily glucose profiles based on sarcopenia in the cognitive impairment group. The figure shows the average ± standard error (SE) glucose level. Average of the highest glucose level: the average of the maximum glucose level of the day during the measurement period. Average of the lowest glucose level: the average of the minimum glucose level of the day during the measurement period. The solid line represents the sarcopenia group, and the dotted line indicates the no sarcopenia group. Fluctuation: the average of the diurnal range from the minimum glucose level to the maximum glucose level. * $p < 0.05$.

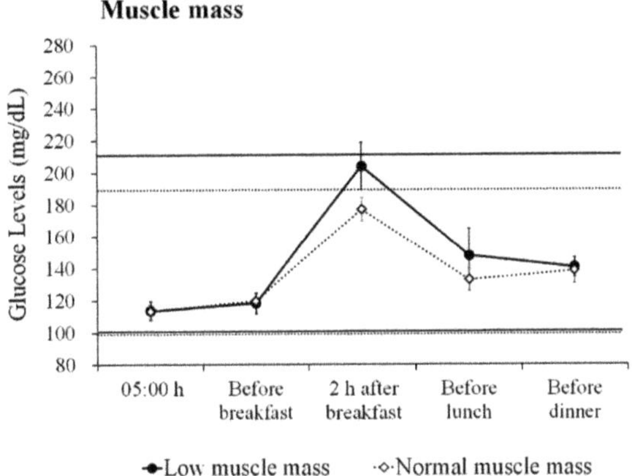

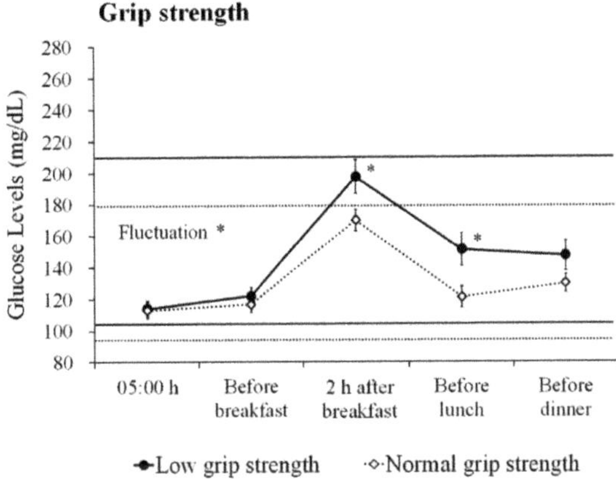

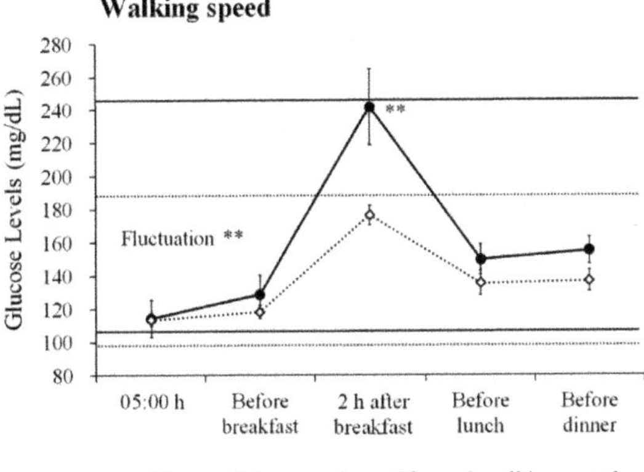

Figure A2. Daily glucose profiles based on the sarcopenia components in the cognitive impairment group. The figures show the average ± SE glucose level based on the components of sarcopenia. ** $p < 0.01$, * $p < 0.05$.

References

1. Cruz-Jentoft, A.J.; Bahat, G.; Bauer, J.; Boirie, Y.; Bruyere, O.; Cederholm, T.; Cooper, C.; Landi, F.; Rolland, Y.; Sayer, A.A.; et al. Sarcopenia: revised European consensus on definition and diagnosis. *Age Ageing* **2018**, *48*, 16–31. [CrossRef] [PubMed]
2. Han, A.; Bokshan, S.L.; Marcaccio, S.E.; DePasse, J.M.; Daniels, A.H. Diagnostic Criteria and Clinical Outcomes in Sarcopenia Research: A Literature Review. *J. Clin. Med.* **2018**, *7*, E70. [CrossRef] [PubMed]
3. Morley, J.E.; Malmstrom, T.K.; Rodriguez-Manas, L.; Sinclair, A.J. Frailty, sarcopenia and diabetes. *J. Am. Med. Dir. Assoc.* **2014**, *15*, 853–859. [CrossRef] [PubMed]
4. Sinclair, A.J.; Abdelhafiz, A.H.; Rodriguez-Manas, L. Frailty and sarcopenia-newly emerging and high impact complications of diabetes. *J. Diabetes Complicat.* **2017**, *31*, 1465–1473. [CrossRef] [PubMed]
5. Yoon, J.W.; Ha, Y.C.; Kim, K.M.; Moon, J.H.; Choi, S.H.; Lim, S.; Park, Y.J.; Lim, J.Y.; Kim, K.W.; Park, K.S.; et al. Hyperglycemia Is Associated with Impaired Muscle Quality in Older Men with Diabetes: The Korean Longitudinal Study on Health and Aging. *Diabetes Metab. J.* **2016**, *40*, 140–146. [CrossRef] [PubMed]
6. Leenders, M.; Verdijk, L.B.; van der Hoeven, L.; Adam, J.J.; van Kranenburg, J.; Nilwik, R.; van Loon, L.J. Patients with type 2 diabetes show a greater decline in muscle mass, muscle strength, and functional capacity with aging. *J. Am. Med. Dir. Assoc.* **2013**, *14*, 585–592. [CrossRef] [PubMed]
7. Kalyani, R.R.; Metter, E.J.; Egan, J.; Golden, S.H.; Ferrucci, L. Hyperglycemia predicts persistently lower muscle strength with aging. *Diabetes Care* **2015**, *38*, 82–90. [CrossRef] [PubMed]
8. Munshi, M.N.; Segal, A.R.; Slyne, C.; Samur, A.A.; Brooks, K.M.; Horton, E.S. Shortfalls of the use of HbA1C-derived eAG in older adults with diabetes. *Diabetes Res. Clin. Pract.* **2015**, *110*, 60–65. [CrossRef] [PubMed]
9. Munshi, M.N.; Slyne, C.; Segal, A.R.; Saul, N.; Lyons, C.; Weinger, K. Liberating A1C goals in older adults may not protect against the risk of hypoglycemia. *J. Diabetes Complicat.* **2017**, *31*, 1197–1199. [CrossRef] [PubMed]
10. Abdelhafiz, A.H.; McNicholas, E.; Sinclair, A.J. Hypoglycemia, frailty and dementia in older people with diabetes: Reciprocal relations and clinical implications. *J. Diabetes Complicat.* **2016**, *30*, 1548–1554. [CrossRef] [PubMed]
11. Burns, J.M.; Johnson, D.K.; Watts, A.; Swerdlow, R.H.; Brooks, W.M. Reduced lean mass in early Alzheimer disease and its association with brain atrophy. *Arch. Neurol.* **2010**, *67*, 428–433. [CrossRef] [PubMed]
12. Fujisawa, C.; Umegaki, H.; Okamoto, K.; Nakashima, H.; Kuzuya, M.; Toba, K.; Sakurai, T. Physical Function Differences Between the Stages From Normal Cognition to Moderate Alzheimer Disease. *J. Am. Med. Dir. Assoc.* **2017**, *18*, e9–e368. [CrossRef] [PubMed]
13. Sugimoto, T.; Nakamura, A.; Kato, T.; Iwata, K.; Saji, N.; Arahata, Y.; Hattori, H.; Bundo, M.; Ito, K.; Niida, S.; et al. Decreased Glucose Metabolism in Medial Prefrontal Areas is Associated with Nutritional Status in Patients with Prodromal and Early Alzheimer's Disease. *J. Alzheimers Dis.* **2017**, *60*, 225–233. [CrossRef] [PubMed]
14. Sugimoto, T.; Ono, R.; Murata, S.; Saji, N.; Matsui, Y.; Niida, S.; Toba, K.; Sakurai, T. Prevalence and associated factors of sarcopenia in elderly subjects with amnestic mild cognitive impairment or Alzheimer disease. *Curr. Alzheimer Res.* **2016**, *13*, 718–726. [CrossRef] [PubMed]
15. McKhann, G.M.; Knopman, D.S.; Chertkow, H.; Hyman, B.T.; Jack, C.R., Jr.; Kawas, C.H.; Klunk, W.E.; Koroshetz, W.J.; Manly, J.J.; Mayeux, R.; et al. The diagnosis of dementia due to Alzheimer's disease: recommendations from the National Institute on Aging-Alzheimer's Association workgroups on diagnostic guidelines for Alzheimer's disease. *Alzheimers Dement.* **2011**, *7*, 263–269. [CrossRef] [PubMed]
16. Petersen, R.C.; Doody, R.; Kurz, A.; Mohs, R.C.; Morris, J.C.; Rabins, P.V.; Ritchie, K.; Rossor, M.; Thal, L.; Winblad, B. Current concepts in mild cognitive impairment. *Arch. Neurol.* **2001**, *58*, 1985–1992. [CrossRef] [PubMed]
17. Gnjidic, D.; Hilmer, S.N.; Blyth, F.M.; Naganathan, V.; Waite, L.; Seibel, M.J.; McLachlan, A.J.; Cumming, R.G.; Handelsman, D.J.; Le Couteur, D.G. Polypharmacy cutoff and outcomes: five or more medicines were used to identify community-dwelling older men at risk of different adverse outcomes. *J. Clin. Epidemiol.* **2012**, *65*, 989–995. [CrossRef] [PubMed]
18. Folstein, M.F.; Folstein, S.E.; McHugh, P.R. "Mini-mental state". A practical method for grading the cognitive state of patients for the clinician. *J. Psychiatr. Res.* **1975**, *12*, 189–198. [CrossRef]

19. Sakurai, T.; Iimuro, S.; Sakamaki, K.; Umegaki, H.; Araki, A.; Ohashi, Y.; Ito, H.; Japanese Elderly Diabetes Intervention Trial Study Group. Risk factors for a 6-year decline in physical disability and functional limitations among elderly people with type 2 diabetes in the Japanese Elderly Diabetes Intervention Trial. *Geriatr. Gerontol. Int.* **2012**, *12*, 117–126. [CrossRef] [PubMed]

20. Monnier, L.; Colette, C.; Dunseath, G.J.; Owens, D.R. The loss of postprandial glycemic control precedes stepwise deterioration of fasting with worsening diabetes. *Diabetes Care* **2007**, *30*, 263–269. [CrossRef] [PubMed]

21. Workgroup on Hypoglycemia, American Diabetes Association. Defining and reporting hypoglycemia in diabetes: a report from the American Diabetes Association Workgroup on Hypoglycemia. *Diabetes Care* **2005**, *28*, 1245–1249. [CrossRef]

22. Chen, L.K.; Liu, L.K.; Woo, J.; Assantachai, P.; Auyeung, T.W.; Bahyah, K.S.; Chou, M.Y.; Chen, L.Y.; Hsu, P.S.; Krairit, O.; et al. Sarcopenia in Asia: consensus report of the Asian Working Group for Sarcopenia. *J. Am. Med. Dir. Assoc.* **2014**, *15*, 95–101. [CrossRef] [PubMed]

23. Pagotto, V.; Silveira, E.A. Methods, diagnostic criteria, cutoff points, and prevalence of sarcopenia among older people. *ScientificWorldJournal* **2014**, *2014*, 231312. [CrossRef] [PubMed]

24. Kawakami, R.; Murakami, H.; Sanada, K.; Tanaka, N.; Sawada, S.S.; Tabata, I.; Higuchi, M.; Miyachi, M. Calf circumference as a surrogate marker of muscle mass for diagnosing sarcopenia in Japanese men and women. *Geriatr. Gerontol. Int.* **2015**, *15*, 969–976. [CrossRef] [PubMed]

25. Okochi, J.; Toba, K.; Takahashi, T.; Matsubayashi, K.; Nishinaga, M.; Takahashi, R.; Ohrui, T. Simple screening test for risk of falls in the elderly. *Geriatr. Gerontol. Int.* **2006**, *6*, 223–227. [CrossRef]

26. Abbatecola, A.M.; Paolisso, G.; Fattoretti, P.; Evans, W.J.; Fiore, V.; Dicioccio, L.; Lattanzio, F. Discovering pathways of sarcopenia in older adults: a role for insulin resistance on mitochondria dysfunction. *J. Nutr. Health Aging* **2011**, *15*, 890–895. [CrossRef] [PubMed]

27. Ferreira, S.T.; Clarke, J.R.; Bomfim, T.R.; De Felice, F.G. Inflammation, defective insulin signaling, and neuronal dysfunction in Alzheimer's disease. *Alzheimers Dement* **2014**, *10*, S76–S83. [CrossRef] [PubMed]

28. Ogawa, S.; Yakabe, M.; Akishita, M. Age-related sarcopenia and its pathophysiological bases. *Inflamm. Regen.* **2016**, *36*, 17. [CrossRef] [PubMed]

29. Asih, P.R.; Tegg, M.L.; Sohrabi, H.; Carruthers, M.; Gandy, S.E.; Saad, F.; Verdile, G.; Ittner, L.M.; Martins, R.N. Multiple Mechanisms Linking Type 2 Diabetes and Alzheimer's Disease: Testosterone as a Modifier. *J. Alzheimers Dis.* **2017**, *59*, 445–466. [CrossRef] [PubMed]

30. Buchman, A.S.; Yu, L.; Wilson, R.S.; Boyle, P.A.; Schneider, J.A.; Bennett, D.A. Brain pathology contributes to simultaneous change in physical frailty and cognition in old age. *J. Gerontol. A Biol. Sci. Med. Sci.* **2014**, *69*, 1536–1544. [CrossRef] [PubMed]

31. Del Campo, N.; Payoux, P.; Djilali, A.; Delrieu, J.; Hoogendijk, E.O.; Rolland, Y.; Cesari, M.; Weiner, M.W.; Andrieu, S.; Vellas, B.; et al. Relationship of regional brain β-amyloid to gait speed. *Neurology* **2016**, *86*, 36–43. [CrossRef] [PubMed]

32. Sato, N.; Morishita, R. The roles of lipid and glucose metabolism in modulation of β-amyloid, tau, and neurodegeneration in the pathogenesis of Alzheimer disease. *Front. Aging Neurosci.* **2015**, *7*, 199. [CrossRef] [PubMed]

33. Vekrellis, K.; Ye, Z.; Qiu, W.Q.; Walsh, D.; Hartley, D.; Chesneau, V.; Rosner, M.R.; Selkoe, D.J. Neurons regulate extracellular levels of amyloid beta-protein via proteolysis by insulin-degrading enzyme. *J. Neurosci.* **2000**, *20*, 1657–1665. [CrossRef] [PubMed]

34. Macauley, S.L.; Stanley, M.; Caesar, E.E.; Yamada, S.A.; Raichle, M.E.; Perez, R.; Mahan, T.E.; Sutphen, C.L.; Holtzman, D.M. Hyperglycemia modulates extracellular amyloid-β concentrations and neuronal activity in vivo. *J. Clin. Investig.* **2015**, *125*, 2463–2467. [CrossRef] [PubMed]

35. Roher, A.E.; Esh, C.L.; Kokjohn, T.A.; Castano, E.M.; Van Vickle, G.D.; Kalback, W.M.; Patton, R.L.; Luehrs, D.C.; Daugs, I.D.; Kuo, Y.M.; et al. Amyloid beta peptides in human plasma and tissues and their significance for Alzheimer's disease. *Alzheimers Dement* **2009**, *5*, 18–29. [CrossRef] [PubMed]

36. Shinohara, M.; Sato, N. Bidirectional interactions between diabetes and Alzheimer's disease. *Neurochem. Int.* **2017**, *108*, 296–302. [CrossRef] [PubMed]

37. Monnier, L.; Mas, E.; Ginet, C.; Michel, F.; Villon, L.; Cristol, J.P.; Colette, C. Activation of oxidative stress by acute glucose fluctuations compared with sustained chronic hyperglycemia in patients with type 2 diabetes. *JAMA* **2006**, *295*, 1681–1687. [CrossRef] [PubMed]

38. Smith-Palmer, J.; Brandle, M.; Trevisan, R.; Orsini Federici, M.; Liabat, S.; Valentine, W. Assessment of the association between glycemic variability and diabetes-related complications in type 1 and type 2 diabetes. *Diabetes Res. Clin. Pract.* **2014**, *105*, 273–284. [CrossRef] [PubMed]

39. Gella, A.; Durany, N. Oxidative stress in Alzheimer disease. *Cell Adh. Migr.* **2009**, *3*, 88–93. [CrossRef] [PubMed]

40. Ogama, N.; Sakurai, T.; Kawashima, S.; Tanikawa, T.; Tokuda, H.; Satake, S.; Miura, H.; Shimizu, A.; Kokubo, M.; Niida, S.; et al. Postprandial Hyperglycemia is Associated with White Matter Hyperintensity and Brain Atrophy in Older Patients with Type 2 Diabetes Mellitus. *Front. Aging Neurosci..* **2018**, *10*, 273. [CrossRef] [PubMed]

41. Ogama, N.; Sakurai, T.; Shimizu, A.; Toba, K. Regional white matter lesions predict falls in patients with amnestic mild cognitive impairment and Alzheimer's disease. *J. Am. Med. Dir. Assoc.* **2014**, *15*, 36–41. [CrossRef] [PubMed]

42. Ogama, N.; Sakurai, T.; Nakai, T.; Niida, S.; Saji, N.; Toba, K.; Umegaki, H.; Kuzuya, M. Impact of frontal white matter hyperintensity on instrumental activities of daily living in elderly women with Alzheimer disease and amnestic mild cognitive impairment. *PLoS ONE* **2017**, *12*, e0172484. [CrossRef] [PubMed]

43. Hirabayashi, N.; Hata, J.; Ohara, T.; Mukai, N.; Nagata, M.; Shibata, M.; Gotoh, S.; Furuta, Y.; Yamashita, F.; Yoshihara, K.; et al. Association Between Diabetes and Hippocampal Atrophy in Elderly Japanese: The Hisayama Study. *Diabetes Care* **2016**, *39*, 1543–1549. [CrossRef] [PubMed]

44. Jackson, A.S.; Janssen, I.; Sui, X.; Church, T.S.; Blair, S.N. Longitudinal changes in body composition associated with healthy ageing: men, aged 20-96 years. *Br. J. Nutr.* **2012**, *107*, 1085–1091. [CrossRef] [PubMed]

45. Kim, K.S.; Park, K.S.; Kim, M.J.; Kim, S.K.; Cho, Y.W.; Park, S.W. Type 2 diabetes is associated with low muscle mass in older adults. *Geriatr. Gerontol. Int.* **2014**, *14*, 115–121. [CrossRef] [PubMed]

46. Kim, K.; Park, S.M. Association of muscle mass and fat mass with insulin resistance and the prevalence of metabolic syndrome in Korean adults: a cross-sectional study. *Sci. Rep.* **2018**, *8*, 2703. [CrossRef] [PubMed]

47. Scott, D.; de Courten, B.; Ebeling, P.R. Sarcopenia: a potential cause and consequence of type 2 diabetes in Australia's ageing population? *Med. J. Aust.* **2016**, *205*, 329–333. [CrossRef] [PubMed]

48. Fried, L.P.; Tangen, C.M.; Walston, J.; Newman, A.B.; Hirsch, C.; Gottdiener, J.; Seeman, T.; Tracy, R.; Kop, W.J.; Burke, G.; et al. Frailty in older adults: evidence for a phenotype. *J. Gerontol. A Biol. Sci. Med. Sci.* **2001**, *56*, M146–M156. [CrossRef] [PubMed]

49. Koch, G.; Belli, L.; Giudice, T.L.; Lorenzo, F.D.; Sancesario, G.M.; Sorge, R.; Bernardini, S.; Martorana, A. Frailty among Alzheimer's disease patients. *CNS Neurol. Disord. Drug Targets* **2013**, *12*, 507–511. [CrossRef] [PubMed]

50. Robertson, D.A.; Savva, G.M.; Kenny, R.A. Frailty and cognitive impairment–a review of the evidence and causal mechanisms. *Ageing Res. Rev.* **2013**, *12*, 840–851. [CrossRef] [PubMed]

51. Wallace, L.; Theou, O.; Rockwood, K.; Andrew, M.K. Relationship between frailty and Alzheimer's disease biomarkers: A scoping review. *Alzheimers Dement.* **2018**, *10*, 394–401. [CrossRef] [PubMed]

52. Walston, J.; Buta, B.; Xue, Q.L. Frailty Screening and Interventions: Considerations for Clinical Practice. *Clin. Geriatr. Med.* **2018**, *34*, 25–38. [CrossRef] [PubMed]

53. Langlois, F.; Vu, T.T.; Chassé, K.; Dupuis, G.; Kergoat, M.J.; Bherer, L. Benefits of physical exercise training on cognition and quality of life in frail older adults. *J. Gerontol. B. Psychol. Sci. Soc. Sci.* **2013**, *68*, 400–404. [CrossRef] [PubMed]

54. Bucci, M.; Huovinen, V.; Guzzardi, M.A.; Koskinen, S.; Raiko, J.R.; Lipponen, H.; Ahsan, S.; Badeau, R.M.; Honka, M.J.; Koffert, J.; et al. Resistance training improves skeletal muscle insulin sensitivity in elderly offspring of overweight and obese mothers. *Diabetologia* **2016**, *59*, 77–86. [CrossRef] [PubMed]

55. Lee, J.; Kim, D.; Kim, C. Resistance Training for Glycemic Control, Muscular Strength, and Lean Body Mass in Old Type 2 Diabetic Patients: A Meta-Analysis. *Diabetes Ther.* **2017**, *8*, 459–473. [CrossRef] [PubMed]

56. Kalyani, R.R.; Varadhan, R.; Weiss, C.O.; Fried, L.P.; Cappola, A.R. Frailty status and altered glucose-insulin dynamics. *J. Gerontol. A Biol. Sci. Med. Sci.* **2012**, *67*, 1300–1306. [CrossRef] [PubMed]

57. Dubois, B.; Feldman, H.H.; Jacova, C.; Hampel, H.; Molinuevo, J.L.; Blennow, K.; DeKosky, S.T.; Gauthier, S.; Selkoe, D.; Bateman, R.; et al. Advancing research diagnostic criteria for Alzheimer's disease: the IWG-2 criteria. *Lancet Neurol.* **2014**, *13*, 614–629. [CrossRef]

58. Jack, C.R., Jr.; Bennett, D.A.; Blennow, K.; Carrillo, M.C.; Dunn, B.; Haeberlein, S.B.; Holtzman, D.M.; Jagust, W.; Jessen, F.; Karlawish, J.; et al. NIA-AA Research Framework: Toward a biological definition of Alzheimer's disease. *Alzheimers Dement.* **2018**, *14*, 535–562. [CrossRef] [PubMed]

59. Bergenstal, R.M. Glycemic Variability and Diabetes Complications: Does It Matter? Simply Put, There Are Better Glycemic Markers! *Diabetes Care* **2015**, *38*, 1615–1621. [CrossRef] [PubMed]

60. Donoghue, O.A.; Dooley, C.; Kenny, R.A. Usual and Dual-Task Walking Speed: Implications for Pedestrians Crossing the Road. *J. Aging Health* **2016**, *28*, 850–862. [CrossRef] [PubMed]

Journal of
Clinical Medicine

Article

Usefulness of Eye Fixation Assessment for Identifying Type 2 Diabetic Subjects at Risk of Dementia

Olga Simó-Servat [1,2,3], Andreea Ciudin [1,2,3], Ángel M. Ortiz-Zúñiga [1,2], Cristina Hernández [1,2,3] and Rafael Simó [1,2,3,*]

[1] Diabetes and Metabolism Research Unit. Vall d'Hebron Research Institute, 08035 Barcelona, Spain;
    olga.simo@vhir.org (O.S.-S.); aciudin@vhebron.net (A.C.); angelmichaelortizzuniga@gmail.com (A.M.O.-Z.);
    cristina.hernandez@vhir.org (C.H.)
[2] Department of Medicine, Universitat Autònoma de Barcelona, 08193 Barcelona, Spain
[3] Centro de Investigación Biomédica en Red de Diabetes y Enfermedades Metabólicas
    Asociadas (CIBERDEM), Instituto de Salud Carlos III (ICSIII), 28029 Madrid, Spain
*   Correspondence: rafael.simo@vhir.org; Tel.: +34-934-894-172

Received: 15 October 2018; Accepted: 1 January 2019; Published: 8 January 2019

Abstract: Type 2 diabetic (T2D) subjects have a significantly higher risk of developing mild cognitive impairment (MCI) and dementia than age-matched non-diabetic individuals. However, the accurate evaluation of cognitive status is based on complex neuropsychological tests, which makes their incorporation into the current standard of care for the T2D population infeasible. Given that the ability to maintain visual gaze on a single location (fixation) is hampered in Alzheimer's disease (AD), the aim of the present study was: (1) To assess whether the evaluation of gaze fixation during fundus-driven microperimetry correlated with cognitive status in T2D subjects; (2) to examine whether the addition of fixational parameters to the assessment of retinal sensitivity increased the predictive value of retinal microperimetry in identifying T2D subjects with MCI. For this purpose, fixation parameters and retinal sensitivity were compared in three age-matched groups of T2D subjects: normocognitive (n = 34), MCI (n = 33), and AD (n = 33). Our results showed that fixation is significantly more unstable in MCI subjects than normocognitive subjects, and even more altered in those affected by AD (ANOVA; $p < 0.01$). Moreover, adding fixation parameters to retinal sensitivity significantly increases the predictive value in identifying those subjects with MCI: ROC (Receiver Operating Characteristic) Area 0.68 with retinal sensitivity alone vs. ROC Area 0.86 when parameters of fixation are added to retinal sensitivity ($p < 0.01$). In conclusion, our results suggest that fixational eye movement parameters assessed by fundus-microperimetry represent a new tool for identifying T2D subjects at risk of dementia.

Keywords: diabetes; dementia; microperimetry

---

1. Introduction

The American Diabetes Association recommends individualizing diabetes treatment by taking into account the cognitive capacity of patients [1]. However, the accurate evaluation of cognitive status is based on complex neuropsychological tests [2], which makes their incorporation into current standards of care for the type 2 diabetic (T2D) population unfeasible. This is an important gap that affects clinical practice because T2D patients have a significantly higher risk of developing Alzheimer's disease (AD) in comparison with age-matched non-diabetic subjects even after adjusting for other risk factors [3–5]. These findings have been recently confirmed in a large nationwide population-based study performed in Taiwan with a follow-up of 10 years, which clearly shows that the risk of developing

AD in the diabetic population is significantly higher than in the non-diabetic population with a hazard ratio (HR) of 1.7 [6]. Similarly, other longitudinal population-based studies in T2D subjects have shown an association with either an accelerated cognitive decline or an increased incidence of dementia [7]. Because of the increase of diabetes prevalence and the concomitant aging of populations, severe cognitive impairment can be considered as an emerging long-time new "diabetic complication" with dramatic consequences for the affected subjects and their families and with a significant impact on healthcare systems. Mild cognitive impairment (MCI), which is also increased in diabetic patients, consists of cognitive impairment on standard tests but no impairment in activities of daily living and represents a transition state between normal cognitive function and dementia. The annual conversion rate from MCI to dementia ranges between 10–30% in the general population [8–10], and we have recently reported that T2D accelerates the conversion to dementia in patients with MCI [11]. Therefore, strategies aimed at identifying T2D patients at risk of dementia are urgently needed. In this regard, it has been estimated that interventions that delay the clinical onset of dementia by 1 year would reduce its prevalence in 2050 by 9 million cases [12].

It could be argued that the benefits of the early diagnosis of dementia are questionable because treatment is still unavailable. However, unrecognized cognitive dysfunction can affect treatment adherence and diabetes self-management resulting in poor glycemic control, an increased frequency of severe hypoglycemic episodes, diabetes-related complications, and hospital admissions [13,14]. In addition, the identification of those T2D subjects affected with MCI would allow the implementation of a more personalized treatment. Hypoglycemic events increase the risk of dementia, therefore treatments that do not increase this risk should be prioritized [15–17]. In addition, T2D patients with cognitive impairment would be more prone to present an impaired self-management of diabetes. Thus, the treatment of diabetes should be simplified in these patients in order to increase treatment adherence.

Interestingly, both AD and diabetic complications share common pathogenic pathways. The impairment of insulin signaling, the presence of low-grade inflammation, and the pathways directly related to chronic hyperglycemia, such as the accumulation of advanced glycation end-products (AGEs) and the increase in oxidative stress play an essential role in the pathogenesis of AD [18]. All these pathways have been implicated in diabetic complications and in particular with diabetic retinopathy (DR). Moreover, several pathways involved in neurodegenerative diseases have been identified in human retinas with early stages of DR [19]. These findings indicate that a common etiology exists between the brain and retinal neurodegeneration in the setting of diabetes, which could explain the increased risk of cognitive impairment that T2D subjects present.

Current strategies to screen for dementia are based on neuropsychological tests such as the Mini-Mental State Examination (MMSE), Montreal Cognitive Assessment (MOCA), and the Digit Symbol Substitution Test (DSST) [20]. Both MMSE and MOCA are time-consuming (10 min each), and DSST, which takes only 2 min, has low specificity [20,21]. Additionally, although these tests have an important role as screening tools, they could be influenced by anxiety or depression which occurs in a significant proportion of ageing T2D subjects [13]. Furthermore, when MCI is suspected, a complete neuropsychological battery is needed, which is cumbersome and time consuming (at least 45 min) [22]. Given the high prevalence of T2D in the older population [23], the detection of MCI subjects based on this battery is unfeasible for daily practice.

In clinical practice, there are no reported phenotypic indicators or reliable examinations for identifying T2D patients at risk of developing dementia. In recent years, growing evidence has shown that retinal neurodegeneration is an early event in the pathogenesis of diabetic retinopathy (DR) [24,25]. The retina is ontogenically a brain-derived tissue and it has been suggested that it may provide an easily accessible and non-invasive way of examining the pathology of the brain [25,26]. Therefore, it seems reasonable to propose that the evaluation of retinal parameters related to neurodegeneration could be useful in identifying those T2D patients at a higher risk of developing cognitive impairment and dementia. In this regard, we have recently demonstrated that retinal sensitivity assessed by microperimetry significantly correlated with neuroimaging parameters and cognitive status [27].

Apart from retinal sensitivity, microperimetry permits us to examine the capacity to maintain visual gaze on a single location (fixation), which has been reported as being hampered in AD and other neurodegenerative diseases [28–32].

Abnormalities in oculomotor function and gaze fixation have been previously reported in subjects affected with AD. In fact, several eye movements like pro-saccades, anti-saccades, and the micro-saccades and saccadic intrusions that occur during fixation are altered in subjects with AD [28–30]. In addition, during fixation, subjects affected with AD present a greater frequency of saccadic intrusions than normal subjects, causing instability in gaze maintenance [29]. Moreover, micro-saccades that occur during normal fixation in AD are notably more oblique compared with normal adults [30]. However, little is known regarding the potential usefulness of the assessment of gaze fixation for identifying T2D patients at risk of AD. This is an important issue because the neural circuits involved in gaze fixation are not the same as those participating in retinal sensitivity and, therefore, complementary information could be obtained.

On this basis, we hypothesize that the assessment of retinal fixation by means of fundus-driven microperimetry could be useful to identify those diabetic subjects with cognitive impairment. For the first time we have analyzed data regarding fixational eye movement parameters assessed by microperimetry with the following aims: (1) To assess whether gaze fixation abnormalities in T2D diabetic subjects are related to cognitive status and, therefore, could be used as a biomarker for identifying T2D subjects at risk of dementia, and (2) to determine whether fixation study is able to increase the diagnostic reliability of retinal sensitivity in identifying subjects at risk of cognitive impairment. Since in diabetic subjects with advanced diabetic retinopathy, both sensitivity and fixation could be altered [33], those subjects with moderate non-proliferative or more advanced stages of diabetic retinopathy were excluded from the study.

2. Materials and Methods

Data regarding fixation parameters obtained from the nested case-control study DIALRET (ClinicalTrial.gov: NCT02360527) were analyzed. A total of 33 patients with AD, 33 with MCI, and 34 with normal cognition in whom a full report of fixation parameters was available were included in the study. All these patients were selected from 214 consecutive type 2 diabetic patients attending a memory clinic (Fundació ACE, Barcelona, Spain). The study was conducted according to the Declaration of Helsinki and was approved by the local Ethics Committee.

The main inclusion criteria were: (a) Age >65 years; (b) type 2 diabetes with a duration >5 years; (c) written informed consent which included accepting participation in brain MRI (Magnetic Resonance Imaging) and PET (Positron Emission Tomography), as well as a potential lumbar puncture.

The main exclusion criteria were: (a) Patients with other neurodegenerative diseases of the brain or retina (i.e., glaucoma) or cerebrovascular diseases (Fazekas scale score $\leq$1) [34]; (b) HbA1c >10% (86 mmol/mol). The exclusion of patients with poor control was because very high blood glucose levels could affect retinal function [35]. (c) Moderate non-proliferative diabetic retinopathy (DR) or more advanced stages of DR according to the International Clinical Diabetic Retinopathy Disease Severity Scale [36]. The exclusion of patients with more advanced DR was based on the fact that severe microvascular impairment could participate in neurodegeneration, and the aim of the study is to assess retinal neurodegeneration regardless of the presence of overt microangiopathy.

All patients underwent complete neuropsychological, neurological, and psychiatric evaluations as previously described [22] including Mini-Mental State Examination (MMSE) and Alzheimer's Disease Assessment Scale-Cognition subscale (ADAS-Cog). Higher values on MMSE and lower values on ADAS-Cog evaluations indicate better cognition. A neuropsychologist and a neurologist from a memory clinic performed the cognitive assessment and the diagnosis of MCI and AD. All patients diagnosed with MCI fulfilled MCI Petersen's diagnosis criteria [8,37]. In addition, a biochemical analysis including HbA1c and a lipid profile was performed.

Retinal sensitivity was evaluated by fundus driven microperimetry (MAIA 3rd generation, Centervue®) after a previous papillary dilation of minimum 4mm. The standard MAIA test covers 10° diameter area with 37 measurement points and a red 1° radius circle was used as the fixation target.

For the evaluation of fixation, the MAIA microperimeter uses high-speed eye trackers (25 Hz). The parameters that evaluate fixation are assessed by tracking eye movements 25 times/Sec and by plotting the resulting distribution over the SLO (Scanning Laser Ophthalmoscope) image. Each movement is represented by a point in the distribution which will constitute the fixation pattern (Figure 1). All fixation positions during the examination are used by the instrument to calculate the fixation indexes P1 and P2, which represent the percentage of fixation points inside a circle of 2 and 4 degrees of diameter from the total of fixation points, respectively. Subjects were classified as having stable fixation if P1 was more than 75%, relatively unstable fixation if P1 was less than 75% and P2 more than 75%, and unstable fixation if both P1 and P2 were below 75%. Microperimetry also provides a more accurate estimation of the fixation pattern using the bivariate contour ellipse area (BCEA). It calculates, by using two orthogonal diameters that describe the extent of the fixation points (in degrees), the area and orientation of an ellipse encompassing a given proportion of the fixation points. Lower BCEA values define better fixation stability. We have selected the BCEA corresponding to 95% of the fixation points [38,39]. For retinal sensitivity and fixation, data corresponding to the right eye were used, which was the first eye explored in all subjects. The duration of the measurements varies from 1 min to 4 min approximately, depending on the subject.

The reliability index is assessed by measuring the number of stimuli reported as seen by the patient when a stimulus is projected onto the optic nerve head (blind spot). A 100% of reliability means that none of the stimulus projected onto the nerve head is reported as seen [39].

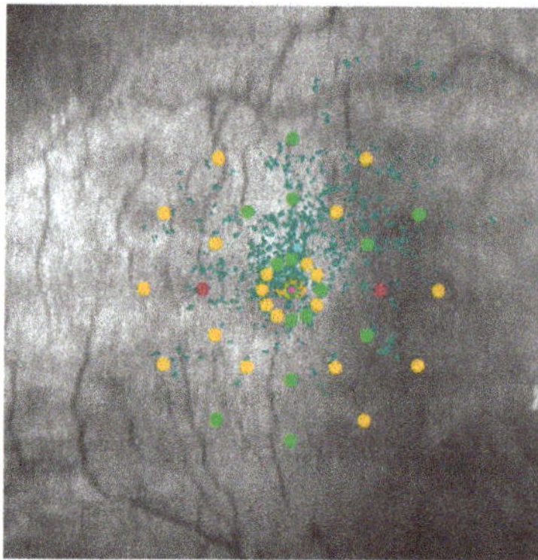

Figure 1. Microperimetry corresponding to a mild cognitive impairment (MCI) subject. The circles in green, yellow, and purple are the 37 measurement points in which retinal sensitivity is assessed. Different colors indicate different retinal sensitivity at each point (green being the best and purple the worst). The blue dots are the fixation points detected during the whole examination (3.5 min).

Statistical Analysis

The categorical variables are presented as percentages. For the quantitative variables, the mean and standard deviation (within the gaps) is displayed, except for BCEA95 in which the median and range are used. Statistical significance was accepted at $p < 0.05$. To assess differences between groups, the Chi-square test for qualitative variables and ANOVA followed by LSD (Least Significant Difference) post-hoc tests for quantitative variables were used. For BCEA95, which does not have a normal distribution (assessed by Shapiro Wilk test, cut-off $p < 0.05$), a non-parametric test was used to compare between groups (Kruskal-Wallis). For the significant differences observed, post hoc pairwise comparison analysis using the Duncan method was performed.

To evaluate the correlation between BCEA95 and ADAS-Cog or MMSE, Spearman's correlation test and regression analysis were performed. Significance was accepted at the level of $p < 0.05$. The Bonferroni correction was used for multiple comparisons.

Logistic regression analysis to predict MCI vs. normocognitive status was performed by using the following variables: Age, sex, retinal sensitivity, fixation indexes P1 and P2, and BCEA95. All possible equations were analysed and the two models that presented a highest AUC (Area Under the Curve) were the following: (1) Retinal sensitivity + P1; (2) retinal sensitivity + P1 + BCEA95. Since the purpose of the study was to evaluate fixation parameters, and as BCEA95 was one of the best fixation measurements, we selected the second model. Multicollinearity was assessed by calculating the Variance Inflation Factor (VIF). ROC curves and the Chi-squared test for ROC area comparison were performed. Statistical analyses were performed with the Stata statistical package.

3. Results

The main clinical characteristics of subjects with type 2 diabetes included in the study are summarized in Table 1. No differences were found among groups in terms of age, sex, hypertension, dyslipidaemia, BMI (Body Mass Index), diabetes duration, or HbA1c. The mean reliability index of the microperimetry test was above 90% in all groups and as expected, retinal sensitivity decreased in the group with MCI and especially in those with AD ($p < 0.01$). The time taken for the examination was longer in AD subjects or MCI ($p < 0.01$). This is because when retinal sensitivity is hampered, the examination takes a longer time since more stimuli and different intensities are presented to the patient. On the other hand, when retinal sensitivity is preserved, only the stimuli with the lowest intensity are presented and the examination takes less time.

Table 1. General characteristics and parameters of retinal sensitivity and fixation.

	AD $n = 33$	MCI $n = 33$	Normocognitive $n = 34$	p
Age (years)	79 (5.57)	77.03 (4.38)	75.53 (6.97)	0.06
Gender (males, %)	42.42	51.52	58.82	0.41
BMI	27.72 (3.79)	28.28 (4.48)	30.28 (4.45)	0.61
Hypertension (%)	78.79	75.76	73.53	0.88
Dyslipidemia (%)	63.64	66.67	71.43	0.81
Diabetes duration (years)	10.45 (6.23)	13.73 (10.64)	10.91 (6.12)	0.22
HbA1c (% of Hb, DCCT *)	6.79 (1.02)	6.99 (1.25)	7.5 (0.95)	0.37
Retinal sensitivity (dB)	16.95 (6.01)	21.05 (4.82)	23.39 (2.47)	<0.00001
Fixation P1 (%)	45.03 (24.89)	58.21 (26.66)	88.52 (12.19)	<0.00001
Fixation P2 (%)	70.2 (20.61)	81.03 (15.60)	96.09 (3.66)	<0.00001
BCEA95 (degrees) (median, range)	46.3 (12.5–187.7)	34.4 (1.2–155.1)	4.55 (0.3–23.5)	<0.00001
Reliability	92.83 (14.49)	97.45 (7.04)	96.59 (13.59)	0.28
Duration of the examination (minutes)	3.1 (1.14)	2.34 (0.83)	1.71 (0.42)	<0.00001

* DCCT: Diabetes Control and Complications Trial was the reference method to calculate the HbA1c.

Fixation was more unstable, in terms of P1 and P2 percentages, in the group with MCI and especially in the group with AD. Therefore, the lowest fixation capacity was found in patients with AD and the highest in patients with normocognition, the differences between the groups being statistically significant ($p < 0.00001$). To illustrate this point, representative examples of the fixation graph, which describes the amplitude of eye movements in degrees versus time, of each group (normocognition, MCI, and AD) are shown in Figure 2. In addition, we also found a significant difference among groups BCEA95 ($p < 0.00001$). In the pairwise comparison analysis, the differences were statistically significant among the three groups as shown in Table 2.

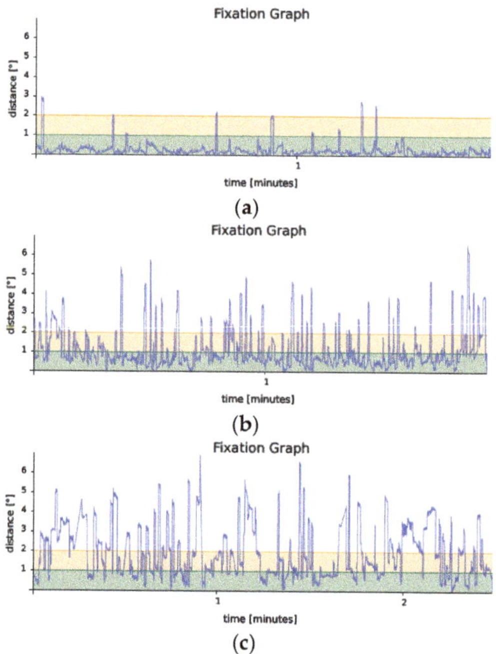

Figure 2. Representative examples of the fixation graph close to the median BCEA95 value of a normocognitive (**a**) (BCEA95 3.1°), MCI (**b**) (BCEA95 34.7°) and AD (**c**) (BCEA95 40.4°) subject. The green and the orange bands represent a circle of 1 and 2 degrees respectively.

Table 2. Pairwise comparison analysis (contrast and 95% confidence interval).

	Normocognitive vs. MCI	MCI vs. AD	Normocognitive vs. AD
Retinal Sensitivity	4.10 (1.78, 6.41)	2.33 (0.08, 4.59)	6.43 (4.00, 8.87)
Fixation P1	13.18 (2.10, 24.26)	30.30 (19.49, 41.11)	43.48 (31.83, 55.14)
Fixation P2	10.83 (3.39, 18.27)	15.06 (7.80, 22.32)	25.89 (18.06, 33.72)
BCEA95	−23.27 (−41.23, −5.31)	−35.11 (−52.93, −17.28)	−58.37 (−77.13, −39.61)

A significant negative correlation was found between BCEA95 and MMSE ($r = -0.5301$, $p < 0.00001$). In addition, BCEA95 also significantly correlated with the ADAS-Cog ($r = 0.4716$, $p < 0.00001$). In other words, a more unstable fixation correlated with lower punctuation in the MMSE test and a higher score in the ADAS-Cog, indicating that fixation instability is associated with a degree of cognitive impairment (Figure 3a,b).

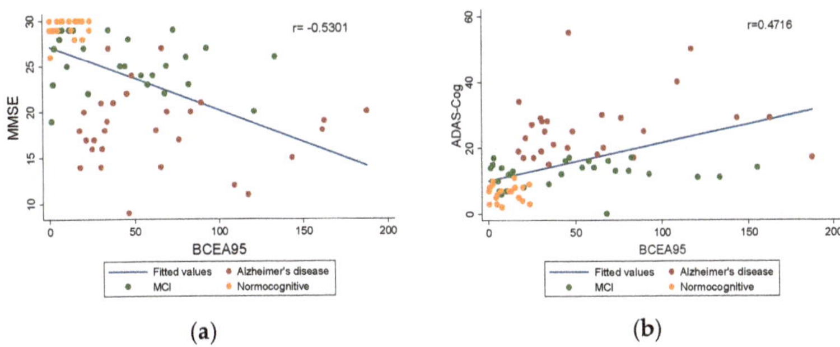

Figure 3. Linear regression graphs between: (**a**) Mini-mental State Examination (MMSE) and BCEA95; (**b**) Alzheimer's Disease Assessment Scale-Cognition subscale (ADAS-Cog) and BCEA95.

Retinal sensitivity, BCEA95, P1, and P2 were also correlated among them, and the coefficients (Spearman correlation) and their significance are displayed in Table 3. Interestingly, when analyzing the same correlation within the normocognitive and MCI groups (excluding those with AD), fixation P1 and P2 values no longer correlate with retinal sensitivity (Pearson correlation coefficients: 0.3199 and 0.311 respectively, *p* value > 0.05).

Table 3. Spearman correlation coefficients (and *p* values*) between retinal sensitivity and fixation parameters.

	Fixation P1	Fixation P2	BCEA95
Retinal Sensitivity	0.41 * <0.001	0.46 * <0.00001	−0.47 * <0.00001
Fixation P1	-	0.93 * <0.00001	−0.93 * <0.00001
Fixation P2	-	-	−0.96 * <0.00001

Finally, a comparison regarding the capacity to discriminate MCI subjects from normocognitive subjects by using retinal sensitivity alone or by adding to the predictive model the parameters of fixation BCEA95 and P1 was performed. The results are summarized in Table 4 and Figure 4. The addition of fixation assessment to the results obtained by measuring retinal sensitivity significantly increased the ROC area (*p* = 0.01), thus enhancing the capacity to discriminate between normocognitive and MCI diabetic subjects. The optimal cut-off based for maximum efficiency (the cut-off point that maximizes the percentage of correct classifications) was 0.494191 with a sensitivity of 72.7% (95% CI: 55.8 to 84.9%), specificity of 87.9% (95% CI: 72.7 to 95.2%), positive predictive value of 85.7% and a negative predictive value of 76.3%. The VIF values (<10) showed that multicollinearity was not present among the predictors (Table 5).

Table 4. ROC Area comparison between retinal sensitivity alone, and retinal sensitivity and fixation parameters (P1 and BCEA95). The outcome is mild cognitive impairment (MCI) (reference = normal cognition).

	ROC Area	Std. Err.	95% CI	*p*-Value
Retinal Sensitivity	0.68	0.07	0.55–0.81	<0.01
Retinal Sensitivity + BCEA95 + Fixation P1	0.86	0.05	0.77–0.95	

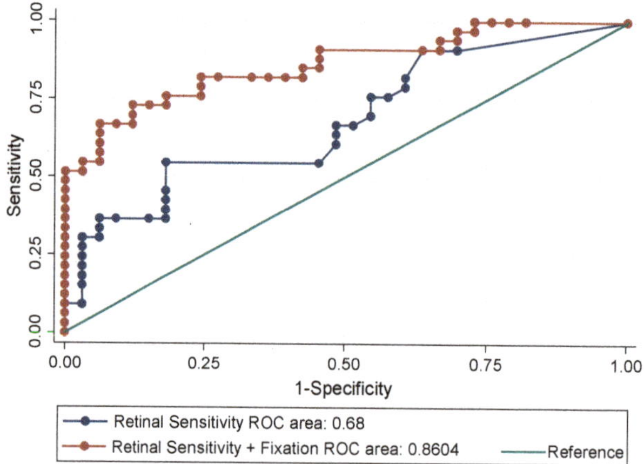

Figure 4. Comparison of two prediction models for MCI versus normal cognition, one based on retinal sensitivity and fixation parameters P1 and BCEA95, and one with retinal sensitivity alone.

Table 5. Multicollinearity assessment: Variance inflation factor (VIF) for each predictor.

	VIF	1/VIF
Retinal Sensitivity	1.15	0.87
BCEA95	4.45	0.22
Fixation P1	4.32	0.23

4. Discussion

Our findings indicate that fixational eye movement parameters assessed by microperimetry are useful for identifying T2D subjects with MCI. Moreover, the addition retinal fixation parameters, and more specifically P1 and BCEA95, to the assessment of retinal sensitivity significantly increase the capability to identify those subjects with MCI. Therefore, the retinal functional assessment of ocular movements as an objective method for identifying T2D patients with MCI opens up new feasible clinical management procedures addressed to testing cognitive status, which could easily be added to the routine ophthalmologic examination for the screening of DR.

Although there is evidence indicating that ocular movements are altered in AD, little is known about the usefulness of the study of ocular movements to identify the prodromal stages of dementia. Kapoula et al. [30] observed that MCI subjects present a major prevalence of oblique micro-saccades during fixation compared with control subjects. Lagun et al. [40] demonstrated that the incorporation of ocular movement information (i.e., duration of fixation, saccade length, and direction) was able to increase the specificity and sensitivity of the measurements of the visual-paired comparison task previously reported. Techniques used to evaluate ocular movements were based on electrodes placed in the region around the eyes and usually incorporated support to restrict head movements [29,41]. New eye-tracking systems have been developed and are usually video-based techniques, which has the advantage of being non-invasive [41–43]. One of their main limitations is head movement, because even small movements may cause large errors in the estimation of a calibrated tracker [42]. Fundus-driven microperimetry is a video-based technique with the advantage of incorporating a chin and forehead bars that restrict head movements. However, the exact distance between the optical stimuli and the eye could vary between subjects due to anatomical reasons and this could represent a limitation when retinal fixation is being assessed.

As previously mentioned, the presence of advanced DR could hamper the microperimetry results. In this subset of subjects, the screening of cognitive impairment should be based on other methods not based on retinal assessment (i.e., neuropsychological tests). However, it should be noted that the vast majority of T2D subjects present no DR or just mild non-proliferative DR. In the present study, given that the T2D patients included did not present signs of DR or other retinal disease, the alterations in fixation detected in those patients with cognitive impairment could be attributed to any neurodegenerative disease rather than an intrinsic eye disease. Marseglia et al. [44] showed that the presence of T2D affects perceptual speed, attention, and primary memory in patients of 70–75 years. They postulate that these domains might be primarily affected by diabetes. It should be noted that fixation eye movements are affected by attention and working memory [28]. For example, micro-saccade rates transiently decrease during an attentional task [45]. Therefore, the evaluation of fixation by fundus-driven microperimetry could be a biomarker of the first neuropsychological abnormalities that occur in diabetes-related cognitive decline. The omnipause neurons (located in the nucleus raphe interpositus of the paramedian pontine reticular formation) or the superior colliculus (SC) are two of the main brain areas involved in fixation. It is possible that these areas are affected early in the neurodegenerative process [46]. During microperimetry, the subjects are asked to maintain their gaze fixated on a central target, while stimuli with different intensities are presented at 37 points in 3 concentric circles of 2, 6, and 10 degrees of diameter [47]. It could be hypothesized that subjects with cognitive impairment are unable to inhibit the saccades triggered by the most eccentric stimuli and maintain their gaze fixated on a central target. Furthermore, certain areas of the cerebral cortex, such as the parietal and frontal cortex (frontal eye fields and the dorsomedial prefrontal cortex), show elevated firing rates during fixation. These areas could contribute to the control of fixation through descending projections to the SC and the omnipause region [46].

We observed statistically significant differences in P1, P2, and BCEA95 among the three groups. Moreover, a correlation was found between BCEA95 and either ADAS-Cog or MMSE punctuations. In addition, when BCEA95 and P1 were added to retinal sensitivity assessment, the capacity to detect MCI by using microperimetry increased significantly. In fact, for a cut-off point of 0.49 of the model that combines retinal sensitivity and fixation parameters (P1 and BCEA95), microperimetry has a sensitivity of 72.7% and a specificity of 87.9% in identifying the subjects with MCI. Furthermore, we did not find collinearity between parameters of retinal sensitivity and fixation. This would mean that this combined test is examining different neural circuits, thus improving the performance of the examination. In this regard, retinal sensitivity relies on the retina and the neural pathway of vision. The first station of the optic tract is the lateral geniculate body of the thalamus, which is the first major visual processing region in the brain and plays a crucial role in relaying information from the retinal ganglion cells to the primary visual cortex [48]. By contrast, the superior colliculus and the parietal and frontal cortex play an essential role in gaze fixation. There is evidence that both the geniculate body and the superior colliculus are affected by AD [49–51]. Therefore, gaze fixation and retinal sensitivity are measuring different neurological pathways and, consequently, can be considered complementary measurements of neurodegeneration, thus enhancing the reliability of the retinal sensitivity assessment alone.

The main limitation of our study is that we have included patients without DR or only those patients with mild non-proliferative DR. Since the presence of more advanced stages of DR, and in particular diabetic macular edema, can alter fixation [52], our results are not representative for the whole diabetic population. Furthermore, longitudinal studies examining retinal fixation in diabetic subjects with and without DR should be performed in order to fully address the utility of fixation assessment by microperimetry and guide its appropriate use.

In conclusion, our results indicate that the assessment of gaze fixation abnormalities in T2D is related to cognitive status and could be a useful tool for identifying patients at risk of developing dementia. In addition, the assessment of retinal sensitivity in combination with parameters of fixation by using microperimetry could be a reliable method for detecting prodromal stages of dementia in the T2D population.

Author Contributions: Conceptualization, O.S., A.C., C.H., R.S.; data curation, O.S., R.S.; methodology, O.S., A.C., A.O., C.H., R.S.; formal analysis, O.S., A.C., C.H.; investigation, O.S., A.C., A.O., C.H., R.S.; resources, R.S.; writing—original draft preparation, O.S.; writing—review and editing, O.S., A.C., A.O., C.H., R.S.; supervision, R.S.; project administration, C.H.; funding acquisition, A.C., C.H., R.S.

Funding: This research was funded by a grant from the European Foundation for the Study of Diabetes (EFSD/Lilly-Mental Health and Diabetes Programme) and Laboratorios Menarini.

Conflicts of Interest: The authors declare no conflict of interest. The funders had no role in the design of the study; in the collection, analyses, or interpretation of data; in the writing of the manuscript, or in the decision to publish the results.

References

1. American Diabetes Association. 6. Glycemic Targets. *Diabetes Care* **2017**, *40* (Suppl. 1), S48–S56. [CrossRef] [PubMed]

2. Albert, M.S.; DeKosky, S.T.; Dickson, D.; Dubois, B.; Feldman, H.H.; Fox, N.C.; Gamst, A.; Holtzman, D.M.; Jagust, W.J.; Petersen, R.C.; et al. The diagnosis of mild cognitive impairment due to Alzheimer's disease: Recommendations from the National Institute on Aging-Alzheimer's Association workgroups on diagnostic guidelines for Alzheimer's disease. *Alzheimers Dement.* **2011**, *7*, 270–279. [CrossRef] [PubMed]

3. Biessels, G.J.; Staekenborg, S.; Brunner, E.; Brayne, C.; Scheltens, P. Risk of dementia in diabetes mellitus: A systematic review. *Lancet Neurol.* **2006**, *5*, 64–74. [CrossRef]

4. Kopf, D.; Frölich, L. Risk of incident Alzheimer's disease in diabetic patients: A systematic review of prospective trials. *J. Alzheimers Dis.* **2009**, *16*, 677–685. [CrossRef] [PubMed]

5. Wang, K.-C.; Woung, L.-C.; Tsai, M.-T.; Liu, C.-C.; Su, Y.-H.; Li, C.-Y. Risk of Alzheimer's disease in relation to diabetes: A population-based cohort study. *Neuroepidemiology* **2012**, *38*, 237–244. [CrossRef]

6. Huang, C.-C.; Chung, C.-M.; Leu, H.-B.; Lin, L.-Y.; Chiu, C.-C.; Hsu, C.-Y.; Chiang, C.-H.; Huang, P.-H.; Chen, T.-J.; Lin, S.-J.; et al. Diabetes mellitus and the risk of Alzheimer's disease: A nationwide population-based study. *PLoS ONE* **2014**, *9*, e87095. [CrossRef] [PubMed]

7. Allen, K.V.; Frier, B.M.; Strachan, M.W.J. The relationship between type 2 diabetes and cognitive dysfunction: Longitudinal studies and their methodological limitations. *Eur. J. Pharmacol.* **2004**, *490*, 169–175. [CrossRef]

8. Petersen, R.C.; Morris, J.C. Mild cognitive impairment as a clinical entity and treatment target. *Arch. Neurol.* **2005**, *62*, 1160–1163, discussion 1167. [CrossRef]

9. Morris, J.C.; Storandt, M.; Miller, J.P.; McKeel, D.W.; Price, J.L.; Rubin, E.H.; Berg, L. Mild cognitive impairment represents early-stage Alzheimer disease. *Arch. Neurol.* **2001**, *58*, 397–405. [CrossRef]

10. Alegret, M.; Cuberas-Borrós, G.; Vinyes-Junqué, G.; Espinosa, A.; Valero, S.; Hernández, I.; Roca, I.; Ruíz, A.; Rosende-Roca, M.; Mauleón, A.; et al. A two-year follow-up of cognitive deficits and brain perfusion in mild cognitive impairment and mild Alzheimer's disease. *J. Alzheimers Dis.* **2012**, *30*, 109–120. [CrossRef]

11. Ciudin, A.; Espinosa, A.; Simó-Servat, O.; Ruiz, A.; Alegret, M.; Hernández, C.; Boada, M.; Simó, R. Type 2 diabetes is an independent risk factor for dementia conversion in patients with mild cognitive impairment. *J. Diabetes Complicat.* **2017**, *31*, 1272–1274. [CrossRef] [PubMed]

12. Hampel, H.; Frank, R.; Broich, K.; Teipel, S.J.; Katz, R.G.; Hardy, J.; Herholz, K.; Bokde, A.L.W.; Jessen, F.; Hoessler, Y.C.; et al. Biomarkers for Alzheimer's disease: Academic, industry and regulatory perspectives. *Nat. Rev. Drug Discov.* **2010**, *9*, 560–574. [CrossRef] [PubMed]

13. Simó, R.; Ciudin, A.; Simó-Servat, O.; Hernández, C. Cognitive impairment and dementia: A new emerging complication of type 2 diabetes-The diabetologist's perspective. *Acta Diabetol.* **2017**, *54*, 417–424. [CrossRef]

14. Xu, W.; Von Strauss, E.; Qiu, C.; Winblad, B.; Fratiglioni, L. Uncontrolled diabetes increases the risk of Alzheimer's disease: A population-based cohort study. *Diabetologia* **2009**, *52*, 1031–1039. [CrossRef] [PubMed]

15. Whitmer, R.A.; Karter, A.J.; Yaffe, K.; Quesenberry, C.P.; Selby, J.V. Hypoglycemic episodes and risk of dementia in older patients with type 2 diabetes mellitus. *JAMA* **2009**, *301*, 1565–1572. [CrossRef] [PubMed]

16. Lin, C.-H.; Sheu, W.H.-H. Hypoglycaemic episodes and risk of dementia in diabetes mellitus: 7-year follow-up study. *J. Intern. Med.* **2013**, *273*, 102–110. [CrossRef] [PubMed]

17. Feinkohl, I.; Aung, P.P.; Keller, M.; Robertson, C.M.; Morling, J.R.; McLachlan, S.; Deary, I.J.; Frier, B.M.; Strachan, M.W.J.; Price, J.F.; et al. Severe hypoglycemia and cognitive decline in older people with type 2 diabetes: The Edinburgh type 2 diabetes study. *Diabetes Care* **2014**, *37*, 507–515. [CrossRef]

18. Simó, R.; Hernández, C. Novel approaches for treating diabetic retinopathy based on recent pathogenic evidence. *Prog. Retin. Eye Res.* **2015**, *48*, 160–180. [CrossRef]

19. Sundstrom, J.M.; Hernández, C.; Weber, S.R.; Zhao, Y.; Dunklebarger, M.; Tiberti, N.; Laremore, T.; Simó-Servat, O.; Garcia-Ramirez, M.; Barber, A.J.; et al. Proteomic analysis of early diabetic retinopathy reveals mediators of neurodegenerative brain diseases. *Investig. Ophthalmol. Vis. Sci.* **2018**, *59*, 2264–2274. [CrossRef]

20. Dong, Y.; Kua, Z.J.; Khoo, E.Y.H.; Koo, E.H.; Merchant, R.A. The Utility of Brief Cognitive Tests for Patients With Type 2 Diabetes Mellitus: A Systematic Review. *J. Am. Med. Dir. Assoc.* **2016**, *17*, 889–895. [CrossRef]

21. Jaeger, J. Digit Symbol Substitution Test: The Case for Sensitivity Over Specificity in Neuropsychological Testing. *J. Clin. Psychopharmacol.* **2018**, *38*, 513–519. [CrossRef] [PubMed]

22. Espinosa, A.; Alegret, M.; Valero, S.; Vinyes-Junqué, G.; Hernández, I.; Mauleón, A.; Rosende-Roca, M.; Ruiz, A.; López, O.; Tárraga, L.; et al. A longitudinal follow-up of 550 mild cognitive impairment patients: Evidence for large conversion to dementia rates and detection of major risk factors involved. *J. Alzheimers Dis.* **2013**, *34*, 769–780. [CrossRef] [PubMed]

23. Corriere, M.; Rooparinesingh, N.; Kalyani, R.R. Epidemiology of diabetes and diabetes complications in the elderly: An emerging public health burden. *Curr. Diabetes Rep.* **2013**, *13*, 805–813. [CrossRef] [PubMed]

24. Simó, R.; Hernández, C. European Consortium for the Early Treatment of Diabetic Retinopathy (EUROCONDOR) Neurodegeneration in the diabetic eye: New insights and therapeutic perspectives. *Trends Endocrinol. Metab.* **2014**, *25*, 23–33. [CrossRef] [PubMed]

25. Simó, R.; Stitt, A.W.; Gardner, T.W. Neurodegeneration in diabetic retinopathy: Does it really matter? *Diabetologia* **2018**, *61*, 1902–1912. [CrossRef] [PubMed]

26. Cheung, C.Y.-L.; Ikram, M.K.; Chen, C.; Wong, T.Y. Imaging retina to study dementia and stroke. *Prog. Retin. Eye Res.* **2017**, *57*, 89–107. [CrossRef] [PubMed]

27. Ciudin, A.; Simó-Servat, O.; Hernández, C.; Arcos, G.; Diego, S.; Sanabria, Á.; Sotolongo, Ó.; Hernández, I.; Boada, M.; Simó, R. Retinal Microperimetry: A New Tool for Identifying Patients With Type 2 Diabetes at Risk for Developing Alzheimer Disease. *Diabetes* **2017**, *66*, 3098–3104. [CrossRef]

28. Molitor, R.J.; Ko, P.C.; Ally, B.A. Eye movements in Alzheimer's disease. *J. Alzheimers Dis.* **2015**, *44*, 1–12. [CrossRef]

29. Fletcher, W.A.; Sharpe, J.A. Saccadic eye movement dysfunction in Alzheimer's disease. *Ann. Neurol.* **1986**, *20*, 464–471. [CrossRef]

30. Kapoula, Z.; Yang, Q.; Otero-Millan, J.; Xiao, S.; Macknik, S.L.; Lang, A.; Verny, M.; Martinez-Conde, S. Distinctive features of microsaccades in Alzheimer's disease and in mild cognitive impairment. *Age (Dordr.)* **2014**, *36*, 535–543. [CrossRef]

31. Bylsma, F.W.; Rasmusson, D.X.; Rebok, G.W.; Keyl, P.M.; Tune, L.; Brandt, J. Changes in visual fixation and saccadic eye movements in Alzheimer's disease. *Int. J. Psychophysiol.* **1995**, *19*, 33–40. [CrossRef]

32. Garbutt, S.; Matlin, A.; Hellmuth, J.; Schenk, A.K.; Johnson, J.K.; Rosen, H.; Dean, D.; Kramer, J.; Neuhaus, J.; Miller, B.L.; et al. Oculomotor function in frontotemporal lobar degeneration, related disorders and Alzheimer's disease. *Brain* **2008**, *131*, 1268–1281. [CrossRef] [PubMed]

33. Midena, E.; Vujosevic, S. Microperimetry in diabetic retinopathy. *Saudi J. Ophthalmol. Off. J. Saudi Ophthalmol. Soc.* **2011**, *25*, 131–135. [CrossRef] [PubMed]

34. Fazekas, F.; Chawluk, J.B.; Alavi, A.; Hurtig, H.I.; Zimmerman, R.A. MR signal abnormalities at 1.5 T in Alzheimer's dementia and normal aging. *AJR Am. J. Roentgenol.* **1987**, *149*, 351–356. [CrossRef] [PubMed]

35. Klemp, K.; Larsen, M.; Sander, B.; Vaag, A.; Brockhoff, P.B.; Lund-Andersen, H. Effect of short-term hyperglycemia on multifocal electroretinogram in diabetic patients without retinopathy. *Investig. Ophthalmol. Vis. Sci.* **2004**, *45*, 3812–3819. [CrossRef] [PubMed]

36. Wilkinson, C.P.; Ferris, F.L.; Klein, R.E.; Lee, P.P.; Agardh, C.D.; Davis, M.; Dills, D.; Kampik, A.; Pararajasegaram, R.; Verdaguer, J.T.; et al. Proposed international clinical diabetic retinopathy and diabetic macular edema disease severity scales. *Ophthalmology* **2003**, *110*, 1677–1682. [CrossRef]

37. Petersen, R.C.; Smith, G.E.; Waring, S.C.; Ivnik, R.J.; Tangalos, E.G.; Kokmen, E. Mild cognitive impairment: Clinical characterization and outcome. *Arch. Neurol.* **1999**, *56*, 303–308. [CrossRef]

38. Molina-Martín, A.; Piñero, D.P.; Pérez-Cambrodí, R.J. Normal values for microperimetry with the MAIA microperimeter: Sensitivity and fixation analysis in healthy adults and children. *Eur. J. Ophthalmol.* **2017**, *27*, 607–613. [CrossRef]

39. Morales, M.U.; Saker, S.; Wilde, C.; Pellizzari, C.; Pallikaris, A.; Notaroberto, N.; Rubinstein, M.; Rui, C.; Limoli, P.; Smolek, M.K.; et al. Reference Clinical Database for Fixation Stability Metrics in Normal Subjects Measured with the MAIA Microperimeter. *Transl. Vis. Sci. Technol.* **2016**, *5*, 6. [CrossRef]

40. Lagun, D.; Manzanares, C.; Zola, S.M.; Buffalo, E.A.; Agichtein, E. Detecting cognitive impairment by eye movement analysis using automatic classification algorithms. *J. Neurosci. Methods* **2011**, *201*, 196–203. [CrossRef]

41. Al-Rahayfeh, A.; Faezipour, M. Eye Tracking and Head Movement Detection: A State-of-Art Survey. *IEEE J. Transl. Eng. Health Med.* **2013**, *1*, 2100212. [CrossRef]

42. Ferhat, O.; Vilariño, F. Low Cost Eye Tracking: The Current Panorama. *Comput. Intell. Neurosci.* **2016**, *2016*, 8680541. [CrossRef] [PubMed]

43. Mele, M.L.; Federici, S. Gaze and eye-tracking solutions for psychological research. *Cogn. Process.* **2012**, *13* (Suppl. 1), 261–265. [CrossRef] [PubMed]

44. Marseglia, A.; Fratiglioni, L.; Laukka, E.J.; Santoni, G.; Pedersen, N.L.; Bäckman, L.; Xu, W. Early Cognitive Deficits in Type 2 Diabetes: A Population-Based Study. *J. Alzheimers Dis.* **2016**, *53*, 1069–1078. [CrossRef] [PubMed]

45. Martinez-Conde, S. Fixational eye movements in normal and pathological vision. *Prog. Brain Res.* **2006**, *154*, 151–176. [PubMed]

46. Krauzlis, R.J.; Goffart, L.; Hafed, Z.M. Neuronal control of fixation and fixational eye movements. *Philos. Trans. R. Soc. Lond. B Biol. Sci.* **2017**, *372*, 20160205. [CrossRef]

47. Rohrschneider, K.; Bültmann, S.; Springer, C. Use of fundus perimetry (microperimetry) to quantify macular sensitivity. *Prog. Retin. Eye Res.* **2008**, *27*, 536–548. [CrossRef]

48. Jones, E.G. A new view of specific and nonspecific thalamocortical connections. *Adv. Neurol.* **1998**, *77*, 49–71, discussion 72–73.

49. Erskine, D.; Taylor, J.P.; Firbank, M.J.; Patterson, L.; Onofrj, M.; O'Brien, J.T.; McKeith, I.G.; Attems, J.; Thomas, A.J.; Morris, C.M.; et al. Changes to the lateral geniculate nucleus in Alzheimer's disease but not dementia with Lewy bodies. *Neuropathol. Appl. Neurobiol.* **2016**, *42*, 366–376. [CrossRef]

50. Dugger, B.N.; Tu, M.; Murray, M.E.; Dickson, D.W. Disease specificity and pathologic progression of tau pathology in brainstem nuclei of Alzheimer's disease and progressive supranuclear palsy. *Neurosci. Lett.* **2011**, *491*, 122–126. [CrossRef]

51. Parvizi, J.; Van Hoesen, G.W.; Damasio, A. The selective vulnerability of brainstem nuclei to Alzheimer's disease. *Ann. Neurol.* **2001**, *49*, 53–66. [CrossRef]

52. Gella, L.; Raman, R.; Pal, S.S.; Ganesan, S.; Sharma, T. Fixation characteristics among subjects with diabetes: SN-DREAMS II, Report No. 5. *Can. J. Ophthalmol.* **2015**, *50*, 302–309. [CrossRef] [PubMed]

Journal of
Clinical Medicine

Article

Risk of Dementia in Older Patients with Type 2 Diabetes on Dipeptidyl-Peptidase IV Inhibitors Versus Sulfonylureas: A Real-World Population-Based Cohort Study

Young-Gun Kim [1,2], Ja Young Jeon [3], Hae Jin Kim [3], Dae Jung Kim [3], Kwan-Woo Lee [3], So Young Moon [4] and Seung Jin Han [3,*]

[1] Department of Medical Sciences, Ajou University Graduate School, Suwon 16499, Korea; ygkim25@gmail.com
[2] Ministry of Health and Welfare, Gyeonggi Provincial Government, Suwon 16444, Korea
[3] Department of Endocrinology and Metabolism, Ajou University School of Medicine, Suwon 16499, Korea; Twinstwins@hanmail.net (J.Y.J.); jinkim@ajou.ac.kr (H.J.K.); djkim@ajou.ac.kr (D.J.K.); LKW65@ajou.ac.kr (K.-W.L.)
[4] Department of Neurology, Ajou University School of Medicine, Suwon 16499, Korea; symoon.bv@gmail.com
* Correspondence: hsj@ajou.ac.kr; Tel.: +82-31-219-5126; Fax: +82-31-219-4497

Received: 12 November 2018; Accepted: 22 December 2018; Published: 28 December 2018

Abstract: Background: Type 2 diabetes is related to an increased risk of dementia. Preclinical studies of dipeptidyl peptidase-IV inhibitors (DPP-4i) for dementia have yielded promising results. Therefore, we investigated the risk of dementia in elderly patients with type 2 diabetes on DPP-4is and sulfonylureas (SU). Methods: Using a claims database called the Korean National Health Insurance Service Senior cohort, new users of DPP-4is and SUs were matched by 1:1 propensity score matching using 49 confounding variables (7552 new DPP-4is users and 7552 new SU users were matched by 1:1 propensity score matching; average age 75.4; mean follow-up period: 1361.9 days). Survival analysis was performed to estimate the risk of dementia. Results: The risk of all-cause dementia was lower in the DPP-4i group compared to the SU group (hazard ratio (HR) 0.66; 95% confidence interval (CI) 0.56–0.78; $p < 0.001$). Particularly, DPP-4i use showed a significantly lower risk of Alzheimer's disease (HR 0.64; 95% CI 0.52–0.79; $p < 0.001$) and a lower risk, albeit non-significant, of vascular dementia compared to SU use (HR 0.66; 95% CI 0.38–1.14; $p = 0.139$). Conclusion: Our findings suggest that DPP-4i use decreases the risk of dementia compared to SU use in elderly patients with type 2 diabetes in a real-world clinical setting.

Keywords: dementia; dipeptidyl-peptidase IV inhibitors; diabetes mellitus; type 2; Alzheimer's disease; dementia; vascular

1. Introduction

Type 2 diabetes and dementia are prevalent in the elderly and have considerable impacts on public health and patient quality of life. Recent estimates suggest that 382 million and 44 million individuals worldwide are affected by type 2 diabetes and dementia, respectively [1,2]. Epidemiological evidence indicates that diabetes is associated with an increased risk of dementia, including Alzheimer's disease and vascular dementia [3,4]. According to a recent meta-analysis of 28 prospective observational studies, patients with diabetes have a 73% higher risk of dementia compared to those without diabetes [5]. Although interventions to prevent and treat the classical macro- and microvascular complications of diabetes have improved, cognitive dysfunction and dementia are emerging as important complications in a rapidly aging society [6].

Type 2 diabetes shares several pathophysiological components with dementia, such as glucotoxicity, insulin resistance, inflammation, and oxidative stress [7]. These similarities suggest that anti-diabetic medications may be effective against dementia. Preclinical and clinical studies have investigated the effects of glucose-lowering agents on dementia and cognitive dysfunction but have reported inconsistent results [8,9].

Dipeptidyl peptidase-4 inhibitors (DPP-4i) are widely used oral hypoglycemic agents associated with a low risk of hypoglycemia and weight gain [10]. DPP-4is improve glucose metabolism by increasing the bioavailability of active glucagon-like peptide-1 by inhibiting its degradation. DPP-4is also have neuroprotective, anti-inflammatory, and anti-atherosclerotic effects. Moreover, DPP-4is attenuated amyloid-β deposition and tau phosphorylation in streptozotocin induced Alzheimer's disease model [11,12]. A DPP-4i also improved memory and learning impairment, brain inflammation, and endothelial dysfunction in a pancreatectomy-induced diabetes model [13].

In a recent cross-sectional study, higher DPP-4 plasma activity was associated with an increased risk of mild cognitive impairment in elderly patients with type 2 diabetes [14]. This suggests that DPP-4is may be effective against cognitive dysfunction in individuals with type 2 diabetes. However, to our knowledge, no clinical study on the effect of DPP-4is on the incidence of dementia in type 2 diabetes has been reported. As long-term use of sulfonylurea (SU) was not associated with an increased risk of Alzheimer's disease in a population-based case-control study [15], we investigated the risk of dementia in older patients on DPP-4is compared with SUs in a population-based cohort study using a national health insurance database.

2. Methods

2.1. Study Design and Data Source

We conducted a population-based retrospective observational cohort study using the Korean National Health Insurance Service Senior cohort (ver. 3.0, 1 January, 2002 to 31 December, 2015), which comprises 550,000 (10%) individuals of the South Korean population >60 years of age as of 2002. The database was created using a stratified random sampling method with 1476 strata and is thus representative of the Korean senior population. It contains information on demographic characteristics, socioeconomic status, and claims, such as diagnosis (International Classification of Diseases, 10th revision (ICD-10) code), drug prescriptions, and medical procedures. Socioeconomic status was indirectly assessed using the annual medical insurance premium determined based on the participant's income and assets, such as property and automobile ownership. Socioeconomic status was defined by dividing medical insurance premiums into 11 quantiles. This study was approved by the Institutional Review Board of Ajou University Hospital (AJIRB-MED-EXP-18-033), which waived the requirement for informed consent because all patient data were de-identified.

2.2. Inclusion and Exclusion Criteria

Patients were included in the cohort if they were aged >60 years with type 2 diabetes and started taking a DPP-4i or SU from 1 November, 2008 to 31 December, 2015, regardless of whether they were taking other hypoglycemic agents (DPP-4is were first approved in Korea on 1 November, 2008). Patients who used both drugs were excluded. A 1-year wash-out period before the first prescription of an SU or DPP-4i enabled identification of new users of each drug type. The prescribed drug and date were defined as the index drug and index date, respectively. Patients who had been diagnosed with type 1 diabetes mellitus or dementia before the index date or who had been prescribed donepezil, memantin, rivastigmine, or galantamine for dementia were excluded. A flowchart of the patient selection process is presented in Figure 1. The follow-up period was calculated from the index date to the first occurrence of study outcomes or the study end date (31 December, 2015).

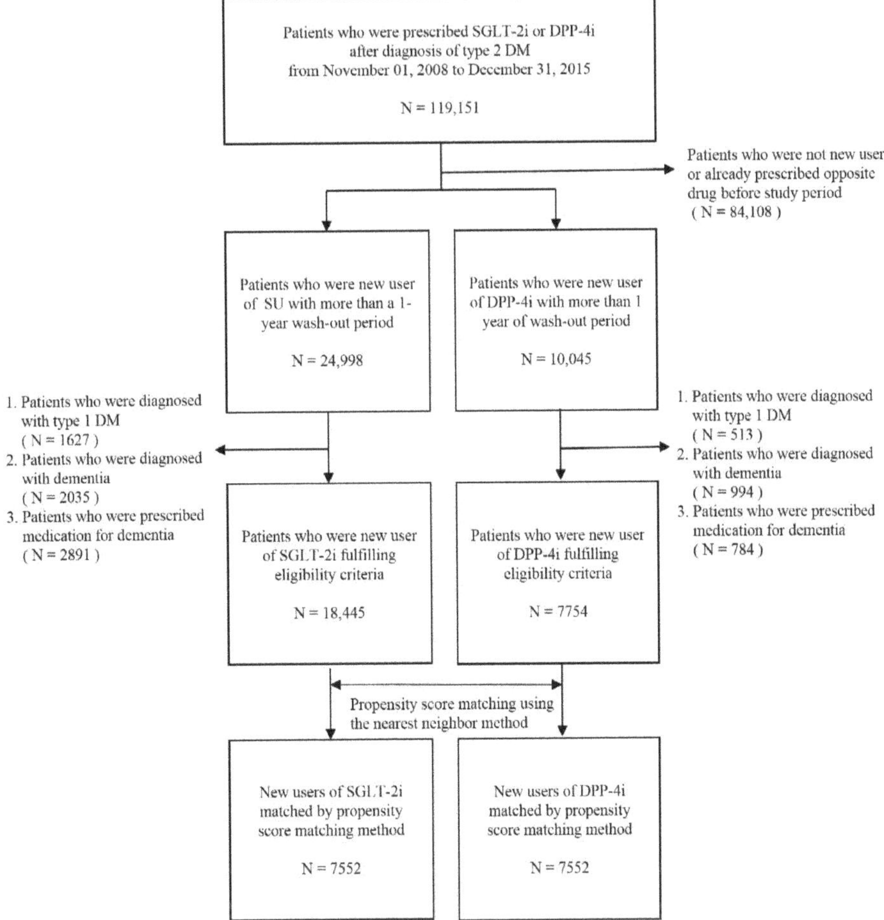

Figure 1. Flow chart of the sample selection process. DM, diabetes mellitus; DPP-4i, dipeptidyl-peptidase IV inhibitor; N, number; SGLT-2i, sodium-glucose co-transporter 2 inhibitor; SU, sulfonylurea.

2.3. Study Outcome and Subgroup Analysis

The primary outcome was the first diagnosis of all-cause dementia (ICD-10 codes: F00, F01, F02, F03, F04, F05, G30, or G31), and the secondary outcomes were the first diagnosis of Alzheimer's disease (F00, G30) or vascular dementia (F01). Subgroup analyses were performed according to sex, age (<75 and ≥75 years), and the presence of DM microvascular or macrovascular complications. DM microvascular complications were defined as at least one of DM nephropathy, neuropathy, or retinopathy, and DM macrovascular complications as at least one of stroke, transient ischemic attack, acute myocardial infarction, other ischemic heart disease, and peripheral artery occlusive disease.

2.4. Statistical Analysis

R software (ver. 3.3.3; R Development Core Team, Vienna, Austria) and SAS (ver. 9.4; SAS Institute, Cary, NC, USA) were used for statistical analyses. Data are expressed as means ± standard deviation. The primary method of statistical adjustment was propensity score matching. Among the patients who met the inclusion/exclusion criteria mentioned above, patients with similar characteristics were selected at a ratio of 1:1 from both groups using propensity score matching. We used the nearest-neighbor technique with a caliper of 0.1 on the probability scale, and replacement of the control was not permitted. The following variables (Table 1): age, sex, socioeconomic status (index date), diagnoses (1 year before the index date), and prescribed drugs (180 days before the index date) were used to calculate propensity scores, and thus those variables were adjusted. Because the claims database does not contain information on the duration of diabetes, we adjusted for several variables that could indirectly reflect disease duration, such as diagnostic codes for DM triopathy, acute myocardial infarction, other ischemic heart diseases, ischemic stroke, hemorrhagic stroke, transient ischemic attack, and peripheral artery occlusive disease, as well as prescriptions for other hypoglycemic agents, including insulin. The quality of correction of confounding variables between the two groups was evaluated as a standardized difference. An absolute standardized difference between groups of <0.1 was considered negligible. After propensity score matching, survival analyses were performed among matched pairs to evaluate the effect of DPP-4is on dementia using the one minus survival probability computed by the Kaplan-Meier approach. As several confounding variables were adjusted for by propensity score matching, univariate Cox regression analysis was performed.

Table 1. Baseline characteristics of the matched pairs.

	SU	DPP-4i	SMD
N	7552	7552	
Age (SD)	75.42 (5.31)	75.39 (4.73)	0.007
Sex (Male, percent)	44.01	43.39	0.013
Socio-economic status (*n*, (%))			0.060
1st to 4th of 11 quantiles	1892 (25.05)	1892 (25.05)	
5th to 8th of 11 quantiles	2301 (30.47)	2378 (31.49)	
9th to 11th of 11 quantiles	3359 (44.48)	3282 (43.46)	
Hypertension	79.61	80.39	0.020
Dyslipidemia	74.13	74.44	0.007
Chronic kidney disease	5.77	5.61	0.007
End-stage renal disease	2.56	2.49	0.004
Any malignancy	12.47	12.27	0.006
Migraine	4.86	4.89	0.001
Asthma	21.93	22.30	0.009
Chronic obstructive pulmonary disease	13.55	13.85	0.009
Connective tissue disease	6.50	6.46	0.002
Atrial fibrillation	4.97	4.67	0.014
Heart failure	8.74	8.91	0.006
Osteoporosis	25.08	25.52	0.010
Cerebrovascular disease			
Ischemic stroke	11.30	11.40	0.003
Hemorrhagic stroke	0.91	0.91	<0.001
Transient ischemic attack	3.42	3.34	0.004
Acute myocardial infarction	2.65	2.78	0.008
Other ischemic heart disease	25.20	25.77	0.013
Other heart disease	18.18	18.68	0.013
Peripheral artery disease	1.44	1.44	<0.001
Microvascular complications of diabetes			
Neuropathy	10.88	10.58	0.010
Nephropathy	5.79	5.75	0.002
Retinopathy	10.05	10.45	0.013

Table 1. *Cont.*

	SU	DPP-4i	SMD
Alcohol use [†]	3.48	3.15	0.018
Tobacco use [†]	0.05	0.07	0.005
Obesity [†]	0.08	0.08	<0.001
Hypoglycemia	2.73	2.32	0.026
Medication use			
Anti-diabetic medicine			
Metformin	93.98	93.95	0.001
Thiazolidinedione	5.27	5.08	0.008
Alpha-glucosidase inhibitor	15.21	15.39	0.005
Meglitinide	8.33	8.10	0.008
SGLT2i	0.54	0.87	0.039
Insulin	37.45	37.01	0.009
Anti-hypertensive agent			
Calcium channel blocker	69.95	70.21	0.005
ACEI	32.56	32.94	0.008
ARB	72.64	72.91	0.006
Beta blocker	47.96	48.34	0.008
Alpha blocker	13.33	13.04	0.009
Diuretics	67.00	66.71	0.006
Aspirin	73.38	73.90	0.012
P2Y12 inhibitor	32.79	32.93	0.003
Warfarin	5.95	5.55	0.017
Other antiplatelet	25.73	25.45	0.006
NOAC	3.30	3.63	0.018
Lipid-lowering agent			
Statin	72.22	73.20	0.022
Fibrate	15.28	15.04	0.007
Ezetimibe	7.79	8.00	0.008

Data are presented as frequencies or means (SD). [†] Confirmed by diagnosis code (International Classification of Diseases, 10th revision). Less than 0.1 (10%) in absolute value of standardized mean difference (SMD) between groups was considered negligible. The mean (SD) standardized difference of all covariates was 1.04% (1.03%). ACEI, angiotensin-converting-enzyme inhibitor; ARB, angiotensin II receptor antagonists; DPP-4i, dipeptidyl peptidase-IV inhibitor; NOAC, novel oral anticoagulant; SD, standard deviation; SGLT2i, sodium-glucose co-transporter 2 inhibitor; SMD, standardized mean difference; SU, sulfonylurea.

3. Results

The cohort comprised 18,445 new SU users and 7754 new DPP-4i users, for a total of 12,833 person-years. After propensity score matching, 7552 pairs remained. The mean follow-up period of the matched pairs was 1361.9 days. Approximately 94, 37, and 15% of patients were already prescribed metformin, insulin, and alpha-glucosidase inhibitors, respectively. Table 1 lists the other baseline characteristics of the matched pairs. The standardized differences of all variables were less than 10%, and the mean standardized difference was 1.04% (1.03%). Thus, the baseline characteristics of the matched pairs were well adjusted.

During the study period, 565 patients had newly developed dementia, among whom 367 had Alzheimer's disease and 54 vascular dementia. When dementia was defined by diagnosis codes, the risk of all-cause dementia was lower in the DPP-4i group compared to the SU group (Figure 2A and Table 2; hazard ratio (HR) 0.66; 95% confidence interval (CI) 0.56–0.78; $p < 0.001$). Additionally, the risk of Alzheimer's disease was significantly lower in the DPP-4i group (Figure 2B, Table 2; HR 0.64; 95% CI 0.52–0.79; $p < 0.001$). The DPP-4i group also had a lower risk, albeit non-significant, of vascular dementia (Figure 2C, Table 2; HR 0.66; 95% CI 0.38–1.14; $p = 0.14$). Furthermore, when dementia was defined using both diagnosis codes and medications, similar trends were observed; that is, the DPP-4i group also had a lower risk of all-cause dementia and Alzheimer's disease (Figure 2D–F and Table 2; HR 0.54; 95% CI 0.40–0.73; $p < 0.001$ for all-cause dementia, HR 0.54; 95% CI 0.39–0.75; $p < 0.001$ for Alzheimer's disease, HR 0.46; 95% CI 0.14–1.46; $p = 0.18$ for vascular dementia).

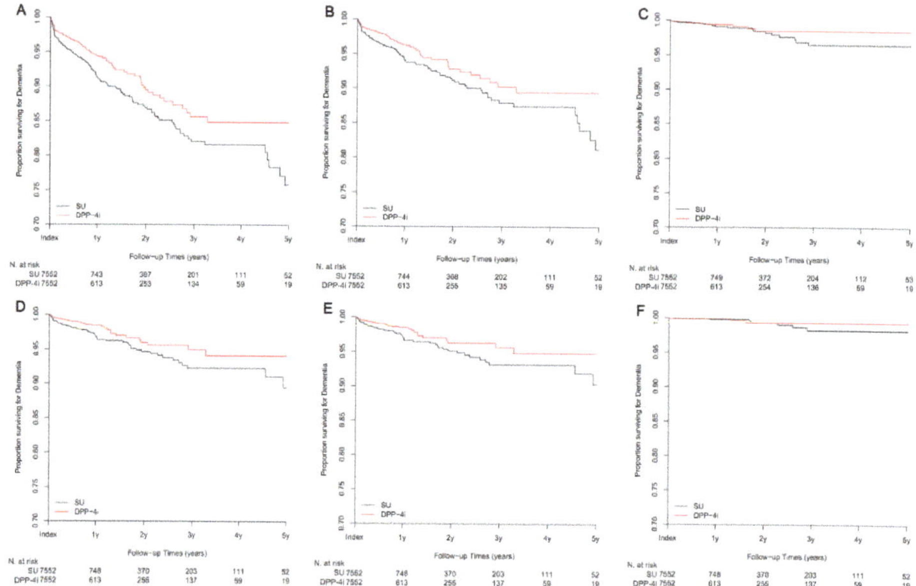

Figure 2. Kaplan-Meier plots for dementia-free survival in new users of DPP-4i and SU. (**A–C**) Dementia was defined by diagnosis codes; all-cause dementia (**A**), Alzheimer's disease (**B**), vascular dementia (**C**). (**D–F**) Dementia was defined by both diagnosis codes and medications; all-cause dementia (**D**), Alzheimer's disease (**E**), and vascular dementia (**F**). DPP-4i, dipeptidyl peptidase-4 inhibitor; *N*, number of patients; SU, sulfonylurea; y, year(s).

Table 2. The risk of dementia in DPP-4i use compared with SU use.

	N	Events	HR	Lower CI	Upper CI	*p*-Value
Event defined with diagnosis codes						
All-cause dementia	15,104	565	0.66	0.56	0.78	<0.001
Alzheimer's disease	15,104	367	0.64	0.52	0.79	<0.001
Vascular dementia	15,104	54	0.66	0.38	1.14	0.14
Event defined with diagnosis codes and medication						
All-cause dementia	15,104	184	0.54	0.40	0.73	<0.001
Alzheimer's disease	15,104	164	0.54	0.39	0.75	<0.001
Vascular dementia	15,104	14	0.46	0.14	1.46	0.18

CI, 95% confidence interval; DPP-4i, dipeptidyl-peptidase IV inhibitor; HR, hazard ratio; *N*, number of patients; SU, sulfonylurea.

Subgroup analyses were performed to determine whether age, sex, and DM complications influenced the protective effect of DPP-4i against dementia (Table 3). DPP-4i use was significantly associated with a lower risk of dementia in males and females. DPP-4i use was associated with a lower risk of dementia in patients aged ≥75 years (HR 0.61; 95% CI 0.50–0.76; *p* < 0.001) but not in those aged <75 years (HR 0.77; 95% CI 0.58–1.03; *p* = 0.08), compared to SU use. Patients without diabetic microvascular complications had a significantly lower HR for dementia in the DPP-4i group compared to the SU group (HR 0.64; 95% CI 0.52–0.78; *p* < 0.001). Among patients with diabetic microvascular complications, DPP-4i use was not significantly associated with an improvement in dementia (HR 0.74; 95% CI 0.53–1.03; *p* = 0.07). However, compared with SU use, DPP-4i use was associated with a lower risk of dementia irrespective of diabetic macrovascular complications.

Table 3. Subgroup analyses according to sex, age, and presence of diabetic microvascular or macrovascular complications.

	N	Events	HR	Lower CI	Upper CI	*p*-Value
Male	6601	202	0.60	0.45	0.80	<0.001
Female	8503	363	0.69	0.56	0.85	<0.001
Patients aged ≥75 years	7662	376	0.61	0.50	0.76	<0.001
Patients aged <75 years	7442	189	0.77	0.58	1.03	0.08
Patients with DM microvascular complication	3418	144	0.74	0.53	1.03	0.07
Patients without DM microvascular complication	11686	421	0.64	0.52	0.78	<0.001
Patients with DM macrovascular complication	5487	227	0.67	0.51	0.87	0.003
Patients without DM macrovascular complication	9617	338	0.65	0.52	0.81	<0.001

CI, 95% confidence interval; HR, hazard ratio; N, number of patients; DM, diabetes mellitus.

4. Discussion

This population-based study demonstrated that use of DPP-4i was associated with a 34% lower risk of all-cause dementia compared with use of SUs in older patients with type 2 diabetes. Indeed, DPP-4i use was related to a significantly lower risk of Alzheimer's disease, but not vascular dementia, compared with SU use.

To our knowledge, this is the first report that DPP-4i use is associated with a lower risk of dementia in older patients with type 2 diabetes. Our cohort was large and representative of the Korean senior population, enabling propensity score-matched analyses. We also used a new-user design with a one-year washout period to reduce the bias inherent in retrospective nonrandomized comparative effectiveness studies.

Insulin resistance and impaired insulin signaling due to chronic hyperglycemia in the brain may induce hyperphosphorylation of tau protein and accumulation of amyloid-β protein, which are hallmarks of Alzheimer's disease [8,16]. In addition, cerebrovascular diseases such as stroke, which are prevalent in diabetes, are closely associated with the development of vascular dementia and the progression of Alzheimer's disease. Because there are interactions between diabetes and dementia and there is no curative treatment for dementia, the effects of antidiabetic medications on cognitive function are of interest.

Our findings support previous reports of a neuroprotective effect of DPP-4is. Research has shown that in human neurons, linagliptin alleviates amyloid-β-induced impaired insulin signaling and neurotoxicity [17]. Long-term sitagliptin treatment attenuated memory impairment and reduced inflammation, nitrosative stress, and amyloid-β protein and amyloid precursor protein accumulation in the brains of transgenic mice with Alzheimer's disease [18]. Vildagliptin and sitagliptin reversed mitochondrial dysfunction in the brain by decreasing mitochondrial reactive oxygen species production and insulin signaling, and improved the learning and memory deficits induced by high-fat-diet consumption [19,20]. In addition, sitagliptin treatment improved memory impairment in mice fed a high-fat diet by enhancing hippocampal neurogenesis and reducing oxidative stress [21]. In a streptozotocin-induced rat model of Alzheimer's disease, saxaglitpin and vildagliptin decreased amyloid-β deposition and tau phosphorylation by increasing hippocampal glucagon-like peptide-1 levels, which reversed the cognitive deficits [11,12]. However, DPP-4is reportedly increases the risk of Alzheimer's disease by aggravating tau phosphorylation and insulin resistance in the hippocampus and primary neurons of OLEF (Otsuka Long-Evans Tokushima Fatty) rats [22].

Few clinical studies have addressed the association between DPP-4is and cognitive function in type 2 diabetic patients. In a prospective pilot study, 10 older patients with type 2 diabetes treated with vildagliptin together with metformin exhibited no cognitive decrements after a 1-year follow up [23]. Furthermore, some previous studies have shown that DPP-4i not only protects against cognitive impairment, but also acts as a cognitive enhancer. Rizzo et al [24]. reported that DPP-4is improved cognitive function compared with SUs, independently of sustained chronic hyperglycemia and glucose variability, in 240 older patients with type 2 diabetes and mild cognitive impairment. In addition, sitagliptin treatment for six months was associated with an increase in the Mini-Mental State Examination score (independent of the change in HbA1c level) compared with metformin treatment in older diabetic patients with or without Alzheimer's disease [25]. These results suggest that DPP-4is could be a cognitive enhancer or protect against cognitive impairment while also functioning as an anti-diabetic agent, which may explain its effects on the risk of dementia. However, these studies had limitations due to a small sample size and short duration of follow-up. The current results are consistent with previous clinical research that reported the beneficial effects of DPP-4is on cognitive function. As our study included a large older population with type 2 diabetes (mean age 75 years) who had a high risk of dementia in a real-world clinical setting, we believe that these findings provide evidence of the protective effects of DPP-4i on the incidence of dementia.

DPP-4i use was associated with a lower risk of Alzheimer's disease, but not vascular dementia, compared with SU use. This finding implies that the efficacy of DPP-4is varies among the types of dementia. Meta-analyses of three large cardiovascular outcome trials of DPP-4i (the SAVOR-TIMI 53, EXAMINE, and TECOS trials) as well as a pooled analysis of small randomized clinical trials showed no significant difference in the risk of stroke between DPP-4i and placebo treatments [26]. Considering these neutral effects of DPP-4i on the risk of stroke, which is a predisposing factor for vascular dementia, DPP-4is may not protect against vascular dementia.

In our subgroup analysis, the association between DPP-4i use and a decreased risk of dementia was not evident in patients aged <75 years or in those with diabetic microvascular complications. Although DPP-4i use was related to a lower risk of dementia in subjects with and without diabetic macrovascular complications, the association was weaker in those with diabetic macrovascular complications. Therefore, the protective effect of DPP-4is against dementia may be greater in older patients and those without diabetic complications.

This study had several limitations. This study was a retrospective analysis, and the claims database lacked information on patient medical histories (most notably, DM duration and body mass index (BMI)), education, lifestyle variables, and laboratory measurements (such as HbA1c); therefore, confounding factors may have influenced the results. Randomized clinical trials on how DPP-4is affects the incidence of dementia are needed to confirm our results. The ongoing CAROLINA-cognition sub-study is exploring whether DPP-4is are superior to SUs in terms of preventing cognitive decline in patients with type 2 diabetes [27]. Additionally, we calculated the incidence of dementia according to diagnosis codes; thus, discrepancies between the medical diagnosis and the diagnosis in the claims data may have reduced the accuracy of the analysis [28]. According to a previous study reporting the accuracy of dementia diagnosis code in Medicare claims data in regard to clinically-diagnosed dementia, the sensitivity and specificity of dementia diagnosis codes in the claims database were 0.85 and 0.89, respectively [29]. When we performed additional survival analyses for dementia defined by both diagnosis codes and prescriptions for dementia, the results showed similar trends. In particular, patients with mild cognitive impairment are less detectable in retrospective observational studies performed using claims databases. Finally, only Koreans were analyzed in this study; therefore, caution should be used when generalizing our results to other ethnicities.

In conclusion, compared with SU use, DPP-4i use was associated with a lower risk of dementia in older Koreans with type 2 diabetes. Further research in other populations using dementia as an endpoint is needed to further assess the neuroprotective effects of DPP-4is.

Author Contributions: Conceptualization, S.J.H.; Data curation, Y.-G.K.; Formal analysis, Y.-G.K.; Funding acquisition, S.J.H.; Investigation, Y.-G.K.; Methodology, Y.-G.K.; Project administration, S.J.H.; Supervision, S.J.H.; Validation, Y.-G.K.; Visualization, Y.-G.K.; Writing—original draft, Y.-G.K. and S.J.H.; Writing—review & editing, J.Y.J., H.J.K., D.J.K., K.-W.L., S.Y.M. and S.J.H.

Funding: This work was supported by the National Research Foundation of Korea (NRF) grant funded by the Korea government (MSIT) (No. NRF-2018R1C1B5044056). The funding entities had no role in the conduct of this study or interpretation of its results

Acknowledgments: This study utilizes the data of the National Health Insurance Service (REQ0000016362) and results are unrelated to the opinion of the National Health Insurance Service.

Conflicts of Interest: The authors declare no conflicts of interest.

References

1. Guariguata, L.; Whiting, D.R.; Hambleton, I.; Beagley, J.; Linnenkamp, U.; Shaw, J.E. Global estimates of diabetes prevalence for 2013 and projections for 2035. *Diabetes Res. Clin. Pract.* **2014**, *103*, 137–149. [CrossRef] [PubMed]
2. World Health Organization. Dementia [article online] 2012. Available online: http://www.who.int/mediacentre/factsheets/fs362/en/ (accessed on 13 February 2015).
3. Profenno, L.A.; Porsteinsson, A.P.; Faraone, S.V. Meta-analysis of Alzheimer's disease risk with obesity, diabetes, and related disorders. *Biol. Psychiatry* **2010**, *67*, 505–512. [CrossRef] [PubMed]
4. Biessels, G.J.; Staekenborg, S.; Brunner, E.; Brayne, C.; Scheltens, P. Risk of dementia in diabetes mellitus: A systematic review. *Lancet Neurol.* **2006**, *5*, 64–74. [CrossRef]
5. Gudala, K.; Bansal, D.; Schifano, F.; Bhansali, A. Diabetes mellitus and risk of dementia: A meta-analysis of prospective observational studies. *J. Diabetes Investig.* **2013**, *4*, 640–650. [CrossRef] [PubMed]
6. Strachan, M.W.; Reynolds, R.M.; Marioni, R.E.; Price, J.F. Cognitive function, dementia and type 2 diabetes mellitus in the elderly. *Nat. Rev. Endocrinol.* **2011**, *7*, 108–114. [CrossRef] [PubMed]
7. Ninomiya, T. Diabetes mellitus and dementia. *Curr. Diab. Rep.* **2014**, *14*, 487. [CrossRef] [PubMed]
8. Femminella, G.D.; Bencivenga, L.; Petraglia, L.; Visaggi, L.; Gioia, L.; Grieco, F.V.; de Lucia, C.; Komici, K.; Corbi, G.; Edison, P.; et al. Antidiabetic Drugs in Alzheimer's Disease: Mechanisms of Action and Future Perspectives. *J. Diabetes Res.* **2017**, *2017*, 7420796. [CrossRef] [PubMed]
9. Biessels, G.J.; Strachan, M.W.; Visseren, F.L.; Kappelle, L.J.; Whitmer, R.A. Dementia and cognitive decline in type 2 diabetes and prediabetic stages: Towards targeted interventions. *Lancet Diabetes Endocrinol.* **2014**, *2*, 246–255. [CrossRef]
10. Phung, O.J.; Scholle, J.M.; Talwar, M.; Coleman, C.I. Effect of noninsulin antidiabetic drugs added to metformin therapy on glycemic control, weight gain, and hypoglycemia in type 2 diabetes. *JAMA* **2010**, *303*, 1410–1418. [CrossRef]
11. Kosaraju, J.; Murthy, V.; Khatwal, R.B.; Dubala, A.; Chinni, S.; Muthureddy Nataraj, S.K.; Basavan, D. Vildagliptin: An anti-diabetes agent ameliorates cognitive deficits and pathology observed in streptozotocin-induced Alzheimer's disease. *J. Pharm. Pharmacol.* **2013**, *65*, 1773–1784. [CrossRef]
12. Kosaraju, J.; Gali, C.C.; Khatwal, R.B.; Dubala, A.; Chinni, S.; Holsinger, R.M.; Madhunapantula, V.S.; Muthureddy Nataraj, S.K.; Basavan, D. Saxagliptin: A dipeptidyl peptidase-4 inhibitor ameliorates streptozotocin induced Alzheimer's disease. *Neuropharmacology* **2013**, *72*, 291–300. [CrossRef] [PubMed]
13. Jain, S.; Sharma, B. Neuroprotective effect of selective DPP-4 inhibitor in experimental vascular dementia. *Physiol. Behav.* **2015**, *152*, 182–193. [CrossRef] [PubMed]
14. Zheng, T.; Qin, L.; Chen, B.; Hu, X.; Zhang, X.; Liu, Y.; Liu, H.; Qin, S.; Li, G.; Li, Q. Association of Plasma DPP4 Activity with Mild Cognitive Impairment in Elderly Patients With Type 2 Diabetes: Results from the GDMD Study in China. *Diabetes Care* **2016**, *39*, 1594–1601. [CrossRef] [PubMed]
15. Imfeld, P.; Bodmer, M.; Jick, S.S.; Meier, C.R. Metformin, other antidiabetic drugs, and risk of Alzheimer's disease: A population-based case-control study. *J. Am. Geriatr. Soc.* **2012**, *60*, 916–921. [CrossRef] [PubMed]
16. de la Monte, S.M. Type 3 diabetes is sporadic Alzheimers disease: Mini-review. *Eur. Neuropsychopharmacol.* **2014**, *24*, 1954–1960. [CrossRef] [PubMed]
17. Kornelius, E.; Lin, C.L.; Chang, H.H.; Li, H.H.; Huang, W.N.; Yang, Y.S.; Lu, Y.L.; Peng, C.H.; Huang, C.N. DPP-4 Inhibitor Linagliptin Attenuates Abeta-induced Cytotoxicity through Activation of AMPK in Neuronal Cells. *CNS Neurosci. Ther.* **2015**, *21*, 549–557. [CrossRef] [PubMed]
18. D'Amico, M.; Di Filippo, C.; Marfella, R.; Abbatecola, A.M.; Ferraraccio, F.; Rossi, F.; Paolisso, G. Long-term inhibition of dipeptidyl peptidase-4 in Alzheimer's prone mice. *Exp. Gerontol.* **2010**, *45*, 202–207. [CrossRef]

19. Pipatpiboon, N.; Pintana, H.; Pratchayasakul, W.; Chattipakorn, N.; Chattipakorn, S.C. DPP4-inhibitor improves neuronal insulin receptor function, brain mitochondrial function and cognitive function in rats with insulin resistance induced by high-fat diet consumption. *Eur. J. Neurosci.* **2013**, *37*, 839–849. [CrossRef]
20. Pintana, H.; Apaijai, N.; Chattipakorn, N.; Chattipakorn, S.C. DPP-4 inhibitors improve cognition and brain mitochondrial function of insulin-resistant rats. *J. Endocrinol.* **2013**, *218*, 1–11. [CrossRef]
21. Gault, V.A.; Lennox, R.; Flatt, P.R. Sitagliptin, a dipeptidyl peptidase-4 inhibitor, improves recognition memory, oxidative stress and hippocampal neurogenesis and upregulates key genes involved in cognitive decline. *Diabetes Obes. Metab.* **2015**, *17*, 403–413. [CrossRef]
22. Kim, D.H.; Huh, J.W.; Jang, M.; Suh, J.H.; Kim, T.W.; Park, J.S.; Yoon, S.Y. Sitagliptin increases tau phosphorylation in the hippocampus of rats with type 2 diabetes and in primary neuron cultures. *Neurobiol. Dis.* **2012**, *46*, 52–58. [CrossRef] [PubMed]
23. Tasci, I.; Naharci, M.I.; Bozoglu, E.; Safer, U.; Aydogdu, A.; Yilmaz, B.F.; Yilmaz, G.; Doruk, H. Cognitive and functional influences of vildagliptin, a DPP-4 inhibitor, added to ongoing metformin therapy in elderly with type 2 diabetes. *Endocr. Metab. Immune Disord. Drug Targets* **2013**, *13*, 256–263. [CrossRef] [PubMed]
24. Rizzo, M.R.; Barbieri, M.; Boccardi, V.; Angellotti, E.; Marfella, R.; Paolisso, G. Dipeptidyl peptidase-4 inhibitors have protective effect on cognitive impairment in aged diabetic patients with mild cognitive impairment. *J. Gerontol. A Biol. Sci. Med. Sci.* **2014**, *69*, 1122–1131. [CrossRef] [PubMed]
25. Isik, A.T.; Soysal, P.; Yay, A.; Usarel, C. The effects of sitagliptin, a DPP-4 inhibitor, on cognitive functions in elderly diabetic patients with or without Alzheimer's disease. *Diabetes Res. Clin. Pract.* **2017**, *123*, 192–198. [CrossRef] [PubMed]
26. Barkas, F.; Elisaf, M.; Tsimihodimos, V.; Milionis, H. Dipeptidyl peptidase-4 inhibitors and protection against stroke: A systematic review and meta-analysis. *Diabetes Metab.* **2017**, *43*, 1–8. [CrossRef] [PubMed]
27. Biessels, G.J.; Janssen, J.; van den Berg, E.; Zinman, B.; Espeland, M.A.; Mattheus, M.; Johansen, O.E.; investigators, C. Rationale and design of the CAROLINA(R)—Cognition substudy: A randomised controlled trial on cognitive outcomes of linagliptin versus glimepiride in patients with type 2 diabetes mellitus. *BMC Neurol.* **2018**, *18*, 7. [CrossRef] [PubMed]
28. Taylor, D.H., Jr.; Fillenbaum, G.G.; Ezell, M.E. The accuracy of medicare claims data in identifying Alzheimer's disease. *J. Clin. Epidemiol.* **2002**, *55*, 929–937. [CrossRef]
29. Taylor, D.H., Jr.; Ostbye, T.; Langa, K.M.; Weir, D.; Plassman, B.L. The accuracy of Medicare claims as an epidemiological tool: The case of dementia revisited. *J. Alzheimer's Dis.* **2009**, *17*, 807–815. [CrossRef]

Journal of
Clinical Medicine

Article

Association between End-Stage Renal Disease and Incident Diabetes Mellitus—A Nationwide Population-Based Cohort Study

Pin-Pin Wu [1,†], Chew-Teng Kor [1,†], Ming-Chia Hsieh [1] and Yao-Peng Hsieh [1,2,3,*]

[1] Department of Internal Medicine, Changhua Christian Hospital, Changhua 500, Taiwan;
    125630@cch.org.tw (P.-P.W.); 179297@cch.org.tw (C.-T.K.); 146446@cch.org.tw (M.-C.H.)
[2] School of Medicine, Kaohsiung Medical University, Kaohsiung 80708, Taiwan
[3] School of Medicine, Chung Shan Medical University, Taichung 40201, Taiwan
* Correspondence: 102407@cch.org.tw; Tel.: +886-4-7238595 (ext. 5889); Fax: +886-4-7277982
† These authors contributed equally to this work.

Received: 24 July 2018; Accepted: 8 October 2018; Published: 11 October 2018

Abstract: Background: Glucose is one of the constituents in hemodialysates and peritoneal dialysates. How the dialysis associates with the incident diabetes mellitus (DM) remains to be assessed. Methods: The claim data of end-stage renal disease (ESRD) patients who initiated dialysis from and a cohort of matched non-dialysis individuals from 2000 to 2013 were retrieved from the Taiwan National Health Insurance Research Database to examine the risk of incident DM among patients on hemodialysis (HD) and peritoneal dialysis (PD). Predictors of incident DM were determined for HD and PD patients using Fine and Gray models to treat death as a competing event, respectively. Results: A total of 2228 patients on dialysis (2092 HD and 136 PD) and 8912 non-dialysis individuals were the study population. The PD and HD patients had 12 and 97 new-onset of DM (incidence rates of 15.98 and 8.69 per 1000 patient-years, respectively), while the comparison cohort had 869 DM events with the incidence rate of 15.88 per 1000 patient-years. The multivariable-adjusted Cox models of Fine and Gray method showed that the dialysis cohort was associated with an adjusted hazard ratio (HR) of 0.49 (95% CI 0.39–0.61, p value < 0.0001) for incident DM compared with the comparison cohort. The adjusted HR of incident DM was 0.46 (95% CI 0.37–0.58, p value < 0.0001) for HD and 0.84 (95% CI 0.47–1.51, p value = 0.56) for PD. Conclusions: ESRD patients were associated with a lower risk of incident DM. HD was associated with a lower risk of incident DM, whereas PD was not.

Keywords: burnt-out diabetes; chronic kidney disease (CKD); dialysis; end-stage renal disease (ESRD); incident diabetes mellitus (DM); insulin resistance

---

1. Introduction

The increasing prevalence of chronic kidney disease (CKD) has a global impact on healthcare management and socioeconomic systems worldwide. Similarly, the prevalence of diabetes mellitus has also been trended high in the general population, especially in the obese and aging [1]. Both of these two entities share many common cardiovascular morbidities and may influence each other in an enhancing manner.

Uremia occurs as a result of enormous retention of various substances when the kidney function is worsening progressively. The acceptable renal replacement therapy includes hemodialysis (HD), peritoneal dialysis (PD), and kidney transplant. Glucose is used as one of the constituents in hemodialysates and peritoneal dialysates [2,3]. Thus, one may expect the higher incidence rate of diabetes mellitus (DM) for dialysis patients owing to the more glucose uptake from the dialysate. However, the epidemiological data on the association of glucose load with incident DM have concluded contradictory results [4,5].

Insulin resistance (IR) is one of the key determinants in the development and progression of DM. IR has existed across all the CKD spectrums and gets exacerbated with the deterioration of renal function as uremia toxins contribute to IR [6,7]. Metabolic acidosis, even of a slight degree, can suppress insulin release and induce IR in CKD [8]. Once non-DM end-stage renal disease (ESRD) patients undergo dialysis treatment, the reduction of uremia toxin and alleviation of metabolic acidosis by dialysis may partially alleviate the degree of IR, thus mitigating the diabetic risk. DM is the leading cause of ESRD worldwide and about 40% of ESRD patients were attributed to diabetic nephropathy [9]. How the dialysis associates with the incident diabetes remained to be assessed. Taiwan has the highest prevalence and incidence rates of ESRD in the world. Therefore, we conducted a retrospective national cohort study using the Taiwan National Insurance Research Database (NHIRD) to compare the risk of incident DM between ESRD patients undergoing PD or HD and non-dialysis patients. In addition, we also determined the risk factors associated with incident DM.

2. Materials and Methods

2.1. Data Source and Study Population

This nationwide retrospective study was conducted using the data retrieved from the Longitudinal Health Insurance Database (LHID), which was randomly selected from the Taiwan National Health Insurance Research Database (NHIRD) and contained the entire claim data for one million beneficiaries. The Taiwan NHIRD was released by the Taiwan National Health Research Institute for scientific research. The Taiwan National Health Insurance (NHI) program has been launched since 1995 and all citizens are enrolled in the program, except prisoners, with a coverage rate of >99%. Therefore, The Taiwan NHIRD can represent the utilization conditions of medical resources for the 23 million residents and is one of the largest databases universally. The NHI adopted the International Classification of Diseases-9th revision, Clinical Modification (ICD-9-CM) for medical payments applications. Patients with ESRD who underwent dialysis treatments will be issued a related catastrophic illness card and the copayment was waived. The Bureau of NHI audits the computerized claim data for medical expenses regularly and those contracted institutions with improper charges or malpractice will face heavy penalties, thus ensuring accurate medical coding. Many high-quality researches have been published using the Taiwan NHIRD [10–12]. This study was exempted from informed consents because the personal identification data were encrypted and transformed in the NHIRD. The Institution Review Board of Changhua Christian Hospital reviewed and approved all the study proposals.

2.2. Study Design

This study was conducted using the inpatient and outpatient claim data from the LHID from 1996 to 2013. Patients who were at the age of 18–100 years and had started maintenance dialysis therapy for at least 90 days between the periods from 2000 to 2013 were enrolled in the dialysis cohort. The date of the first ESRD diagnosis was referred to as the index date. Patients who had undergone dialysis for ESRD from 1996 to 1999, received kidney transplant before the index date, type 2 DM before the index date or type 1 DM throughout the entire period, or incomplete demographics were excluded. Dialysis patients were further categorized as PD and HD groups according to their initial dialysis modality. The reference non-dialysis cohort was recruited from the same dataset with four controls matched to each one dialysis patient by age, gender, and the index year after excluding enrollees with CKD, ESRD, or renal transplant throughout the study period or DM before the index date. Follow-up data for the cohorts was reviewed from the index date until the date of incident DM, the end of 2013 or censored due to death, whichever occurred first.

2.3. Definition of ESRD, DM and Other Comorbidities

The inpatient and outpatient reimbursement data from LHID were linked to define the baseline demographic features and clinical conditions for both cohorts. Individuals were having any comorbidity if they fulfilled the following rules: One diagnostic code at discharge or at least two diagnostic codes in outpatient claims. To minimize the accidental miscoding in the outpatient reimbursement data, the diagnosis from outpatient encounters also required that the first and last diagnoses within one year were at least 30 days apart. ESRD on dialysis was ascertained from the catastrophic illness certificates with the code 585. DM was diagnosed only when there were both DM codes (ICD-9 code 250) and the use of anti-diabetic agents.

Aside from demographic data (e.g., sex, age, and residency area), we also collected information about comorbidities and drug treatments from the LHID. The baseline comorbidities included hypertension (ICD-9 codes 401–405), ischemic heart disease (ICD-9 codes 410–414), congestive heart failure (CHF) (ICD-9 code 428), cerebrovascular disease (ICD-9 codes 430–438), rheumatoid disease (ICD-9 codes 446.5, 710.0–710.4, 714.0–714.2, 714.8, 725.x), gout (ICD-9 code 274), and chronic obstructive pulmonary disease (ICD-9 codes 491, 492, 496). We also obtained information on pharmacotherapy regarding statins, anti-hypertensive drugs, analgesics, and glucocorticoids for multivariate adjustment.

3. Statistical Analysis

Summary descriptive data were shown as mean ± standard deviation (SD) and frequency with percentage for continuous and categorical covariates, respectively. The distributions of variables between the case cohort and matched reference cohort were compared using Student's test, or Mann-Whitney test and the Chi-square test or Fisher's exact test, as appropriate.

The outcome of interest in this study was incident DM event. The incidence rate was calculated as the number of new-onset DM cases divided by the follow-up time and expressed as the number of events per 1000 person-years for the dialysis and comparison groups. As death, a competing risk for the development of incident DM, was censored, we ran the Fine and Gray competing risk models with hazard ratio (HR) and 95% confidence interval (CI) to compare the DM risk between the cohorts. We also compared the risk of mortality (non-DM death), death without DM, between the cohorts. The association was further examined by stepwise cause-specific Cox proportional hazard models with SLENTRY = 0.15 and SLSTAY = 0.15. The HRs of incident DM for dialysis cohort versus comparison cohort were adjusted for all the baseline variables. The propensity score was calculated by logistic regression analysis and was used to adjust in the Cox regression model to reduce bias from unmeasured confounding factors. Four levels of sensitivity tests was performed as (1) adjusted for propensity score; (2) kidney transplant as a censored covariate; (3) change of renal replacement therapy as a time-dependent covariate; and (4) change of dialysis mode or kidney transplant as a censored event. The associations were also assessed in subgroups stratified by gender, age, income, and the number of medical visits in one year after study entry. Risk factors for incident DM were determined in the entire study population, comparison cohort and ESRD dialysis cohort, respectively. Statistical analyses were done using IBM SPSS Statistics for Windows, Version 22.0 (IBM Corp., Armonk, NY, USA). A two-sided *p*-value was set to < 0.05 with statistical significance.

4. Results

4.1. Patients' Baseline Characteristics

The enrollment flowchart was depicted in Figure 1, demonstrating that after the matching processes a total of 2228 dialysis cases (2092 HD; 136 PD) and 8912 control cases were recruited from 2000 to 2013 for the analysis. The differences of the baseline patient characteristics between the dialysis and comparison cohorts were compared in the Table 1. As expected, the ESRD dialysis cohort had more medical visits, higher prevalence of most comorbidities, and higher proportion of patients taking medication than the comparison cohort.

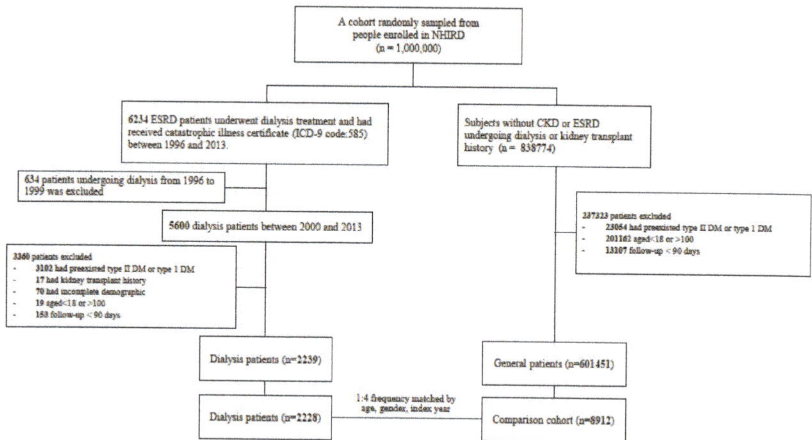

Figure 1. Flowchart of patient selection processes for the end-stage renal disease (ESRD) cohort and the comparison cohort.

4.2. Prevalence of Incident DM and Mortality Rate in the Dialysis and Comparison Cohorts

During the follow-up period of around five–six years, the dialysis cohort had lower rate of DM and higher death rate compared to the comparison cohort (4.89% vs. 9.75%, p value < 0.001 for DM; 30.12% vs. 14.03% for death, p value < 0.001, Table 1). The cumulative incidence rate of DM was significantly lower in the dialysis cohort than in the comparison cohort, while the mortality rate was higher in the dialysis cohort than in the comparison cohort (Figures 2 and 3; log-rank test, p value < 0.001 and p value < 0.001, respectively). The PD and HD patients had 12 and 97 DM events (incidence rates of 15.98 and 8.69 per 1000 patient-years, respectively), while the comparison cohort had 869 DM events with the incidence rate of 15.88 per 1000 patient-years (Table 2). The multivariable-adjusted Cox models of Fine and Gray method showed that dialysis cohort was associated with an adjusted HR of 0.49 (95% CI 0.39–0.61, p value < 0.0001) for incident DM compared with the comparison cohort. The reduced diabetic risk in dialysis cohort was attributed to the lower diabetogenic effect of HD (adjusted HR 0.46, p value < 0.0001) rather than that of PD (adjusted HR 0.84, p value = 0.56). The significant associations were also consistently found when running stepwise Cox models. Regarding mortality, PD, and HD cohorts had significantly higher mortality risk in both Cox models.

4.3. Sensitivity Analysis

Table 3 showed the results of the four sensitivity analyses to corroborate the previous models.

In all assessments, a significant negative correlation between dialysis patients and DM events further confirmed the finding that dialysis is a protective factor in the development of DM. HD was significantly associated with a reduced risk of incident DM compared to the comparison cohort, while the risk between PD and comparison cohort did not differ in the four sensitivity analyses. Regarding non-DM death, dialysis cohort was associated with higher mortality risk than the comparison cohort.

4.4. Association of Dialysis with Incident DM Stratified by Sex, Age, Year of Enrollment, Numbers of Medical Visits, and Economic Incomes

Dialysis cohort was associated with a lower risk of incident DM than the comparison cohort in all the subgroup analyses, while a significantly higher mortality risk was seen in the dialysis cohort (Table 4).

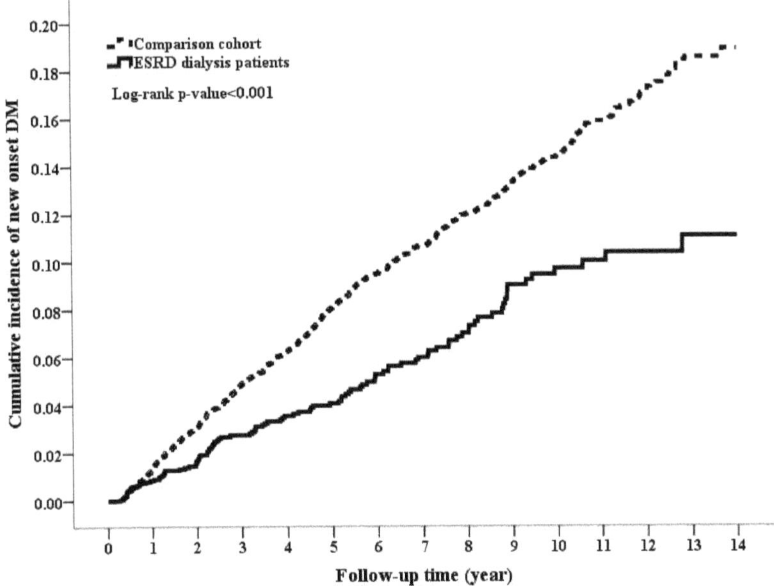

Figure 2. Cumulative incidence rate of diabetes mellitus between the end-stage renal disease (ESRD) cohort and the comparison cohort (*p*-value < 0.001, Log-rank test).

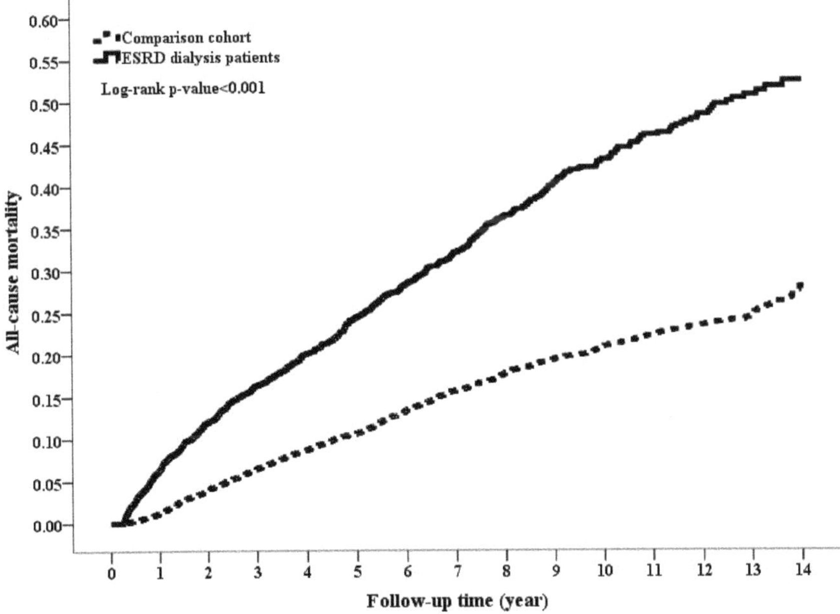

Figure 3. Cumulative incidence rate of mortality between the end-stage renal disease (ESRD) cohort and the comparison cohort (*p*-value < 0.001, Log-rank test).

Table 1. Baseline characteristics and clinical outcomes between the dialysis and comparison cohorts.

No. of Patients	Frequency Matched by Age, Gender, Index Year (1:4)		*p*-Value
	ESRD Dialysis Patient **2228**	**Comparison Cohort** **8912**	
Demographics			
Age	60.29 ± 16.43	60.29 ± 16.43	1.000
Gender, Male	1096 (49.19%)	4384 (49.19%)	1.000
The year of diagnosis			
2000	223 (10.01%)	892 (10.01%)	1.000
2001	122 (5.48%)	488 (5.48%)	1.000
2002	138 (6.19%)	552 (6.19%)	1.000
2003	139 (6.24%)	556 (6.24%)	1.000
2004	159 (7.14%)	636 (7.14%)	1.000
2005	168 (7.54%)	672 (7.54%)	1.000
2006	176 (7.9%)	704 (7.9%)	1.000
2007	159 (7.14%)	636 (7.14%)	1.000
2008	167 (7.5%)	668 (7.5%)	1.000
2009	157 (7.05%)	628 (7.05%)	1.000
2010	172 (7.72%)	688 (7.72%)	1.000
2011	162 (7.27%)	648 (7.27%)	1.000
2012	174 (7.81%)	696 (7.81%)	1.000
2013	112 (5.03%)	448 (5.03%)	1.000
Geographic location			
Northern Taiwan	958 (43%)	3846 (43.16%)	0.912
Central Taiwan	428 (19.21%)	1643 (18.44%)	0.418
Southern Taiwan	796 (35.73%)	3208 (36%)	0.832
Eastern Taiwan and islands	46 (2.06%)	215 (2.41%)	0.372
Monthly income, NTD	14,054.51 ± 13,083.05	14,383.13 ± 13,818.97	0.310
Number of medical visits in 1 year after study entry	35.62 ± 19.28	26.34 ± 18.42	<0.001
Pre-existing comorbidities			
Hypertension	1757 (78.86%)	6986 (78.39%)	0.628
Gout	473 (21.23%)	1337 (15%)	<0.001
Ischemic heart disease	400 (17.95%)	1373 (15.41%)	0.003
Congestive heart failure	430 (19.3%)	835 (9.37%)	<0.001
Cerebrovascular disease	181 (8.12%)	691 (7.75%)	0.561
Chronic obstructive pulmonary disease	297 (13.33%)	1051 (11.79%)	0.047
Rheumatoid disease	83 (3.73%)	280 (3.14%)	0.165
Charlson's comorbidity index score	3.95 ± 1.84	2.55 ± 1.84	<0.001
Long-term medication use			
Anti-hypertensive drugs	1563 (70.15%)	5419 (60.81%)	<0.001
ACEIs/ARBs	1085 (48.7%)	3418 (38.35%)	<0.001
Diuretics	873 (39.18%)	2306 (25.88%)	<0.001
Beta-blockers	946 (42.46%)	3219 (36.12%)	<0.0001
NSAIDs	231 (10.37%)	737 (8.27%)	0.002
Analgesic drugs other than NSAIDs	320 (14.36%)	875 (9.82%)	<0.001
Statin	437 (19.61%)	1186 (13.31%)	<0.001
Corticosteroid	245 (11%)	499 (5.6%)	<0.001
Outcome			
New-onset DM	109 (4.89%)	869 (9.75%)	<0.001
Death	671 (30.12%)	1250 (14.03%)	<0.001
Follow-up time (years)	5.35 ± 3.92	6.14 ± 3.95	<0.001

Values are expressed as mean ± SD or number (percentage). Abbreviations: ACE inhibitor, angiotensin-converting enzyme inhibitor; ARB, angiotensin II receptor blocker; NSAID, Non-Steroidal Anti-Inflammatory Drug; NTD, New Taiwan Dollar.

4.5. Risk Factors for Incident DM among Dialysis Patients

We further determine the contributing factors to incident DM in the study population (Table 5). Among the entire study population, the factors contributing to the occurrence of incident DM consisted of hypertension, gout, and the user of statins with the presence of hypertension having the greatest risk being adjusted HR of 1.95 (95% CI 1.62–2.35, *p* value < 0.0001). The risk factors were not exactly the same among the whole study population, dialysis population and comparison population. In dialysis cohort, age was the only contributing factor for incident DM with an adjusted HR of 1.02 (1.01–1.04).

Table 2. Incidence and risk of new-onset DM and mortality between ESRD dialysis cohort and the comparison cohort.

	Events (n/N)	Incidence Rate [a]	cHR (95% CI)	p Value	aHR [b] (95% CI)	p Value	aHR [c] (95% CI)	p Value
New-onset DM								
Cohorts								
Comparison cohort	869/8912	15.88 (14.83–16.94)	1.00	—	1.00	—	1.00	—
ESRD dialysis cohort	109/2228	9.15 (7.43–10.87)	0.50 (0.41–0.61)	<0.0001	0.49 (0.39–0.61)	<0.0001	0.56 (0.46–0.69)	<0.0001
Categorized dialysis type at start								
Comparison cohort	869/8912	15.88 (14.83–16.94)	1.00	—	1.00	—	1.00	—
ESRD dialysis cohort								
PD	12/136	15.98 (6.94–25.03)	0.94 (0.53–1.66)	0.82	0.84 (0.47–1.51)	0.56	0.93 (0.52–1.64)	0.8
HD	97/2092	8.69 (6.96–10.42)	0.47 (0.38–0.58)	<0.0001	0.46 (0.37–0.58)	<0.0001	0.54 (0.44–0.66)	<0.0001
Mortality								
Cohorts								
Comparison cohort	1250/8912	22.85 (21.58–24.11)	1.00	—	1.00	—	1.00	—
ESRD dialysis cohort	671/2228	56.33 (52.06–60.59)	2.54 (2.31–2.79)	<0.0001	2.32 (2.07–2.60)	<0.0001	2.25 (2.00–2.47)	<0.0001
Categorized dialysis type at start								
Comparison cohort	1250/8912	22.85 (21.58–24.11)	1.00	—	1.00	—	1.00	—
ESRD dialysis cohort								
PD	29/136	38.62 (24.57–52.68)	1.67 (1.16–2.40)	0.006	2.37 (1.61–3.51)	<0.0001	2.50 (1.72–3.63)	<0.0001
HD	642/2092	57.52 (53.07–61.97)	2.60 (2.36–2.86)	<0.0001	2.32 (2.07–2.60)	<0.0001	2.21 (1.99–2.46)	<0.0001

Abbreviations: ESRD, end-stage renal disease; aHR, adjusted hazard ratio; cHR, crude hazard ratio; CI, confidence interval. [a] per 1000 person-years. [b] Adjusted for all variables in Table 1 by Cox proportional hazard model with Fine and Grey's method to consider death as a competing risk. [c] Adjusted for all variables in Table 1 by stepwise Cox proportional hazard model (SLENTRY = 0.15 and SLSTAY = 0.15) with cause-specified method to consider death as a competing risk.

Table 3. Sensitivity analysis.

	New-Onset DM				Non-DM Death			
	aHR [a] (95% CI)	p Value	aHR [b] (95% CI)	p Value	aHR [a] (95% CI)	p Value	aHR [b] (95% CI)	p Value
Adjusted for the propensity score								
Comparison cohort	1.00	—	1.00	—	1.00	—	1.00	—
ESRD dialysis cohort	0.53 (0.43–0.65)	<0.0001	0.57 (0.46–0.7)	<0.0001	1.85 (1.66–2.06)	<0.0001	1.78 (1.6–1.98)	<0.0001
PD	0.98 (0.55–1.74)	0.94	0.99 (0.56–1.75)	0.96	1.89 (1.7–2.11)	<0.0001	1.81 (1.63–2.01)	<0.0001
HD	0.50 (0.40–0.63)	<0.0001	0.54 (0.43–0.67)	<0.0001	1.24 (0.85–1.81)	0.27	1.29 (0.89–1.86)	0.18
Kidney transplant as a censored covariate								
Comparison cohort	1.00	—	1.00	—	1.00	—	1.00	—
ESRD dialysis cohort	0.50 (0.40–0.63)	<0.0001	0.53 (0.43–0.65)	<0.0001	2.32 (2.07–2.59)	<0.0001	2.24 (2.02–2.49)	<0.0001
PD	0.87 (0.46–1.64)	0.66	0.89 (0.47–1.65)	0.7	2.30 (1.54–3.44)	<0.0001	2.46 (1.67–3.61)	<0.0001
HD	0.47 (0.37–0.60)	<0.0001	0.50 (0.40–0.63)	<0.0001	2.31 (2.06–2.58)	<0.0001	2.22 (1.99–2.47)	<0.0001
Change of renal replacement therapy as time-dependent covariates								
Comparison cohort	1.00	—	1.00	—	1.00	—	1.00	—
ESRD dialysis cohort	0.54 (0.43–0.67)	<0.0001	0.53 (0.43–0.66)	<0.0001	2.31 (2.06–2.58)	<0.0001	2.23 (2–2.47)	<0.0001
PD	0.96 (0.54–1.71)	0.88	0.93 (0.51–1.69)	0.81	2.49 (1.65–3.77)	<0.0001	2.47 (1.63–3.75)	<0.0001
HD	0.51 (0.40–0.64)	<0.0001	0.50 (0.40–0.62)	<0.0001	2.30 (2.06–2.57)	<0.0001	2.22 (1.99–2.47)	<0.0001
Change of dialysis mode or kidney transplant as censored covariates								
Comparison cohort	1.00	—	1.00	—	1.00	—	1.00	—
ESRD dialysis cohort	0.57 (0.45–0.71)	<0.0001	0.61 (0.5–0.75)	<0.0001	2.43 (2.18–2.72)	<0.0001	2.36 (2.13–2.62)	<0.0001
PD	1.13 (0.6–2.14)	0.7	1.37 (0.73–2.56)	0.33	3.8 (2.5–5.79)	<0.0001	4.39 (3.02–6.38)	<0.0001
HD	0.54 (0.43–0.68)	<0.0001	0.59 (0.47–0.73)	<0.0001	2.4 (2.14–2.68)	<0.0001	2.32 (2.08–2.57)	<0.0001

Abbreviations: aHR, adjusted hazard ratio; CI, confidence interval; HD, hemodialysis; PD, peritoneal dialysis. [a] Adjusted for all variables in Table 1 by Cox proportional hazard model with Fine and Grey's method to consider death as a competing risk. [b] Adjusted for all variables in Table 1 by stepwise Cox proportional hazard model (SLENTRY = 0.15 and SLSTAY = 0.15) with cause-specified method to consider death as a competing risk.

Table 4. Subgroup analyses of risk for DM and mortality between the ESRD and control cohorts. [a]

Subgroup		Comparison Subjects			ESRD Subjects			ESRD Subjects vs. Comparison Subjects					
								Risk of DM			Risk of Non-DM Death		
		N	DM	Non-DM Death	N	DM	Non-DM Death	aHR (95% CI)	p Value	$P_{interaction}$	aHR (95% CI)	p Value	$P_{interaction}$
Gender													
	Female	4528	470	531	1132	61	302	0.48 (0.36–0.64)	<0.001	0.76	2.44 (2.05–2.91)	<0.0001	0.86
	Male	4384	399	719	1096	48	369	0.50 (0.36–0.71)	<0.001		2.25 (1.94–2.62)	<0.0001	
Age, years													
	<65	4968	563	371	1242	65	223	0.43 (0.33–0.58)	<0.001	0.28	1.82 (1.49–2.23)	<0.001	0.76
	≥65	3944	306	879	986	44	448	0.63 (0.44–0.89)	0.010		2.40 (2.09–2.76)	<0.001	
Year of index													
	2000–2005	3796	573	695	949	82	340	0.61 (0.47–0.79)	<0.001	0.03	2.03 (1.73–2.39)	<0.0001	<0.001
	2006–2013	5116	296	555	1279	27	331	0.39 (0.26–0.60)	<0.001		2.79 (2.37–3.28)	<0.0001	
Income (New Taiwan dollars)													
	<15,840	4740	412	852	1177	52	403	0.59 (0.43–0.82)	0.001	0.62	2.18 (1.89–2.51)	<0.001	0.04
	≥15,840	4172	457	398	1051	57	268	0.49 (0.36–0.66)	<0.001		2.58 (2.13–3.13)	<0.001	
Number of medical visits in 1 year after study entry													
	<24	4812	471	587	643	24	213	0.39 (0.25–0.60)	<0.001	0.26	4.09 (3.39–4.94)	<0.001	<0.001
	≥24	4100	398	663	1585	85	458	0.59 (0.46–0.76)	<0.001		1.83 (1.59–2.09)	<0.001	

Abbreviations: aHR, adjusted hazard ratio; CI, confidence interval; [a] Adjusted for all variables in Table 1.

Table 5. The risk factors associated with of new onset DM [a].

Risk Factor	Study Patients (N = 11140)		Comparison Cohort (N = 8912)		ESRD Dialysis Cohort (N = 2228)	
	aHR [b] (95% CI)	p Value	aHR [b] (95% CI)	p Value	aHR [b] (95% CI)	p Value
ESRD dialysis	0.56 (0.46–0.69)	<0.0001	—	—	—	—
Age	—	—	—	—	1.02 (1.01–1.04)	0.0009
Gender, Male	0.9 (0.79–1.02)	0.11	—	—	—	—
Geographic location						
Northern Taiwan	—	—	1 (reference)		—	—
Central Taiwan	—	—	1.15 (0.95–1.39)	0.14	—	—
Southern Taiwan	—	—	1.23 (1.05–1.43)	0.008	—	—
Eastern Taiwan and islands	—	—	1.02 (0.65–1.61)	0.92	—	—
Monthly income	—	—	—	—	1.13 (0.98–1.29)	0.09
Hypertension	1.95 (1.62–2.35)	<0.0001	2.12 (1.74–2.59)	<0.0001		
Gout	1.36 (1.15–1.6)	0.0002	1.32 (1.11–1.56)	0.0015	1.44 (0.93–2.23)	0.1
Analgesic drugs other than NSAIDs	0.81 (0.63–1.04)	0.09	0.78 (0.6–1.03)	0.08	—	—
statin	1.27 (1.06–1.52)	0.009	1.34 (1.11–1.62)	0.0026	—	—
Anti-hypertensive drugs	—	—	—	—	0.71 (0.47–1.06)	0.09

Abbreviations: aHR, adjusted hazard ratio; CI, confidence interval. [a] only the variables with a p value < 0.15 were listed. [b] Adjusted for all variables in Table 1 by Cox proportional hazard model with Fine and Grey's method to consider death as a competing risk.

5. Discussion

To the best of our knowledge, our study is the first one using a representative nationwide data, to quantify the incidence rate of DM in patients undergoing dialysis compared with a non-dialysis reference cohort. Strikingly, ESRD patients were associated with a lower risk of incident DM (aHR 0.49; 95% CI 0.39–0.61, $p < 0.0001$), which was attributed to HD (aHR 0.46; 95% CI 0.37–0.58, $p < 0.0001$), not to PD (aHR 0.84, 95% CI 0.47–1.51, $p = 0.56$). The contributors to incident DM in the entire study cohort included hypertension, gout, and the use of statins.

Our findings were consolidated by the following reasons. First, we attempted to mitigate the impact of different patient's characteristics distributions on the measured outcomes in the study cohorts by adjusting the propensity scores. Second, all established confounding factors for DM, including age, gender, comorbidities, and some pharmacotherapies were adjusted in competing- risk Cox models, and thus these covariates could not explain the reduced risk of incident DM in relation with dialysis. Third, the number of medical visits was also adjusted in the multivariate Cox models, so the detection bias, possibly caused by more frequent visits in dialysis patients, was minimized. Fourth, in our study, ESRD is, as expected, to associate to higher mortality risk, but surprisingly relate to lower incident DM risk when compared with non-ESRD. One may attribute those observations of our study to the higher death rates occurring in the ESRD patients, thus preventing them from developing DM. In order to resolve this issue, Fine and Gray method of competing risk analysis was chosen in our investigation, which more substantiated our findings.

The practice of dialysis procedures is very complex and various aspects related to dialysis per se and patients have different or even oppose effects on the pathogenesis of DM. Most of the available data suggested the liability of ESRD patients to develop DM. For example, intake of foods with high glycemic index is believed to predispose to postprandial hyperglycemia and higher insulin levels. The results from epidemiological studies on glycemic load in association with incident DM were contradictory. Villegas et al. reported that high intake of foods with a high glycemic index and glycemic load, especially rice, may increase the risk of type 2 DM in Chinese women, but this association was not found in the Whitehall II study [4,5]. Regardless, two meta-analyses, one of prospective cohort studies and the other of retrospective studies, reported a positive association of both dietary glycemic index and glycemic load and risk of type 2 DM [13,14]. The glucose content in the dialysate is another source of caloric supply in the dialysis patients. Based on our findings, the association between ESRD and incident DM could not be solely explained by glucose load.

Tremendous contributors to the development of DM have been identified with IR being one of the most important factors. IR exists in every stage of CKD and the etiologies are multifactorial, including vitamin D deficiency, erythropoietin deficiency, uremia milieu, inflammation, and hyperparathyroidism [15]. In addition, many comorbid factors, not related to CKD itself, contributing to IR, include old age, obesity, dyslipidemia, and hypertension, and so forth. IR has been shown to increase in ESRD patients, but dialysis treatment was reportedly capable of alleviating resistance. Using the euglycemia hyperinsulinaemic glucose clamp technique, Kabayashi et al. reported that both HD and PD can improve IR observed in uremia milieu [6]. Similar findings were also documented by Satirapoj et al. who reported decreased IR after five weeks of HD and PD in the same patient group [16]. However, no significant difference was found between predialysis and dialysis groups in a later investigation [17]. Therefore, once CKD patients do not develop DM before reaching ESRD, dialysis treatments may have positive effects on IR, thus reducing the incident DM risk.

Of the risk factors for type 2 DM, increasing body weight has been reported to one of the most important contributors to impaired glucose tolerance or even type 2 DM. Body mass index (BMI) was the greatest contributor among the three covariates (age, race/ethnicity, BMI) to the increase in diabetes prevalence after adjustments in a study of five NHANES involving 23,932 participants aged 20 to 74 years [18]. The nutritional parameters showed a longitudinal decline with dialysis vintage in dialysis patients. In an analysis of 17,022 patients commencing PD or HD, the BMI trajectory changed in a non-linear fashion, where mean BMI initially decreased, followed by increment and then stabilization at three years [19]. Afterwards, it dropped gradually. The protein-energy wasting (PEW) in ESRD patients on dialysis is caused by uremic toxins, inadequate dietary intake due to anorexia, and inflammation, and is closely associated with mortality [20]. Therefore, PEW or malnutrition may contribute to both lower incidence rate of DM and higher mortality rate. However, we did not have the BMI values for the present study.

Patients with DM and CKD have been found to require reduced or even discontinued of antidiabetic medication with the progression of CKD, initiation of dialysis therapy and gradual loss of residual renal function with time [21]. "Burnt-out diabetes" refers to the situation where diabetic patients can attain appropriate glycemic control, e.g., HbA1C < 6%, without any antidiabetic agents. A great number of mechanisms contributing to this phenomenon consist of loss of dietary intake due to diabetic gastroparesis and uremia, diminished renal and hepatic clearance of insulin with prolongation of insulin half-life, impaired renal gluconeogenesis, protein-energy wasting, disrupted counter-regulation of hypoglycemia, imposed dietary restriction, and hypoglycemia effects of dialysis [22,23]. Burnt-out diabetes was reported in 20.7% and 5.4% of Japanese diabetic patients undergoing HD by using glycated hemoglobin and glycated albumin, respectively [21]. Therefore, considering these effects underlying "burnt-out diabetes" on glucose homeostasis among non-diabetic patients at the initiation of dialysis, dialysis procedure per se might prevent those from developing DM.

In dialysis modalities (HD or PD), no consistent conclusion was drawn on which one is preferable to the other in terms of clinical outcomes. Typically, uremic toxins and excessive fluid have been removed intermittently in HD and continuously in PD. While higher inflammation was observed in oxidative stress and IR in ESRD, the influence of PD and HD on them were not exactly the same. Initiation of dialysis can improve IR and glucose tolerance. However, glucose load and absorption was also different. Glucose load is continuous in the PD patients throughout the day, whereas approximately 15–25 g of glucose may be removed during HD with the net absorption of glucose determined by the concentration of glucose-containing hemodialysates [24,25]. The glucose concentrations used to achieve appropriate ultrafiltration range from 1360 to 3860 md/dL and the glucose load delivered by PD can be as much as 10% to 30% of a patient's total energy intake [26,27]. Burnt-out diabetes was only evidenced in HD, not in PD. Therefore, the effects of PD and HD on the occurrence of incident DM were different and the reduced risk of DM was only documented for HD.

CKD was shown to complicate IR, mainly due to post-receptor defect of insulin action, and may be prone to the development of incident DM in some reports. The aim of our study was to address whether the dialysis per se would increase or reduce the incidence of DM, so those pre-dialysis CKD patients developing DM in their CKD period ahead of ESRD requiring dialysis were excluded from our case cohort. We proposed that the remaining ESRD patients without developing DM in the CKD period may have some protective factors in the environmental and genetic aspects that mitigate the pro-diabetic tendency. Those protective factors may continue or even amplify once they progress to ESRD. This can partially explain why ESRD was associated with a lower risk of incident DM, although their influence cannot be comprehensively addressed in this study.

As with other nationwide studies using Taiwan NHRID, there are several limitations in our study. First, some known contributing factors to incident DM, such as body mass index, physical activity, family history of DM, smoking, alcohol consumption, quantity and quality of sleep and dietary patterns, were not available in the NHRID. Those factors usually lead to cardiovascular comorbidities, which were controlled in our study, so we believe the bias caused by un-adjustment of those factors could be minimized. Second, four types of peritoneal dialysis solution (PDS) have been available in Taiwan: (1) The conventional glucose-based PDS; (2) neutral- pH PDS with low concentration of glucose degradation products; (3) icodextrin-based PDS; and (4) amino acid- containing PDS [28]. Many studies reported that biocompatible PDS led to less inflammation as compared to bio-incompatible PDS. Long-term online hemodiafiltration with ultrapure solution reportedly caused less inflammation via enhanced clearance of middle molecules than low-flux hemodialysis [29]. We cannot compare the influence of the four types of PDS and these two HD modes on glucose homeostasis due to technical limitations. Third, the number of patients on PD was relatively small, which may lead to the over- or underestimation of incident diabetic risk with PD. Therefore, the risk of diabetes in PD should be interpreted with caution. Fourth, the lack of laboratory data in our study is one limitation of NHRID. However, in order to avoid mis-coding problems, the use of DM's ICD-9 code and anti-diabetic drugs is a prerequisite for diagnosing DM to improve diagnostic accuracy.

In conclusion, this study highlights the significant reduction in the risk of developing diabetes in patients undergoing dialysis and this association was attributed to HD rather than PD. In addition to Cox models of Fine and Gray's method, stepwise cause-specific hazard models and sensitivity tests further corroborated this paradoxical relationship. Large-scale prospective studies on the underlying mechanisms of this phenomenon are urgent.

Author Contributions: Conceptualization, P.-P.W.; Formal analysis, C.-T.K.; Investigation, M.-C.H.; Writing—original draft, Y.-P.H.

Funding: This study was funded by grants 107-CCH-HCR-033 from the Changhua Christian Hospital Research Foundation.

Acknowledgments: None.

Conflicts of Interest: The authors declare no conflicts of interest.

References

1. Wild, S.; Roglic, G.; Green, A.; Sicree, R.; King, H. Global prevalence of diabetes: Estimates for the year 2000 and projections for 2030. *Diabetes Care* **2004**, *27*, 1047–1053. [CrossRef] [PubMed]
2. Raimann, J.G.; Kruse, A.; Thijssen, S.; Kuntsevich, V.; Dabel, P.; Bachar, M.; Diaz-Buxo, J.A.; Levin, N.W.; Kotanko, P. Metabolic effects of dialyzate glucose in chronic hemodialysis: Results from a prospective, randomized crossover trial. *Nephrol. Dial. Transplant.* **2012**, *27*, 1559–1568. [CrossRef] [PubMed]
3. Wideröe, T.E.; Smeby, L.C.; Myking, O.L. Plasma concentrations and transperitoneal transport of native insulin and C-peptide in patients on continuous ambulatory peritoneal dialysis. *Kidney Int.* **1984**, *25*, 82–87. [CrossRef] [PubMed]
4. Villegas, R.; Liu, S.; Gao, Y.T.; Yang, G.; Li, H.; Zheng, W.; Shu, X.O. Prospective study of dietary carbohydrates, glycemic index, glycemic load, and incidence of type 2 diabetes mellitus in middle-aged Chinese women. *Arch. Intern. Med.* **2007**, *167*, 2310–2316. [CrossRef] [PubMed]
5. Mosdøl, A.; Witte, D.R.; Frost, G.; Marmot, M.G.; Brunner, E.J. Dietary glycemic index and glycemic load are associated with high-density-lipoprotein cholesterol at baseline but not with increased risk of diabetes in the Whitehall II study. *Am. J. Clin. Nutr.* **2007**, *86*, 988–994. [CrossRef] [PubMed]
6. Kobayashi, S.; Maejima, S.; Ikeda, T.; Nagase, M. Impact of dialysis therapy on insulin resistance in end-stage renal disease: Comparison of haemodialysis and continuous ambulatory peritoneal dialysis. *Nephrol. Dial. Transplant.* **2000**, *15*, 65–70. [CrossRef] [PubMed]
7. Nerpin, E.; Risérus, U.; Ingelsson, E.; Sundström, J.; Jobs, M.; Larsson, A.; Basu, S.; Arnlöv, J. Insulin sensitivity measured with euglycemic clamp is independently associated with glomerular filtration rate in a community-based cohort. *Diabetes Care* **2008**, *31*, 1550–1555. [CrossRef] [PubMed]
8. Souto, G.; Donapetry, C.; Calviño, J.; Adeva, M.M. Metabolic acidosis-induced insulin resistance and cardiovascular risk. *Metab. Syndr. Relat. Disord.* **2011**, *9*, 247–253. [CrossRef] [PubMed]
9. Collins, A.J.; Foley, R.N.; Chavers, B.; Gilbertson, D.; Herzog, C.; Johansen, K.; Kasiske, B.; Kutner, N.; Liu, J.; St Peter, W.; et al. United States renal data system 2011 annual data report: Atlas of chronic kidney disease & end-stage renal disease in the United States. *Am. J. Kidney Dis.* **2012**, *59*, e1-420.
10. Chen, Y.C.; Lin, H.Y.; Li, C.Y.; Lee, M.S.; Su, Y.C. A nationwide cohort study suggests that hepatitis C virus infection is associated with increased risk of chronic kidney disease. *Kidney Int.* **2014**, *85*, 1200–1207. [CrossRef] [PubMed]
11. Wu, C.Y.; Chen, Y.J.; Ho, H.J.; Hsu, Y.C.; Kuo, K.N.; Wu, M.S.; Lin, J.T. Association between nucleoside analogues and risk of hepatitis B virus-related hepatocellular carcinoma recurrence following liver resection. *JAMA* **2012**, *308*, 1906–1914. [CrossRef] [PubMed]
12. Kuo, H.W.; Tsai, S.S.; Tiao, M.M.; Yang, C.Y. Epidemiological features of CKD in Taiwan. *Am. J. Kidney Dis.* **2007**, *49*, 46–55. [CrossRef] [PubMed]

13. Dong, J.Y.; Zhang, L.; Zhang, Y.H.; Qin, L.Q. Dietary glycaemic index and glycaemic load in relation to the risk of type 2 diabetes: A meta-analysis of prospective cohort studies. *Br. J. Nutr.* **2011**, *106*, 1649–1654. [CrossRef] [PubMed]

14. Barclay, A.W.; Petocz, P.; McMillan-Price, J.; Flood, V.M.; Prvan, T.; Mitchell, P.; Brand-Miller, J.C. Glycemic index, glycemic load, and chronic disease risk—A meta-analysis of observational studies. *Am. J. Clin. Nutr.* **2008**, *87*, 627–637. [CrossRef] [PubMed]

15. De Boer, I.H.; Zelnick, L.; Afkarian, M.; Ayers, E.; Curtin, L.; Himmelfarb, J.; Ikizler, T.A.; Kahn, S.E.; Kestenbaum, B.; Utzschneider, K. Impaired glucose and insulin homeostasis in moderate-severe CKD. *J. Am. Soc. Nephrol.* **2016**, *27*, 2861–2871. [CrossRef] [PubMed]

16. Satirapoj, B.; Supasyndh, O.; Phantana Angkul, P.; Ruangkanchanasetr, P.; Nata, N.; Chaiprasert, A.; Kanjanakul, I.; Choovichian, P. Insulin resistance in dialysis versus nondialysis end stage renal disease patients without diabetes. *J. Med. Assoc. Thai.* **2011**, *94*, S87–S93. [PubMed]

17. Akalın, N.; Köroğlu, M.; Harmankaya, Ö.; Akay, H.; Kumbasar, B. Comparison of insulin resistance in the various stages of chronic kidney disease and inflammation. *Ren. Fail.* **2015**, *37*, 237–240. [CrossRef] [PubMed]

18. Menke, A.; Rust, K.F.; Fradkin, J.; Cheng, Y.J.; Cowie, C.C. Associations between trends in race/ethnicity, aging, and body mass index with diabetes prevalence in the United States: A series of cross-sectional studies. *Ann. Intern. Med.* **2014**, *161*, 328. [CrossRef] [PubMed]

19. Dong, J.; Yang, Z.K.; Chen, Y. Older age, higher body mass index and inflammation increase the risk for new-onset diabetes and impaired glucose tolerance in patients on peritoneal dialysis. *Perit. Dial. Int.* **2016**, *36*, 277–283. [CrossRef] [PubMed]

20. Obi, Y.; Qader, H.; Kovesdy, C.P.; Kalantar-Zadeh, K. Latest consensus and update on protein-energy wasting in chronic kidney disease. *Curr. Opin. Clin. Nutr. Metab. Care* **2015**, *18*, 254–262. [CrossRef] [PubMed]

21. Abe, M.; Hamano, T.; Hoshino, J.; Wada, A.; Inaba, M.; Nakai, S.; Masakane, I. Is there a "burnt-out diabetes" phenomenon in patients on hemodialysis? *Diabetes Res. Clin. Pract.* **2017**, *130*, 211–220. [CrossRef] [PubMed]

22. Abe, M.; Kalanter-Zadeh, K. Haemodialysis-induced hypoglycaemia and glycaemic disarrays. *Nat. Rev. Nephrol.* **2015**, *11*, 302–313. [CrossRef] [PubMed]

23. Park, J.; Lertdumrongluk, P.; Molnar, M.Z.; Kovesdy, C.P.; Kalantar-Zadeh, K. Glycemic control in diabetic dialysis patients and the burnt-out diabetes phenomenon. *Curr. Diabetes Rep.* **2012**, *12*, 432–439. [CrossRef] [PubMed]

24. Liu, Y.; Xiao, X.; Qin, D.P.; Tan, R.S.; Zhong, X.S.; Zhou, D.Y.; Liu, Y.; Xiong, X.; Zheng, Y.Y. Comparison of intradialytic parenteral nutrition with glucose or amino acid mixtures in maintenance hemodialysis patients. *Nutrients* **2016**, *8*, 220. [CrossRef] [PubMed]

25. Wathen, R.L.; Keshaviah, P.; Hommeyer, P.; Cadwell, K.; Comty, C.M. The metabolic effects of hemodialysis with and without glucose in the dialysate. *Am. J. Clin. Nutr.* **1978**, *31*, 1870–1875. [CrossRef] [PubMed]

26. Duong, U.; Mehrotra, R.; Molnar, M.Z.; Noori, N.; Kovesdy, C.P.; Nissenson, A.R.; Kalantar-Zadeh, K. Glycemic control and survival in peritoneal dialysis patients with diabetes mellitus. *Clin. J. Am. Soc. Nephrol.* **2011**, *6*, 1041–1048. [CrossRef] [PubMed]

27. Grodstein, G.P.; Blumenkrantz, M.J.; Kopple, J.D.; Moran, J.K.; Coburn, J.W. Glucose absorption during continuous ambulatory peritoneal dialysis. *Kidney Int.* **1981**, *19*, 564–567. [CrossRef] [PubMed]

28. Velloso, M.S.; Otoni, A.; de Paula Sabino, A.; de Castro, W.V.; Pinto, S.W.; Marinho, M.A.; Rios, D.R. Peritoneal dialysis and inflammation. *Clin. Chim. Acta* **2014**, *430*, 109–114. [CrossRef] [PubMed]

29. Den Hoedt, C.H.; Bots, M.L.; Grooteman, M.P.; van der Weerd, N.C.; Mazairac, A.H.; Penne, E.L.; Levesque, R.; ter Wee, P.M.; Nubé, M.J.; Blankestijn, P.J.; et al. Online hemodiafiltration reduces systemic inflammation compared to low-flux hemodialysis. *Kidney Int.* **2014**, *86*, 423–432. [CrossRef] [PubMed]

Journal of
Clinical Medicine

Review

The New Era for Reno-Cardiovascular Treatment in Type 2 Diabetes

Clara García-Carro [1,†], Ander Vergara [1,†], Irene Agraz [1,2], Conxita Jacobs-Cachá [1,2,*],
Eugenia Espinel [1,2], Daniel Seron [1,2] and María José Soler [1,2,*]

1   Nephrology Research Group, Vall d'Hebron Research Institute (VHIR), Nephrology Department,
    Hospital Universitari Vall d'Hebron, Universitat Autònoma de Barcelona, 08035 Barcelona, Spain;
    clara.garcia@vhebron.net (C.G.-C.); vergara.ander@gmail.com (A.V.); iagraz@vhebron.net (I.A.);
    eespinel@vhebron.net (E.E.); dseron@vhebron.net (D.S.)
2   Red de Investigación Renal (REDINREN), Instituto Carlos IIIFEDER, 28029 Madrid, Spain
*   Correspondence: conxita.jacobs@vhir.org (C.J.-C.); mjsoler01@gmail.com (M.J.S.);
    Tel.: +34-934-89-30-00 (C.J.-C. & M.J.S.)
†   These authors contributed equally to this work.

Received: 9 May 2019; Accepted: 11 June 2019; Published: 17 June 2019

Abstract: Diabetic kidney disease (DKD) is the leading cause of end-stage renal disease in the developed world. Until 2016, the only treatment that was clearly demonstrated to delay the DKD was the renin-angiotensin system blockade, either by angiotensin-converting enzyme inhibitors or angiotensin receptor blockers. However, this strategy only partially covered the DKD progression. Thus, new strategies for reno-cardiovascular protection in type 2 diabetic patients are urgently needed. In the last few years, hypoglycaemic drugs, such as sodium-glucose co-transporter 2 inhibitors and glucagon-like peptide-1 receptor agonists, demonstrated a cardioprotective effect, mainly in terms of decreasing hospitalization for heart failure and cardiovascular death in type 2 diabetic patients. In addition, these drugs also demonstrated a clear renoprotective effect by delaying DKD progression and decreasing albuminuria. Another hypoglycaemic drug class, dipeptidyl peptidase 4 inhibitors, has been approved for its use in patients with advanced chronic kidney disease, avoiding, in part, the need for insulinization in this group of DKD patients. Studies in diabetic and non-diabetic experimental models suggest that these drugs may exert their reno-cardiovascular protective effect by glucose and non-glucose dependent mechanisms. This review focuses on newly demonstrated strategies that have shown reno-cardiovascular benefits in type 2 diabetes and that may change diabetes management algorithms.

Keywords: diabetes; diabetic kidney disease; reno-cardiovascular protection; sodium-glucose co-transporter 2 inhibitors; glucagon-like peptide-1 receptor agonists; dipeptidyl peptidase 4 inhibitors

---

1. Introduction

Diabetic kidney disease (DKD) is the first cause of chronic kidney disease, leading to premature death and end-stage renal disease (ESRD) in the developed and developing world. In response, multiple potential therapeutic agents have been tested, focusing on the treatment of hyperglycaemia and hypertension, mainly directed at the renin-angiotensin system blockade [1–4]. However, these therapies only partially delay the progression of DKD to ESRD, so there is an urgent need for additional effective treatments. In this context, sodium-glucose co-transporter 2 (SGLT2) inhibitors and the glucagon-like peptide-1 receptor agonists (GLP-1RAs) have recently emerged as new potential strategies for both diabetic type 2 and 1 patients [5–7].

The SGLT2 is expressed in the proximal tubule of the kidneys and is responsible for 90% of renal glucose reabsorption. SGLT2 inhibitors promote the urinary excretion of glucose and, consequently, lower blood glucose levels. Interestingly, these drugs scarcely provoke hypoglycaemia, as their effect is not related to beta cell function or modifications in insulin sensitivity [8]. SGLT2 inhibitors were first approved by the U.S. Food and Drug Administration (FDA) in 2013 for patients with type 2 diabetes, and the first study that demonstrated their beneficial effects in terms of delaying DKD progression was published in 2016 [5]. Glucagon-like peptide-1 (GLP-1) is an endogenous incretin peptide released from intestinal L cells in response to ingested nutrients. GLP-1 is rapidly inactivated by the enzyme dipeptidyl-peptidase-4 (DPP-4) and cleared by the kidneys. GLP-1 stimulates pancreatic insulin synthesis and insulin secretion in a glucose-dependent manner, slows gastric emptying, inhibits glucagon release, and promotes satiety. GLP-1 receptor agonists are structurally similar to GLP-1 but resist dipeptidyl peptidase 4 (DDP-4) degradation [8]. Exenatide was the first GLP-1RA approved by the FDA in 2005 for patients with type 2 diabetes, and in 2016, liraglutide was the first that demonstrated beneficial effects in terms of decreasing albuminuria [9]. DPP-4 inhibitors are a class of oral hypoglycaemics that block the DPP-4 enzyme and subsequently neutralise several incretin peptides, including the glucose-dependent insulinotropic polypeptide (GIP) and GLP-1 [8]. Thus, DPP-4 inhibitors increase GLP-1 and reduce blood glucose by inhibiting glucagon release and stimulating insulin secretion. Sitagliptin was the first DPP-4 inhibitor approved by the FDA in 2006, followed by linagliptin, saxagliptin, and alogliptin. More recently, in 2019, linagliptin demonstrated beneficial effects in reducing the progression of albuminuria in DKD [10].

In this review, we describe the new strategies for reno-cardiovascular protection in patients with type 2 diabetes and their potential mechanisms. We cover the most important studies focused on the renoprotection exerted by SGLT2 inhibitors, GLP-1RAs, and DPP-4 inhibitors in DKD patients.

2. Classical Pharmacological Reno-Cardiovascular Approaches in Diabetes

Patients with diabetes have a higher prevalence of cardiovascular morbidity and mortality compared to the general population [11]. It is well known that diabetes is associated with accelerated atherosclerosis, affecting the coronaries, which increases the risk for myocardial infarction, heart failure, and may cause diabetic cardiomyopathy independent of coronary artery disease, hypertension, and valvular complications [12]. According to some authors, during the early stages of diabetes, there is an increase in plasma renin activity, mean arterial pressure, and renal vascular resistance [13], suggesting that renin-angiotensin-aldosterone system (RAAS) activation plays a major role in the development of cardiovascular disease (CVD) [14]. Therefore, angiotensin-converting enzyme inhibitors (ACEi) and angiotensin II receptor blockers (ARB) have been, for many years, the first line therapy for secondary CVD prevention in patients with diabetes [15].

In the late 80's to early 90's, an ACEi, enalapril, demonstrated its effectiveness in reducing mortality in patients with heart failure [16,17]. Later, the heart outcomes prevention evaluation (HOPE) trial included subjects with high cardiovascular risk, such as diabetic patients, and demonstrated that another ACEi, ramipril, significantly reduced the rates of death, myocardial infarction, and stroke in patients with vascular disease or diabetes [14]. Similarly, the losartan intervention for endpoint reduction (LIFE) trial showed that Losartan was more effective than Atenolol in reducing cardiovascular morbidity and mortality in patients with hypertension, diabetes, and ventricular hypertrophy [17]. In 2001, two seminal studies demonstrated the nephroprotective effect of RAAS blockades in patients with type 2 diabetes [2,3]. A subsequent metanalysis supported the use of ACEi in patients who have diabetic kidney disease (DKD) with significant albuminuria [18]; however, the beneficial effect of RAAS inhibition as a primary prevention in diabetic patients has not been demonstrated. The beneficial effect of RAAS blockade in patients with advanced chronic kidney disease (estimated glomerular filtration rate (eGFR) < 30 mL/min/1.73 m^2) is unknown, given the fact that those patients have been systematically excluded from clinical trials [1,19].

After the previously mentioned results, and based on preliminary studies demonstrating an added effect on decreasing albuminuria with a dual RAAS blockade [18], later research was focused on studying the effect of the combination of ACEi and ARB in high risk DKD patients [20]. During this period, the ongoing telmisartan alone and in combination with ramipril global endpoint trial. (ONTARGET) evaluated whether the combination of an ACEi (ramipril) with an ARB (telmisartan) was better than the full dose of either drug. This study showed that there was no superiority of the ACEi versus the ARB and that the dual blockade did not confer greater cardiovascular protection. Moreover, the combination of ACEi and ARB increased the risk of adverse events, namely hyperkalemia, hypotensive symptoms, and the over declined eGFR, more than monotherapy [20]. Similar results were obtained from the aliskiren trial in type 2 diabetes using cardio-renal endpoints (ALTITUDE), which compared the effect of a direct renin inhibitor aliskiren to placebo in high-risk type 2 diabetic patients on top of an ACEi or an ARB. The ALTITUDE trial, just like the ONTARGET study, demonstrated that the simultaneous administration of aliskiren with an ACE inhibitor or an ARB should be avoided. The study was halted early because an increased incidence of hypotension, hyperkalemia, renal complications, and non-fatal stroke (HR = 1.25; 95% CI 0.98–1.60; $p = 0.07$) was observed in the aliskiren arm during the follow-up of approximately 2.7 years [21]. Consequently, Novartis (Basel, Switzerland) immediately suspended all promotional and educational programs related to aliskiren and its combinations.

3. Reno-Cardiovascular Protection of SGLT2 Inhibition

SGLT2 inhibitors enhance renal glucose excretion by inhibiting renal glucose reabsorption in the renal proximal tubule. Consequently, SGLT2 inhibitors reduce plasma glucose in an insulin-independent manner and improve insulin resistance in diabetes [22]. SGLT2 inhibitors were shown to reduce glycated haemoglobin (HbA1c) by approximately 0.6%–1.2%, with a lower rate of hypoglycaemia [23]. Four recent major trials have shown that SGLT2 inhibitors are superior to other anti-diabetic medications in the prevention of cardiovascular events and renal protection [5,24–26]. The empagliflozin cardiovascular outcome event trial in type 2 diabetes mellitus patients (EMPA REG OUTCOME) was the first clinical trial examining the effects of empagliflozin compared to placebo on cardiovascular morbimortality in patients with type 2 diabetes and at a high risk for cardiovascular events [27]. EMPA REG included 7020 patients, all of them with established cardiovascular disease and an eGFR $\geq$ 30 mL/min/1.73 m^2, randomly assigned to receive empagliflozin at -10 or 25 mg or a placebo on top of standard care. Death from cardiovascular causes, non-fatal myocardial infarction, or non-fatal stroke was decreased in the empagliflozin group (10.5%) compared to the placebo group (12.1%) (HR 0.86; 95% CI 0.74–0.99; $p = 0.04$). Interestingly, this benefit was higher in older patients (>65 years) and with HbA1c $\leq$ 8.5%. Empagliflozin more significantly decreased cardiovascular death (3.7% vs. 5.9%) and death from any cause (5.7% vs. 8.3%) compared to placebo (HR 0.62; 95% CI 0.49–0.77; $p < 0.001$ and HR 0.68; 95% CI 0.57–0.82, $p < 0.001$), as well as hospitalization for heart failure (HR 0.65; 95% CI 0.50–0.85; $p = 0.002$). HbA1c levels were similar in both groups. In contrast, in previous studies, empagliflozin was associated with a decrease in HbA1c levels in patients with type 2 diabetes [28,29]. In the EMPA REG study, patients receiving empagliflozin showed a decrease in weight, waist circumference, uric acid level, blood pressure, and increased cholesterol levels. 2250 out of 7020 patients (32%) included in the EMPA REG OUTCOME trial had chronic kidney disease (eGFR < 60 mL/min/1.73 m^2 and/or macroalbuminuria) [5], and 1896 patients (27%) presented microalbuminuria and eGFR $\geq$ 60 mL/min/1.73 m^2. Treatment with empagliflozin showed benefits in terms of incident or worsening nephropathy (12.7% vs. 18.8% in placebo group) (HR 0.61; 95% CI 0.53–0.70; $p < 0.001$) and progression to macroalbuminuria (11.2% vs. 16.2%) (HR 0.62; 95% CI 0.54–0.72; $p < 0.001$). Patients receiving empagliflozin demonstrated a decrease in the doubling of serum creatinine compared to the placebo group (HR 0.56; 95% CI 0.39–0.79; $p < 0.001$), as well as a lower rate of renal replacement therapy initiation (0.3% vs. 0.6%) (HR 0.45; 95% CI 0.21–0.97; $p = 0.04$) (see Table 1). eGFR decreased in the empagliflozin group within the first weeks of treatment, but at the end of follow up, eGFR remained stable with empagliflozin compared to a decrease in the placebo group [30].

The Canvas Program studied the cardiovascular and renal effects of canagliflozin (100 mg or 300 mg) versus placebo in 10,142 type 2 diabetic patients with a previous history of cardiovascular disease or two or more cardiovascular risk factors and eGFR > 30 mL/min/1.73 m^2. The mean eGFR was 76.5 ± 20.5 mL/min/1.73 m^2; 22.6% presented microalbuminuria and 7.6% presented macroalbuminuria. Canagliflozin administration significantly decreased HbA1c levels, as well as blood pressure and body weight. In the canagliflozin group, deaths from cardiovascular cause, non-fatal myocardial infarction, and non-fatal stroke were significantly decreased by 14% (26.9 vs. 31.5 patients, with an event per 1000 patient-years) [24]. Regarding renal outcomes, canagliflozin reduced the risk of the composite outcome of a sustained 40% reduction in eGFR, renal replacement therapy initiation, and death from renal causes. Canagliflozin reduces albuminuria progression (HR 0.73; 95% CI 0.67–0.69), and the regression of albuminuria occurred more frequently in canagliflozin treated patients (HR 1.7; 95% CI 1.51–1.91).

The dapagliflozin effect on cardiovascular events (DECLARE-TIMI 58) evaluated the effects of dapagliflozin (SGLT2 inhibitor) versus placebo on cardiovascular and renal outcomes in patients who had, or were at risk for, cardiovascular disease. Eligible patients were older than 40 years old, had type 2 diabetes, a history of cardiovascular disease or more than two classical cardiovascular risk factors, and an eGFR ≥ 60 mL/min/1.73 m^2. The study included 17,160 patients, 40.6% with cardiovascular disease and 59.4% with multiple cardiovascular risk factors. The mean eGFR was 85.2 mL/min/1.73 m^2 and only 7% of patients had an eGFR <60 mL/min/1.73 m^2. Dapagliflozin significantly decreased HbA1c levels, body weight, and blood pressure [25]. There were no differences in cardiovascular death, myocardial infarction, and ischemic stroke between the two groups. Dapagliflozin reduced the risk of this composite outcome by 17% due to a significantly lower rate of hospitalization for heart failure in the dapagliflozin group (HR 0.73; 95% CI 061–0.88). It also reduced the composite renal outcome (a decrease of 40% or more in eGFR, end-stage renal disease, or death from renal or cardiovascular cause) risk by 24% (HR 0.76; 95% CI 0.67–0.87). Interestingly, the renoprotection observed with dapagliflozin was independent of the presence of established CVD [25].

The renoprotective effect of SGLT2 inhibition has become evident in the last 3 years. However, until now its use has been limited to patients with eGFR > 45 mL/min/1.73 m^2 (Table 2). A recent paper published by Vlado Perkovic clearly demonstrated the beneficial effect of canagliflozin in patients with moderate–advanced DKD (CREDENCE trial) [26]. This trial studied the effect of canagliflozin compared to placebo in patients with eGFR = 30–90 mL/min/1.73 m^2 and a urinary albumin-to-creatinine ratio > 300–5000 mg/g creatinine, receiving a stable dose of an ACE or ARB. It included 4401 patients with a mean eGFR of 56.2 mL/min/1.73 m^2 and a mean albumin-to creatinine ratio of 927 mg/g. Of note, 31% of patients had an eGFR < 45 mL/min/1.73 m^2. The trial was halted early because the number of primary outcome events in the placebo group required to trigger analysis was reached sooner than estimated. Canagliflozin decreased the risk of end-stage kidney disease, doubling of the serum creatinine level, or renal or cardiovascular death by 30% compared to placebo [26]. Patients in the canagliflozin group also showed a 30% reduction in the risk of cardiovascular death or hospitalization for heart failure (HR 0.69; 95% CI 0.57–0.83; *p* < 0.001), a 20% reduction of cardiovascular death, myocardial infarction or stroke (HR 0.80; 95% CI 0.67–0.95), and a 29% reduction of hospitalization for heart failure (HR 0.61; 95% CI 0.47–0.80). Canagliflozin group showed a decreased eGFR slope compared with the placebo group (−3.19 ± 0.15 vs. −4.71 ± 0.15 mL/min/1.73 m^2 per year), which means a difference of 1.52 mL/min/1.73 m^2 per year. Rates of adverse events were similar in both groups. In conclusion, the use of canagliflozin in type 2 diabetic patients with established kidney disease on top of renin-angiotensin system blockade was safe and decreased the risk of kidney failure and cardiovascular events [26].

Table 1. Summary of renal outcomes in controlled randomized trials with sodium-glucose co-transporter 2 (SGLT2) inhibitors and glucagon-like peptide-1 (GLP 1) receptor agonists.

Pharmacological Class	SGLT2 Inhibitors				GLP-1 Receptor Agonists	
Trials	EMPA-REG OUTCOME	CANVAS Program	DECLARE-TIMI 58	CREDENCE	LEADER	SUSTAIN-6
Antidiabetic agent	Empagliflozin	Canagliflozin	Dapagliflozin	Canagliflozin	Liraglutide	Semaglutide
Median follow-up (years)	3.1	2.4	4.2	2.6	3.84	2.1
Number of patients (n) (active vs. placebo)	4687 vs. 2333	5795 vs. 4347	8582 vs. 8578	2202 vs. 2199	4668 vs. 4672	1648 vs. 1649
% Patients with moderate-to-severe renal disease [a]	25.9%	NR	7%	59.8%	23.1%	28.5%
Hazard Ratio (95% CI)						
Composite renal outcome	0.61 (0.53–0.70) [b]	0.60 (0.47–0.77) [c]	0.76 (0.67–0.87) [d]	0.70 (0.59–0.82) [e]	0.78 (0.67–0.92) [b]	0.64 (0.46–0.88) [b]
New onset of persistent macroalbuminuria	0.62 (0.54–0.72)	NR	NR	NR	0.62 (0.54–0.72)	NR
Persistent doubling of serum creatinine	0.56 (0.39–0.79) [f]	NR	NR	0.60 (0.48–0.76) [g]	0.89 (0.67–1.19) [f]	NR
Initiation of renal-replacement therapy	0.45 (0.21–0.97)	NR	NR	0.74 (0.55–1.00)	0.87 (0.61–1.24)	NR
Death due to renal disease	NA	NR	NR	NA	1.59 (0.52–4.87)	NR

[a] Estimated glomerular filtration rate (eGFR) of less than 60 mL/min/1.73 m^2. Different primary composite outcomes: [b] new-onset of persistent macroalbuminuria, persistent doubling of serum creatinine and an eGFR less than 45 mL/min/1.73 m^2, need for renal-replacement therapy in the absence of a reversible cause or death due to renal disease; [c] 40% reduction in the eGFR, need for renal-replacement therapy in the absence of a reversible cause or death due to renal disease; [d] ≥40% decrease in eGFR to less than 60 mL/min/1.73 m^2, new end-stage renal disease or death from renal or cardiovascular causes; [e] end-stage kidney disease, persistent doubling of serum creatinine level or death from renal or cardiovascular causes. [f] persistent doubling of serum creatinine and an eGFR less than 45 mL/min/1.73 m^2; [g] persistent doubling of serum creatinine level. NA: not applicable because there are less than 10 events reported. NR: not reported.

Table 2. New antidiabetic drugs, dipeptidyl peptidase 4 (DDP4) inhibitors, GLP-1 receptors agonists and SGLT2 inhibitors dose and drug indications according chronic kidney disease stage.

	Chronic Kidney Diseases Stages (mL/min/1.73 m^2)					
	1 ($\geq$90)	2 (60–89)	3a (45–59)	3b (30–44)	4 (15–29)	5 (<15)
DPP4 inhibitors (oral)						
Sitagliptin	25–100 mg/day	No dose adjustment		50 mg/day	25 mg/day	
Vildagliptin	50–100 mg/day	No dose adjustment		50 mg/day		
Saxagliptin	2.5–5 mg/day	No dose adjustment		2.5 mg/day	Avoid in Dialysis	
Linagliptin	5 mg/day	No dose adjustment				
Alogliptin	6.5–25 mg/day	No dose adjustment		12.5 mg/day	6.5 mg/day	
GLP-1 receptor agonists (subcutaneously)						
Exetanide	10µg/12 hours	No dose adjustment			Avoid	
Exetanide LP	2 mg/week	No dose adjustment			Avoid	
Lixisetanide	20µg/day	No dose adjustment			Avoid	
Albiglutide	30–50 mg/week	No dose adjustment			Avoid	
Dulaglutide	0.75–1.5 mg/week	No dose adjustment			Avoid	
Liraglutide	1.2–1.8 mg/day	No dose adjustment			Avoid	
Semaglutide	0.25–1 mg/week	No dose adjustment			Avoid	
SGLT2 inhibitors (oral)						
Empagliflozin	10–25 mg/day	No dose adjustment			Avoid	
Canagliflozin	100–300 mg/day	No dose adjustment			Avoid	
Dapagliflozin	5–10 mg/day	No dose adjustment			Avoid	

4. Reno-Cardiovascular Protection of GLP1 Receptor Agonists

GLP-1RAs are glucagon-like peptide-1 (GLP-1) molecule analogues. GLP-1 is an incretin secreted by intestinal enteroendocrine L-cells in response to food intake, which increases insulin secretion in a glucose-dependent manner [31]. GLP-1 Ras have been shown to improve glycaemic control and reduce glycated haemoglobin (HbA1c) by approximately 1%–1.5% in short-term treatments with a lower rate of hypoglycaemia [9,32]. The reduction of hypoglycaemic events is important in diabetic patients with chronic kidney disease (CKD), where GLP-1RAs have proven to be safe while many other antihyperglycemic drugs require dose adjustment as insulin or are simply contraindicated (Table 2). These new drugs have also been shown to reduce weight during treatment, which seems to be a class effect and an interesting outcome when it comes to treating overweight type 2 diabetic patients [9,32–34]. However, their most significant impact is that GLP-1RAs were demonstrated to reduce major adverse cardiovascular events (MACEs), liraglutide, semaglutide, and albiglutide [9,33,35], and delay DKD progression liraglutide, semaglutide, and dulaglutide [6,32,33].

In the liraglutide effect and action in diabetes: evaluation of CV outcome results (LEADER) trial, Liraglutide was shown to decrease three-point MACEs and death from cardiovascular causes in a 3.8 year follow-up when compared to the placebo added to standard care (HR 0.87; 95% CI 0.78–0.97; $p < 0.001$) [9]. Moreover, death from any cause was lower in the Liraglutide group and, although there were no statistically significant differences in the occurrence of myocardial infarction or stroke between both groups, there was a trend toward a reduced incidence of both events. In the subgroup analyses, these protective effects were more evident in patients with kidney disease and an eGFR rate below 60 mL/min/1.73 m^2, as well as in patients with cardiovascular disease at baseline. In concordance, semaglutide also reduced the MACE composite outcome when compared to placebo during a follow-up period of 2.1 years in the trial to evaluate cardiovascular and other long-term outcomes with semaglutide in subjects with type 2 diabetes (SUSTAIN-6) (HR 0.74; 95% CI 0.58–0.95; $p < 0.001$) [33]. There was a significant reduction in the incidence of non-fatal stroke and a non-significant trend for a lower incidence of non-fatal myocardial infarction in the semaglutide group. However, and in contrast to the results displayed in the LEADER trial, the rates of death from cardiovascular causes, and death from any cause, were similar in both groups. This could be, in part, explained by the larger number of patients recruited in LEADER and the longer observation time. In the randomized, double-blind, placebo-controlled trial of the effect of albiglutide on major cardiovascular events in patients with type 2 diabetes mellitus (HARMONY), albiglutide also reduced the incidence of three-point MACEs compared to placebo during a follow-up of 1.6 years (HR 0.78; 95% CI 0.68–0.90; $p < 0.0001$) [35]. This decrease was mainly driven by a reduction in the incidence of myocardial infarction, but rates of stroke and death from cardiovascular causes were similar in both treatment groups.

Conversely, other GLP-1RAs did not demonstrate clear cardiovascular benefits. In the evaluation of lixisenatide in acute coronary syndrome (ELIXA) trial, lixisenatide exhibited non-inferiority when compared to a placebo added to standard care during a 25 month follow-up in patients with established cardiovascular disease, but did not reduce the incidence of the composite cardiovascular outcome [36]. Nonetheless, all patients included in the ELIXA were at high cardiovascular risk, with a recent previous history of myocardial infarction or unstable angina, while 81.2% of patients in LEADER had established cardiovascular disease, without previous ischemic heart disease [9], and 60.5% in SUSTAIN-6 [33], and 71% of patients in HARMONY [35] had a previous history of ischemic heart disease. Another trial, the exenatide study of cardiovascular event lowering (EXSCEL), showed a non-significant trend to reduce the incidence of three-point MACEs compared to placebo, where 73.1% of the trial population had previous cardiovascular disease [34].

When it comes to evaluating the renoprotective effects of GLP-1RAs, trials usually examine a composite outcome of new-onset persistent macroalbuminuria, persistent doubling of serum creatinine, end-stage renal disease, or death due to renal causes. A secondary renal outcome analysis of LEADER showed a reduced incidence of this composite outcome in patients receiving liraglutide (HR 0.78; 95% CI 0.67–0.92; $p = 0.003$) [6]. This reduction was mainly driven by a lower incidence of persistent macroalbuminuria, but there were no differences in the doubling of serum creatinine, end-stage renal disease, or death due to renal causes between the groups. New-onset microalbuminuria incidence was lower in the treated group. The effect of Liraglutide appeared to be independent in a subgroup analysis by baseline albuminuria or eGFR. A slightly slower decline of eGFR was also observed in the treated group. In the SUSTAIN-6 trial, semaglutide also reduced the secondary composite renal outcomes of new or worsening nephropathy (HR 0.64; 95% CI 0.46–0.88; $p = 0.005$) [33]. Recently, secondary renal outcomes analysis of the dulaglutide versus insulin glargine in patients with type 2 diabetes and moderate-to-severe chronic kidney disease (AWARD-7) trial revealed that dulaglutide therapy in patients with moderate-to-severe CKD (stages 3 and 4) reduced the glomerular filtration rate decline in the short-term when compared to insulin glargine during a 1 year follow-up [32]. This finding was especially significant in participants who seemed to have a more severe disease with a baseline urine albumin-to-creatinine ratio (UACR) higher than 300 mg/g. In addition, albuminuria reduction was more pronounced in this subgroup (AWARD-7). Interestingly, the AWARD-7 is an randomized clinical trial (RCT), where dulaglutide was compared to insulin glargine, and both treatments were combined with insulin lispro. There were little to no differences in glycaemic control between groups, which may indicate that protective renal effects could be mediated by other mechanisms related to this pharmacological class [32]. Whether cardiovascular protection is a class effect or a specific outcome of certain GLP-1RAs is still controversial and future trials like the dulaglutide and cardiovascular outcomes in type 2 diabetes (REWIND) will give us more information on this matter. Furthermore, although it seems that GLP-1RAs exert renoprotective effects, specific trials should be designed to evaluate this aspect.

5. Combination of SGLT2 Inhibitors and GLP1 Receptor Agonists in Diabetes

Despite their promising cardiovascular and renal protective effects, there are still very few trials that verify the positive outcomes of SGLT2 inhibitors and GLP-1RAs used in combination. In the randomized controlled trial 104-week results—once-weekly exenatide plus once-daily dapagliflozin vs. once-weekly exetanide or dapagliflozin alone (DURATION-8), patients with type 2 diabetes inadequately controlled by metformin were randomly assigned to receive exenatide plus dapagliflozin, exenatide alone, and dapagliflozin alone [37]. During a 28 week follow-up, there was a significant reduction of HbA1c in the combined therapy, compared to exenatide or dapagliflozin, and a considerably higher proportion of patients achieved a glycated haemoglobin ≤7%. Exenatide plus dapagliflozin also exhibited a significant weight reduction compared to dapagliflozin or exenatide alone, and systolic blood pressure (SBP) reduction was slightly higher in the combined treatment. It is worth mentioning that no episodes of hypoglycaemia were described and only one case of ketoacidosis was reported in the exenatide group. In the dulaglutide as add-on therapy to SGLT2 inhibitors in patients with inadequately controlled type 2 diabetes (AWARD-10) trial, two different doses of dulaglutide (0.75 or 1.5 mg) or placebo were assigned to patients previously receiving SGLT2 inhibitors with or without metformin [38]. Dapagliflozin and empagliflozin were the most used SGLT2 inhibitors during the 24 week follow-up. A higher proportion of patients in both dulaglutide groups also achieved the goal of HbA1c ≤ 7%. Only the dulaglutide dose of 1.5 mg provided a significant weight and SBP reduction when compared to placebo. Similar to DURATION-8, the reported adverse effects were mainly gastrointestinal and were more frequently described with high doses of dulaglutide. There were no differences in hypoglycemic events between groups, and the described rates were low. In another recent trial, the superior efficacy of insulin degludec/liraglutide vs. insulin glargine (IGlar U100) as add-on to SGLT2 inhibitors ± oral antidiabetic drug therapy in patients with type 2 diabetes (DUAL IX) trial, patients already receiving a SGLT2 inhibitor with or without other oral antidiabetic drugs were randomized to be treated with an injectable combination of insulin degludec and liraglutide (IDegLira) or insulin glargine (IGlar) alone [39]. After 26 weeks of treatment, the group receiving IDegLira had a significant reduction in HbA1c compared to IGlar (treatment difference −0.36%; 95% CI −0.5, −0.21; $p < 0.0001$). However, no changes in body weight were found with the IDegLira treatment. Hypoglycaemic episodes, although were rarely described, were more frequent when compared to the rates displayed in DURATION-8 and AWARD-10, which may be related to the addition of insulin.

Considering the evidence, the combination of the SGLT2 inhibitor and GLP-1RAs has proven to be safe and well tolerated. The rates of adverse effects are similar to those described in monotherapy and of a mild to moderate intensity. Hypoglycaemic episodes were barely described, although no trials have evaluated their safety in populations with established chronic kidney disease, where there is a higher risk of antidiabetic treatment related hypoglycaemia. Greater reductions of HbA1c, body weight, and blood pressure were observed in combined treatments. Despite these results, cardiovascular outcomes and/or DKD progression with add-on therapy have not been analyzed in the previous trials. Therefore, more studies are needed to evaluate these events.

6. Potential Nephroprotective Mechanisms of SGLT2 Inhibitors and GLP1 Receptor Agonists

Blood glucose level and the reduced blood pressure produced by SGLT2 inhibitors and GLP-1RAs have undeniable beneficial effects in terms of cardio-renal protection. However, other independent pathways have also been described and may potentially be relevant to explain their renoprotective effects. Both drugs have natriuretic properties that produce hemodynamic effects on the kidneys. In healthy subjects, glucose and sodium (Na^+), among other metabolites and electrolytes, are reabsorbed by the tubular cells, mainly in the proximal portion (98% of the total glucose and 67% of the Na^+). Glucose is absorbed together with Na^+ by sodium-glucose co-transporters 1 and 2 (SGLT1 and SGLT2) located at the apical membrane of the tubular cells. Although both transporters are expressed in the tubular cells, the major part of the glucose (90%) is reabsorbed by SGLT2 [40,41]. Na^+ is further reabsorbed by other transporters, of which the Na^+/H^+ exchanger isoform 3 is the most important (NHE3) [42]. Glucose is transported to the blood stream by the glucose transporters (GLUT) -mainly GLUT2- [40,41] and Na^+ via the Na^+/K^+ ATPase [42], both located at the basolateral membrane of the tubular cells. This active transport of glucose and Na^+ contributes importantly to glucose homeostasis and to the maintenance of the intraglomerular tone. In diabetic patients, due to glomerular hyperfiltration, glucose and Na^+ levels are increased in the lumen of the tubules. In response, the tubular reabsorbing mechanisms are upregulated both by an increase in the activity or expression of the transporters mentioned above [43] and by a translocation of GLUT2 to the apical membrane [44,45]. Therefore, glucose and Na^+ blood levels rise, contributing to hyperglycaemia and hypertension. In addition, the enhanced reabsorption in the proximal tubule, decreases Na^+ flow to the distal tubule and activates the macula densa. Both SGLT2 inhibitors and GLP-1RAs increase natriuresis, leading to the restoration of the tubuloglomerular feedback, which results in afferent arteriole vasoconstriction and, finally, in a reduction of the intraglomerular pressure. SGLT2 inhibitors have a more important natriuretic effect than GLP-1RAs, most probably related to the direct blockade of the tubular Na^+ uptake mediated by SGLT2 and/or the impairment of the Na^+/H^+ exchanger isoform 3 (NHE3) activity, which is glucose-dependent and seems to be indirectly blocked by SGLT2 inhibition [46] (see Figure 1). It has recently been described that treatment with empagliflozin decreases NHE3 expression in the kidneys of diabetic otsuka long-evans tokushima Fatty (OLETF) rats, as well as the expression of Na^+-K^+-$2Cl^-$ cotransporters and epithelial Na^+ channels, when compared to untreated littermates, suggesting that SGLT2 modulates the Na^+ reabsorption of several tubular transporters [47]. GLP-1RAs only seem to decrease NHE3 activity [48–50], a fact that would explain the difference in the natriuretic potential of these two drugs [51]. It is unclear whether GLP-1RAs inhibit NHE3 activity through interaction with the GLP-1 receptor or extrarenal mechanisms, such as the renin–angiotensin system or atrial natriuretic peptide modulation [52,53].

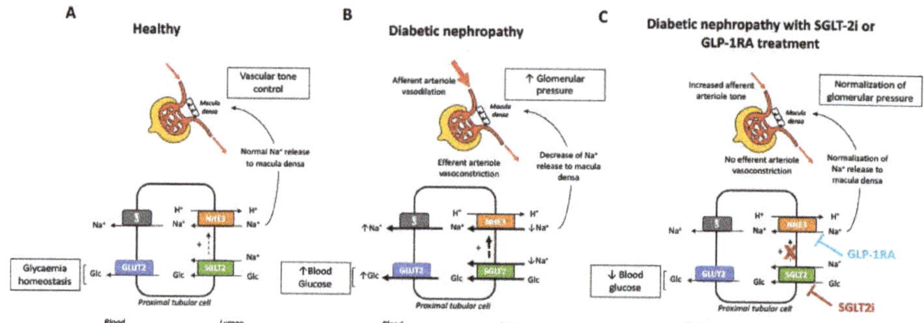

Figure 1. Mechanisms of glucose reabsorption and intraglomerular pressure control in healthy individuals and in diabetic patients. (**A**) In healthy individuals, approximately 90% of glucose is reabsorbed by SGLT2 that is located at the apical membrane of the proximal tubular cells and transported back to the blood stream by the basolateral glucose transporter 2 (GLUT2) transporter. Na$^+$ is reabsorbed by the apical transporters SGLT2 (with glucose) and the Na$^+$/H$^+$ exchanger isoform 3 (NHE3) and returned to the circulation via several basolateral Na$^+$ transporters: sodium bicarbonate transporters, Na$^+$ channels, and the Na$^+$/K$^+$ ATPase. Tubular glucose and Na$^+$ reabsorption mechanisms contribute to glucose homeostasis and glomerular tone control thanks to the tubuloglomerular feedback controlled by the macula densa. (**B**) In diabetic patients, glucose and Na$^+$ reabsorption mechanisms are increased secondary to the hyperfiltration. This fact contributes importantly to the hyperglycaemia and raises the intraglomerular pressure. (**C**) Both SGLT2i and GLP-1RAs produce natriuresis that leads to a decrease of the glomerular pressure. In the case of SGLT2 inhibitors, the natriuretic effect is due to the direct blockade of SGLT2 and the collateral inhibition of NHE3, which has an SGLT2 dependent activity. Regarding GLP-1RAs, these drugs impair only NHE3 activity by an unknown mechanism. Moreover, SGLT2 inhibition contributes to blood glucose level control. ζ: Basolateral Na$^+$ transporters.

Interestingly, SGLT2 inhibitors and GLP-1RAs probably have other effects on the kidneys, independent of the glycaemic control and the natriuresis produced by both drugs. As mentioned before, SGLT2 is clearly expressed in the tubular compartment [54]. However, the glomerular cells can also express SGLT2 under protein overload conditions [55], suggesting that this transporter can be upregulated in a non-diabetic context. In addition, SGLT2 is overexpressed in human tubular cells in a culture (HK-2) treated with transforming growth factor beta 1 (TGF-β1). In this model, the effects of TGF-β1 are reverted by empagliflozin most probably by nuclear factor kappa B/toll-like receptor 4 (NF-κB/TLR4) pathway inhibition [56]. Moreover, non-diabetic murine and rat models of kidney injury have shown that treatment with SGLT2 inhibitors decreases kidney fibrosis and inflammation markers [55,57,58]. In a similar way, GLP-1RAs seem to have beneficial effects on the kidneys beyond natriuresis. In a high-fat-diet-induced obesity mice model with renal impairment, liraglutide treatment likely protected subjects from kidney injury through lipid and mitochondrial metabolism regulation via the sirtuin/AMP-activated protein kinase/peroxisome proliferator-activated receptor gamma coactivator 1-alpha (Sirt1/AMPK/PGC1α) pathways [59]. The results obtained in diabetic and non-diabetic experimental models using SGLT2 inhibitors and GLP-1RAs clearly highlight the possibility that these drugs have a direct protective effect on the kidneys.

7. Current Role of DPP-4 Inhibitors in Patients with Type 2 Diabetes and CKD

The indications for Dipeptidyl Peptidase 4 Inhibitors in patients with type 2 diabetes and CKD have been analyzed over the last 10 years. In early 2010, diverse studies involving a small number of patients demonstrated the safety and efficacy of an adjusted dose of vildagliptin in patients with advanced CKD in hemodialysis and peritoneal dialysis [60,61]. In agreement with these studies, clinical trials performed with alogliptin and sitagliptin demonstrated the safety and efficacy of these DPP-4 inhibitors in type 2 ESRD patients receiving dialysis [62,63]. Later studies by Laakso et al. showed that no dose adjustment was needed when linagliptin was used in patients with moderate to severe renal impairment [64]. Drug dosage adjustments based on renal function of the currently available DPP-4 inhibitors are depicted in Table 2. Three important randomized clinical trials assessed the protective effect of DPP-4 inhibitors on renal functions. The first published study, the saxagliptin and cardiovascular outcomes in patients with type 2 diabetes mellitus (SAVOR-TIMI 53) trial, demonstrated that treatment with saxagliptin was associated with a reduction in UACR compared with placebo (median observation time, 2.1 years) in patients with T2D at a high cardiovascular risk with diverse baseline renal characteristics, including a substantial number of patients with renal dysfunction and/or albuminuria. Interestingly, decreased UACR in saxagliptin-treated patients seemed to be independent of its effect on glycemia. This was observed in patients with normoalbuminuria, microalbuminuria, and macroalbuminuria, irrespective of their eGFR at baseline. However, the positive effect of saxagliptin on UACR was not accompanied by a reduction of other renal outcomes [65]. The second study, the linagliptin and its effects on hyperglycaemia and albuminuria in patients with type 2 diabetes and renal dysfunction (MARLINA-T2D) trial, was aimed at investigating the glycaemic and renal effects of linagliptin added to the standard-of-care in patients with type 2 diabetes and albuminuria [66]. Previous studies by the same group retrospectively included four completed studies with 217 patients with type 2 diabetes and prevalent albuminuria, randomized to either linagliptin 5 mg/day ($n = 162$) or placebo ($n = 55$). They found that linagliptin significantly reduced albuminuria from baseline by 28%, compared to placebo, after 24 weeks of treatment. Of note is that an effect on albuminuria was already seen after 12 weeks of treatment [67]. In the MARLINA trial, linagliptin improved glycaemic control in type 2 diabetes patients and those with early stages of diabetic kidney disease but did not improve renal damage, as estimated using the surrogate endpoint of albuminuria, although significantly more of the participants in the linagliptin group than those in the placebo group experienced a meaningful improvement in albuminuria [66]. The last recently published clinical trial, the cardiovascular and renal microvascular outcome study with linagliptin (CARMELINA) study, aimed to test the long-term (median observation time, 2.2 years) effect of linagliptin on hard renal outcomes. This study included adults with type 2 diabetes and high cardiovascular and renal risk (74% of patients had prevalent chronic kidney disease, 43% had an eGFR below 45 mL/min/1.73 m^2, and 15.2% had an eGFR below 30 mL/min/1.73 m^2). Linagliptin reduced the progression of the albuminuria category (i.e., a change from normoalbuminuria to microalbuminuria/macroalbuminuria or a change from microalbuminuria to macroalbuminuria) by 14% (HR 0.86; 95% CI 0.78–0.95; $p = 0.003$), although the composite renal endpoints of sustained ESRD, death due to renal failure, or a sustained decrease of 50% or more were not different [10].

8. Conclusions

Diabetic kidney disease is the leading cause of premature death and end-stage renal disease in the developed world. Until 2016, the only treatment that demonstrated to be able to attenuate DKD was a renin-angiotensin system blockade, either by ACEi or ARB. However, the partial effectiveness of these agents means that new therapeutic strategies are still needed to delay or prevent progression to ESRD. In recent years, the reno-cardiovascular safety profile of SGLT2 inhibitors and GLP-1RAs has been demonstrated [6,9,68–70]. These new drug classes offer reno-cardiovascular protective effects in diabetic patients. Thus, cardiologist and nephrologists should consider the administration of SGLT2 inhibitors or GLP-1RAs for renovascular protection in their type 2 diabetic patients. Recent studies published by endocrinologists, nephrologists and/or cardiologist all recommend the use of SGLT2 inhibitors or GLP-1RAs as a second line treatment in type 2 diabetic patients when it is not contraindicated [71–74].

Author Contributions: C.G.-C., I.A., A.V., and M.J.S. wrote the manuscript and designed the tables. C.J.-C. wrote the manuscript and designed the figure. M.J.S., D.S. and E.E. reviewed and edited the manuscript.

Funding: The authors are current recipients of research grants from the FONDO DE INVESTIGACIÓN SANITARIA-FEDER, ISCIII, PI17/00257, and REDINREN, RD16/0009/0030.

Acknowledgments: The authors are current recipients of research grants from the FONDO DE INVESTIGACIÓN SANITARIA-FEDER, ISCIII, PI17/00257, and REDINREN, RD16/0009/0030. A.V. performed this work for the basis of his thesis at the Department de Medicina of Universitat Autònoma de Barcelona (UAB).

Conflicts of Interest: M.J.S. reports conflicts of interest with NovoNordisk, Janssen, Boehringer, Eli Lilly, AstraZeneca, and Esteve.

References

1. Guideline Development Group; Bilo, H.; Coentrao, L.; Couchoud, C.; Covic, A.; De Sutter, J.; Drechsler, C.; Gnudi, L.; Goldsmith, D.; Heaf, J.; et al. Clinical Practice Guideline on management of patients with diabetes and chronic kidney disease stage 3b or higher (eGFR < 45 mL/min). *Nephrol. Dial. Transplant.* **2015**, *30*, 1–142.
2. Lewis, E.J.; Hunsicker, L.G.; Clarke, W.R.; Berl, T.; Pohl, M.A.; Lewis, J.B.; Ritz, E.; Atkins, R.C.; Rohde, R.; Raz, I. Renoprotective effect of the angiotensin-receptor antagonist irbesartan in patients with nephropathy due to type 2 diabetes. *N. Engl. J. Med.* **2001**, *345*, 851–860. [CrossRef] [PubMed]
3. Brenner, B.M.; Cooper, M.E.; de Zeeuw, D.; Keane, W.F.; Mitch, W.E.; Parving, H.H.; Remuzzi, G.; Snapinn, S.M.; Zhang, Z.; Shahinfar, S. Effects of losartan on renal and cardiovascular outcomes in patients with type 2 diabetes and nephropathy. *N. Engl. J. Med.* **2001**, *345*, 861–869. [CrossRef] [PubMed]
4. Anguiano, L.; Riera, M.; Pascual, J.; Soler, M.J. Endothelin Blockade in Diabetic Kidney Disease. *J. Clin. Med.* **2015**, *4*, 1171–1192. [CrossRef] [PubMed]
5. Wanner, C.; Inzucchi, S.E.; Lachin, J.M.; Fitchett, D.; von Eynatten, M.; Mattheus, M.; Johansen, O.E.; Woerle, H.J.; Broedl, U.C.; Zinman, B.; et al. Empagliflozin and Progression of Kidney Disease in Type 2 Diabetes. *N. Engl. J. Med.* **2016**, *375*, 323–334. [CrossRef] [PubMed]
6. Mann, J.F.E.; Ørsted, D.D.; Brown-Frandsen, K.; Marso, S.P.; Poulter, N.R.; Rasmussen, S.; Tornøe, K.; Zinman, B.; Buse, J.B.; LEADER Steering Committee and Investigators. Liraglutide and Renal Outcomes in Type 2 Diabetes. *N. Engl. J. Med.* **2017**, *377*, 839–848. [CrossRef] [PubMed]
7. Zelniker, T.A.; Wiviott, S.D.; Raz, I.; Im, K.; Goodrich, E.L.; Bonaca, M.P.; Mosenzon, O.; Kato, E.T.; Cahn, A.; Furtado, R.H.M.; et al. SGLT2 inhibitors for primary and secondary prevention of cardiovascular and renal outcomes in type 2 diabetes: A systematic review and meta-analysis of cardiovascular outcome trials. *Lancet* **2019**, *393*, 31–39. [CrossRef]
8. Malham, S.B.; Herrick, C.J. New Pharmacologic Agents for Diabetes Treatment. *Mo Med.* **2016**, *113*, 361–366.
9. Marso, S.P.; Daniels, G.H.; Brown-Frandsen, K.; Kristensen, P.; Mann, J.F.E.; Nauck, M.A.; Nissen, S.E.; Pocock, S.; Poulter, N.R.; Ravn, L.S.; et al. Liraglutide and Cardiovascular Outcomes in Type 2 Diabetes. *N. Engl. J. Med.* **2016**, *375*, 311–322. [CrossRef]
10. Rosenstock, J.; Perkovic, V.; Johansen, O.E.; Cooper, M.E.; Kahn, S.E.; Marx, N.; Alexander, J.H.; Pencina, M.; Toto, R.D.; Wanner, C.; et al. Effect of Linagliptin vs Placebo on Major Cardiovascular Events in Adults with Type 2 Diabetes and High Cardiovascular and Renal Risk. *JAMA* **2019**, *321*, 69–79. [CrossRef]
11. Grundy, S.M.; Benjamin, I.J.; Burke, G.L.; Chait, A.; Eckel, R.H.; Howard, B.V.; Mitch, W.; Smith, S.C.; Sowers, J.R. Diabetes and cardiovascular disease: A statement for healthcare professionals from the American Heart Association. *Circulation* **1999**, *100*, 1134–1146. [CrossRef] [PubMed]
12. Rubler, S.; Dlugash, J.; Yuceoglu, Y.Z.; Kumral, T.; Branwood, A.W.; Grishman, A. New type of cardiomyopathy associated with diabetic glomerulosclerosis. *Am. J. Cardiol.* **1972**, *30*, 595–602. [CrossRef]
13. Miller, J.A.; Floras, J.S.; Zinman, B.; Skorecki, K.L.; Logan, A.G. Effect of hyperglycaemia on arterial pressure, plasma renin activity and renal function in early diabetes. *Clin. Sci.* **1996**, *90*, 189–195. [CrossRef] [PubMed]
14. Heart Outcomes Prevention Evaluation Study Investigators; Yusuf, S.; Sleight, P.; Pogue, J.; Bosch, J.; Davies, R.; Dagenais, G. Effects of an Angiotensin-Converting–Enzyme Inhibitor, Ramipril, on Cardiovascular Events in High-Risk Patients. *N. Engl. J. Med.* **2000**, *342*, 145–153.
15. Matsusaka, H.; Kinugawa, S.; Ide, T.; Matsushima, S.; Shiomi, T.; Kubota, T.; Sunagawa, K.; Tsutsui, H. Angiotensin II type 1 receptor blocker attenuates exacerbated left ventricular remodeling and failure in diabetes-associated myocardial infarction. *J. Cardiovasc. Pharmacol.* **2006**, *48*, 95–102. [CrossRef]
16. CONSENSUS Trial Study Group. Effects of Enalapril on Mortality in Severe Congestive Heart Failure. *N. Engl. J. Med.* **1987**, *316*, 1429–1435. [CrossRef]

17. SOLVD Investigators; Yusuf, S.; Pitt, B.; Davis, C.E.; Hood, W.B.; Cohn, J.N. Effect of Enalapril on Survival in Patients with Reduced Left Ventricular Ejection Fractions and Congestive Heart Failure. *N. Engl. J. Med.* **1991**, *325*, 293–302.

18. Delanaye, P.; Scheen, A.J. Preventing and treating kidney disease in patients with type 2 diabetes. *Expert Opin. Pharmacother.* **2019**, *20*, 277–294. [CrossRef]

19. Zoccali, C.; Blankestijn, P.J.; Bruchfeld, A.; Capasso, G.; Fliser, D.; Fouque, D.; Goumenos, D.; Ketteler, M.; Massy, Z.; Rychlık, I.; et al. Children of a lesser god: Exclusion of chronic kidney disease patients from clinical trials. *Nephrol. Dial. Transplant.* **2019**. [CrossRef]

20. Yusuf, S.; Teo, K.K.; Pogue, J.; Dyal, L.; Copland, I.; Schumacher, H.; Dagenais, G.; Sleight, P.; Anderson, C. Telmisartan, ramipril, or both in patients at high risk for vascular events. *N. Engl. J. Med.* **2008**, *358*, 1547–1559.

21. Parving, H.-H.; Brenner, B.M.; McMurray, J.J.V.; de Zeeuw, D.; Haffner, S.M.; Solomon, S.D.; Chaturvedi, N.; Persson, F.; Desai, A.S.; Nicolaides, M.; et al. Cardiorenal end points in a trial of aliskiren for type 2 diabetes. *N. Engl. J. Med.* **2012**, *367*, 2204–2213. [CrossRef] [PubMed]

22. Isaji, M. SGLT2 inhibitors: Molecular design and potential differences in effect. *Kidney Int. Suppl.* **2011**, *79*, S14–S19. [CrossRef] [PubMed]

23. Bashier, A.; Khalifa, A.A.; Rashid, F.; Abdelgadir, E.I.; Al Qaysi, A.A.; Ali, R.; Eltinay, A.; Nafach, J.; Alsayyah, F.; Alawadi, F. Efficacy and Safety of SGLT2 Inhibitors in Reducing Glycated Hemoglobin and Weight in Emirati Patients with Type 2 Diabetes. *J. Clin. Med. Res.* **2017**, *9*, 499–507. [CrossRef] [PubMed]

24. Neal, B.; Perkovic, V.; Mahaffey, K.W.; de Zeeuw, D.; Fulcher, G.; Erondu, N.; Shaw, W.; Law, G.; Desai, M.; Matthews, D.R.; et al. Canagliflozin and Cardiovascular and Renal Events in Type 2 Diabetes. *N. Engl. J. Med.* **2017**, *377*, 644–657. [CrossRef] [PubMed]

25. Wiviott, S.D.; Raz, I.; Bonaca, M.P.; Mosenzon, O.; Kato, E.T.; Cahn, A.; Silverman, M.G.; Zelniker, T.A.; Kuder, J.F.; Murphy, S.A.; et al. Dapagliflozin and Cardiovascular Outcomes in Type 2 Diabetes. *N. Engl. J. Med.* **2018**, *380*, 347–357. [CrossRef] [PubMed]

26. Perkovic, V.; Jardine, M.J.; Neal, B.; Bompoint, S.; Heerspink, H.J.L.; Charytan, D.M.; Edwards, R.; Agarwal, R.; Bakris, G.; Bull, S.; et al. Canagliflozin and Renal Outcomes in Type 2 Diabetes and Nephropathy. *N. Engl. J. Med.* **2019**. [CrossRef] [PubMed]

27. Zinman, B.; Wanner, C.; Lachin, J.M.; Fitchett, D.; Bluhmki, E.; Hantel, S.; Mattheus, M.; Devins, T.; Johansen, O.E.; Woerle, H.J.; et al. Empagliflozin, Cardiovascular Outcomes, and Mortality in Type 2 Diabetes. *N. Engl. J. Med.* **2015**, *373*, 2117–2128. [CrossRef]

28. Haring, H.-U.; Merker, L.; Seewaldt-Becker, E.; Weimer, M.; Meinicke, T.; Woerle, H.J.; Broedl, U.C.; EMPA-REG METSU Trial Investigators. Empagliflozin as Add-on to Metformin Plus Sulfonylurea in Patients with Type 2 Diabetes: A 24-week, randomized, double-blind, placebo-controlled trial. *Diabetes Care* **2013**, *36*, 3396–3404. [CrossRef]

29. Kovacs, C.S.; Seshiah, V.; Swallow, R.; Jones, R.; Rattunde, H.; Woerle, H.J.; Broedl, U.C.; EMPA-REG PIO™ trial investigators. Empagliflozin improves glycaemic and weight control as add-on therapy to pioglitazone or pioglitazone plus metformin in patients with type 2 diabetes: A 24-week, randomized, placebo-controlled trial. *Diabetes Obes. Metab.* **2014**, *16*, 147–158. [CrossRef]

30. Wanner, C.; Heerspink, H.J.L.; Zinman, B.; Inzucchi, S.E.; Koitka-Weber, A.; Mattheus, M.; Hantel, S.; Woerle, H.-J.; Broedl, U.C.; von Eynatten, M.; et al. Empagliflozin and Kidney Function Decline in Patients with Type 2 Diabetes: A Slope Analysis from the EMPA-REG OUTCOME Trial. *J. Am. Soc. Nephrol.* **2018**, *29*, 2755–2769. [CrossRef]

31. Paternoster, S.; Falasca, M. Dissecting the Physiology and Pathophysiology of Glucagon-Like Peptide-1. *Front. Endocrinol.* **2018**, *9*, 584. [CrossRef] [PubMed]

32. Tuttle, K.R.; Lakshmanan, M.C.; Rayner, B.; Busch, R.S.; Zimmermann, A.G.; Woodward, D.B.; Botros, F.T. Dulaglutide versus insulin glargine in patients with type 2 diabetes and moderate-to-severe chronic kidney disease (AWARD-7): A multicentre, open-label, randomised trial. *Lancet Diabetes Endocrinol.* **2018**, *6*, 605–617. [CrossRef]

33. Marso, S.P.; Bain, S.C.; Consoli, A.; Eliaschewitz, F.G.; Jódar, E.; Leiter, L.A.; Lingvay, I.; Rosenstock, J.; Seufert, J.; Warren, M.L.; et al. Semaglutide and Cardiovascular Outcomes in Patients with Type 2 Diabetes. *N. Engl. J. Med.* **2016**, *375*, 1834–1844. [CrossRef] [PubMed]

34. Holman, R.R.; Bethel, M.A.; Mentz, R.J.; Thompson, V.P.; Lokhnygina, Y.; Buse, J.B.; Chan, J.C.; Choi, J.; Gustavson, S.M.; Iqbal, N.; et al. Effects of Once-Weekly Exenatide on Cardiovascular Outcomes in Type 2 Diabetes. *N. Engl. J. Med.* **2017**, *377*, 1228–1239. [CrossRef] [PubMed]

35. Hernandez, A.F.; Green, J.B.; Janmohamed, S.; D'Agostino, R.B.; Granger, C.B.; Jones, N.P.; Leiter, L.A.; Rosenberg, A.E.; Sigmon, K.N.; Somerville, M.C.; et al. Albiglutide and cardiovascular outcomes in patients with type 2 diabetes and cardiovascular disease (Harmony Outcomes): A double-blind, randomised placebo-controlled trial. *Lancet* **2018**, *392*, 1519–1529. [CrossRef]

36. Pfeffer, M.A.; Claggett, B.; Diaz, R.; Dickstein, K.; Gerstein, H.C.; Køber, L.V.; Lawson, F.C.; Ping, L.; Wei, X.; Lewis, E.F.; et al. Lixisenatide in Patients with Type 2 Diabetes and Acute Coronary Syndrome. *N. Engl. J. Med.* **2015**, *373*, 2247–2257. [CrossRef] [PubMed]

37. Frías, J.P.; Guja, C.; Hardy, E.; Ahmed, A.; Dong, F.; Öhman, P.; Jabbour, S.A. Exenatide once weekly plus dapagliflozin once daily versus exenatide or dapagliflozin alone in patients with type 2 diabetes inadequately controlled with metformin monotherapy (DURATION-8): A 28 week, multicentre, double-blind, phase 3, randomised control. *Lancet. Diabetes Endocrinol.* **2016**, *4*, 1004–1016. [CrossRef]

38. Ludvik, B.; Frías, J.P.; Tinahones, F.J.; Wainstein, J.; Jiang, H.; Robertson, K.E.; García-Pérez, L.-E.; Woodward, D.B.; Milicevic, Z. Dulaglutide as add-on therapy to SGLT2 inhibitors in patients with inadequately controlled type 2 diabetes (AWARD-10): A 24-week, randomised, double-blind, placebo-controlled trial. *Lancet. Diabetes Endocrinol.* **2018**, *6*, 370–381. [CrossRef]

39. Philis-Tsimikas, A.; Billings, L.K.; Busch, R.; Portillo, C.M.; Sahay, R.; Halladin, N.; Eggert, S.; Begtrup, K.; Harris, S. Superior efficacy of insulin degludec/liraglutide versus insulin glargine U100 as add-on to sodium-glucose co-transporter-2 inhibitor therapy: A randomized clinical trial in people with uncontrolled type 2 diabetes. *Diabetes Obes. Metab.* **2019**, *21*, 1399–1408. [CrossRef] [PubMed]

40. Szablewski, L. Distribution of glucose transporters in renal diseases. *J. Biomed. Sci.* **2017**, *24*, 64. [CrossRef] [PubMed]

41. Navale, A.M.; Paranjape, A.N. Glucose transporters: Physiological and pathological roles. *Biophys. Rev.* **2016**, *8*, 5. [CrossRef] [PubMed]

42. Wang, X.; Armando, I.; Upadhyay, K.; Pascua, A.; Jose, P.A. The regulation of proximal tubular salt transport in hypertension: An update. *Curr. Opin. Nephrol. Hypertens.* **2009**, *18*, 412–420. [CrossRef] [PubMed]

43. Umino, H.; Hasegawa, K.; Minakuchi, H.; Muraoka, H.; Kawaguchi, T.; Kanda, T.; Tokuyama, H.; Wakino, S.; Itoh, H. High Basolateral Glucose Increases Sodium-Glucose Cotransporter 2 and Reduces Sirtuin-1 in Renal Tubules through Glucose Transporter-2 Detection. *Sci. Rep.* **2018**, *8*, 6791. [CrossRef]

44. Cohen, M.; Kitsberg, D.; Tsytkin, S.; Shulman, M.; Aroeti, B.; Nahmias, Y. Live imaging of GLUT2 glucose-dependent trafficking and its inhibition in polarized epithelial cysts. *Open Biol.* **2014**, *4*, 140091. [CrossRef] [PubMed]

45. Hinden, L.; Udi, S.; Drori, A.; Gammal, A.; Nemirovski, A.; Hadar, R.; Baraghithy, S.; Permyakova, A.; Geron, M.; Cohen, M.; et al. Modulation of Renal GLUT2 by the Cannabinoid-1 Receptor: Implications for the Treatment of Diabetic Nephropathy. *J. Am. Soc. Nephrol. JASN* **2018**, *29*, 434–448. [CrossRef] [PubMed]

46. Pessoa, T.D.; Campos, L.C.G.; Carraro-Lacroix, L.; Girardi, A.C.C.; Malnic, G. Functional role of glucose metabolism, osmotic stress, and sodium-glucose cotransporter isoform-mediated transport on Na+/H+ exchanger isoform 3 activity in the renal proximal tubule. *J. Am. Soc. Nephrol. JASN* **2014**, *25*, 2028–2039. [CrossRef] [PubMed]

47. Chung, S.; Kim, S.; Son, M.; Kim, M.; Koh, E.S.; Shin, S.J.; Ko, S.-H.; Kim, H.-S. Empagliflozin Contributes to Polyuria via Regulation of Sodium Transporters and Water Channels in Diabetic Rat Kidneys. *Front. Physiol.* **2019**, *10*, 271. [CrossRef]

48. Carraro-Lacroix, L.R.; Malnic, G.; Girardi, A.C.C. Regulation of Na+/H+ exchanger NHE3 by glucagon-like peptide 1 receptor agonist exendin-4 in renal proximal tubule cells. *Am. J. Physiol. Ren. Physiol.* **2009**, *297*, F1647–F1655. [CrossRef]

49. Farah, L.X.S.; Valentini, V.; Pessoa, T.D.; Malnic, G.; McDonough, A.A.; Girardi, A.C.C. The physiological role of glucagon-like peptide-1 in the regulation of renal function. *Am. J. Physiol. Ren. Physiol.* **2016**, *310*, F123–F127. [CrossRef]

50. Crajoinas, R.O.; Oricchio, F.T.; Pessoa, T.D.; Pacheco, B.P.M.; Lessa, L.M.A.; Malnic, G.; Girardi, A.C.C. Mechanisms mediating the diuretic and natriuretic actions of the incretin hormone glucagon-like peptide-1. *Am. J. Physiol. Ren. Physiol.* **2011**, *301*, F355–F363. [CrossRef]

51. Packer, M. Contrasting effects on the risk of macrovascular and microvascular events of antihyperglycemic drugs that enhance sodium excretion and lower blood pressure. *Diabet. Med.* **2018**, *35*, 707–713. [CrossRef] [PubMed]

52. Skov, J. Effects of GLP-1 in the Kidney. *Rev. Endocr. Metab. Disord.* **2014**, *15*, 197–207. [CrossRef] [PubMed]

53. Andersen, A.; Lund, A.; Knop, F.K.; Vilsbøll, T. Glucagon-like peptide 1 in health and disease. *Nat. Rev. Endocrinol.* **2018**, *14*, 390–403. [CrossRef] [PubMed]

54. Wang, X.X.; Levi, J.; Luo, Y.; Myakala, K.; Herman-Edelstein, M.; Qiu, L.; Wang, D.; Peng, Y.; Grenz, A.; Lucia, S.; et al. SGLT2 Protein Expression Is Increased in Human Diabetic Nephropathy. *J. Biol. Chem.* **2017**, *292*, 5335–5348. [CrossRef] [PubMed]

55. Cassis, P.; Locatelli, M.; Cerullo, D.; Corna, D.; Buelli, S.; Zanchi, C.; Villa, S.; Morigi, M.; Remuzzi, G.; Benigni, A.; et al. SGLT2 inhibitor dapagliflozin limits podocyte damage in proteinuric nondiabetic nephropathy. *JCI Insight* **2018**, *3*, 15. [CrossRef] [PubMed]

56. Panchapakesan, U.; Pegg, K.; Gross, S.; Komala, M.G.; Mudaliar, H.; Forbes, J.; Pollock, C.; Mather, A. Effects of SGLT2 Inhibition in Human Kidney Proximal Tubular Cells—Renoprotection in Diabetic Nephropathy? *PLoS ONE* **2013**, *8*, e54442. [CrossRef]

57. Abbas, N.A.T.; El. Salem, A.; Awad, M.M. Empagliflozin, SGLT2inhibitor, attenuates renal fibrosis in rats exposed to unilateral ureteric obstruction: Potential role of klotho expression. *Naunyn-Schmiedeberg's Arch. Pharmacol.* **2018**, *391*, 1347–1360. [CrossRef]

58. Zhang, Y.; Nakano, D.; Guan, Y.; Hitomi, H.; Uemura, A.; Masaki, T.; Kobara, H.; Sugaya, T.; Nishiyama, A. A sodium-glucose cotransporter 2 inhibitor attenuates renal capillary injury and fibrosis by a vascular endothelial growth factor–dependent pathway after renal injury in mice. *Kidney Int.* **2018**, *94*, 524–535. [CrossRef]

59. Wang, C.; Li, L.; Liu, S.; Liao, G.; Li, L.; Chen, Y.; Cheng, J.; Lu, Y.; Liu, J. GLP-1 receptor agonist ameliorates obesity-induced chronic kidney injury via restoring renal metabolism homeostasis. *PLoS ONE* **2018**, *13*, e0193473. [CrossRef]

60. Ito, M.; Abe, M.; Okada, K.; Sasaki, H.; Maruyama, N.; Tsuchida, M.; Higuchi, T.; Kikuchi, F.; Soma, M. The dipeptidyl peptidase-4 (DPP-4) inhibitor vildagliptin improves glycemic control in type 2 diabetic patients undergoing hemodialysis. *Endocr. J.* **2011**, *58*, 979–987. [CrossRef]

61. Ito, H.; Mifune, M.; Matsuyama, E.; Furusho, M.; Omoto, T.; Shinozaki, M.; Nishio, S.; Antoku, S.; Abe, M.; Togane, M.; et al. Vildagliptin is Effective for Glycemic Control in Diabetic Patients Undergoing either Hemodialysis or Peritoneal Dialysis. *Diabetes Ther.* **2013**, *4*, 321–329. [CrossRef] [PubMed]

62. Fujii, Y.; Abe, M.; Higuchi, T.; Mizuno, M.; Suzuki, H.; Matsumoto, S.; Ito, M.; Maruyama, N.; Okada, K.; Soma, M. The dipeptidyl peptidase-4 inhibitor alogliptin improves glycemic control in type 2 diabetic patients undergoing hemodialysis. *Expert Opin. Pharmacother.* **2013**, *14*, 259–267. [CrossRef] [PubMed]

63. Arjona Ferreira, J.C.; Corry, D.; Mogensen, C.E.; Sloan, L.; Xu, L.; Golm, G.T.; Gonzalez, E.J.; Davies, M.J.; Kaufman, K.D.; Goldstein, B.J. Efficacy and safety of sitagliptin in patients with type 2 diabetes and ESRD receiving dialysis: A 54-week randomized trial. *Am. J. Kidney Dis. Off. J. Natl. Kidney Found.* **2013**, *61*, 579–587. [CrossRef] [PubMed]

64. Laakso, M.; Rosenstock, J.; Groop, P.-H.; Barnett, A.H.; Gallwitz, B.; Hehnke, U.; Tamminen, I.; Patel, S.; von Eynatten, M.; Woerle, H.-J. Treatment with the Dipeptidyl Peptidase-4 Inhibitor Linagliptin or Placebo Followed by Glimepiride in Patients with Type 2 Diabetes With Moderate to Severe Renal Impairment: A 52-Week, Randomized, Double-Blind Clinical Trial: Figure 1. *Diabetes Care* **2015**, *38*, e15–e17. [CrossRef] [PubMed]

65. Udell, J.A.; Bhatt, D.L.; Braunwald, E.; Cavender, M.A.; Mosenzon, O.; Steg, P.G.; Davidson, J.A.; Nicolau, J.C.; Corbalan, R.; Hirshberg, B.; et al. Saxagliptin and Cardiovascular Outcomes in Patients with Type 2 Diabetes Mellitus and Moderate or Severe Renal Impairment: Observations From the SAVOR-TIMI 53 Trial. *Diabetes Care* **2014**, *38*, dc141850. [CrossRef] [PubMed]

66. Groop, P.-H.; Cooper, M.E.; Perkovic, V.; Hocher, B.; Kanasaki, K.; Haneda, M.; Schernthaner, G.; Sharma, K.; Stanton, R.C.; Toto, R.; et al. Linagliptin and its effects on hyperglycaemia and albuminuria in patients with type 2 diabetes and renal dysfunction: The randomized MARLINA-T2D trial. *Diabetes Obes. Metab.* **2017**, *19*, 1610–1619. [CrossRef] [PubMed]

67. Groop, P.-H.; Cooper, M.E.; Perkovic, V.; Emser, A.; Woerle, H.-J.; von Eynatten, M. Linagliptin lowers albuminuria on top of recommended standard treatment in patients with type 2 diabetes and renal dysfunction. *Diabetes Care* **2013**, *36*, 3460–3468. [CrossRef]
68. Fioretto, P.; Zambon, A.; Rossato, M.; Busetto, L.; Vettor, R. SGLT2 Inhibitors and the Diabetic Kidney. *Diabetes Care* **2016**, *39*, S165–S171. [CrossRef]
69. DeFronzo, R.A.; Norton, L.; Abdul-Ghani, M. Renal, metabolic and cardiovascular considerations of SGLT2 inhibition. *Nat. Rev. Nephrol.* **2017**, *13*, 11–26. [CrossRef]
70. Heerspink, H.J.L.; Perkins, B.A.; Fitchett, D.H.; Husain, M.; Cherney, D.Z.I. Sodium Glucose Cotransporter 2 Inhibitors in the Treatment of Diabetes Mellitus: Cardiovascular and Kidney Effects, Potential Mechanisms, and Clinical Applications. *Circulation* **2016**, *134*, 752–772. [CrossRef]
71. Davies, M.J.; D'Alessio, D.A.; Fradkin, J.; Kernan, W.N.; Mathieu, C.; Mingrone, G.; Rossing, P.; Tsapas, A.; Wexler, D.J.; Buse, J.B. Management of Hyperglycemia in Type 2 Diabetes, 2018. A Consensus Report by the American Diabetes Association (ADA) and the European Association for the Study of Diabetes (EASD). *Diabetes Care* **2018**, *41*, 2669–2701. [CrossRef] [PubMed]
72. Sarafidis, P.; Ferro, C.J.; Morales, E.; Ortiz, A.; Malyszko, J.; Hojs, R.; Khazim, K.; Ekart, R.; Valdivielso, J.; Fouque, D.; et al. SGLT-2 inhibitors and GLP-1 receptor agonists for nephroprotection and cardioprotection in patients with diabetes mellitus and chronic kidney disease. A consensus statement by the EURECA-m and the DIABESITY working groups of the ERA-EDTA. *Nephrol. Dial. Transplant.* **2019**, *34*, 208–230. [CrossRef] [PubMed]
73. Seferović, P.M.; Petrie, M.C.; Filippatos, G.S.; Anker, S.D.; Rosano, G.; Bauersachs, J.; Paulus, W.J.; Komajda, M.; Cosentino, F.; de Boer, R.A.; et al. Type 2 diabetes mellitus and heart failure: A position statement from the Heart Failure Association of the European Society of Cardiology. *Eur. J. Heart Fail.* **2018**, *20*, 853–872. [CrossRef] [PubMed]
74. Vardeny, O.; Vaduganathan, M. Practical Guide to Prescribing Sodium-Glucose Cotransporter 2 Inhibitors for Cardiologists. *JACC Heart Fail.* **2019**, *7*, 169–172. [CrossRef] [PubMed]

MDPI

St. Alban-Anlage 66

4052 Basel

Switzerland

Tel. +41 61 683 77 34

Fax +41 61 302 89 18

www.mdpi.com

Journal of Clinical Medicine Editorial Office

E-mail: jcm@mdpi.com

www.mdpi.com/journal/jcm